NEW RESEARCH ON SILICON - STRUCTURE, PROPERTIES, TECHNOLOGY

Edited by **Vitalyi Igorevich Talanin**

New Research on Silicon - Structure, Properties, Technology
http://dx.doi.org/10.5772/65609
Edited by Vitalyi Igorevich Talanin

Contributors

Dawei Luo, Maurizio Casalino, Guilherme Gaspar, Antoine Autruffe, José Mário Pó, Wei Li, Fuwan Gan, Hao Li, Haiyang Huang, Yogesh Sharma, Néstor David Espinosa-Torres, José Álvaro David Hernández-De-La-Luz, Javier Martínez-Juárez, Seigo Ito, Arkady Skvortsov, Jaime Andres Perez Taborda, Marisol Martin Gonzalez, Olga Caballero Calero, Didac Vega, Angel Rodriguez, Azrul Azlan Hamzah, Norhafizah Burham, Jumril Yunas, Burhanuddin Yeop Majlis, Vitalyi Igorevich Talanin, Igor Evgenievich Talanin, V.I. Lashko

Published by InTech
Janeza Trdine 9, 51000 Rijeka, Croatia

Publishing Process Manager Mirena Calmic
Technical Editor SPi Global
Cover InTech Design team

First published May, 2017
Printed in Croatia
Legal deposit, Croatia: National and University Library in Zagreb

Additional hard copies can be obtained from orders@intechopen.com

New Research on Silicon - Structure, Properties, Technology, Edited by Vitalyi Igorevich Talanin
p. cm.
Print ISBN 978-953-51-3159-5
Online ISBN 978-953-51-3160-1

For EU product safety concerns:
IN TECH d.o.o., Prolaz Marije Krucifikse Kozulić 3, 51000 Rijeka, Croatia,
info@intechopen.com or visit our website at intechopen.com.

Meet the editor

Dr. V. I. Talanin was born in 1977 into a Russian noble family in Zaporozhye, USSR. He completed his MSc degree (2000) in Physics and Technology of Materials and Components in Electronic Technique, MSc degree (2006) in Software Engineering, and PhD degree (2002) in Physics of Semiconductors and Dielectrics. His PhD dissertation was on the mechanisms of grown-in microdefect formation in dislocation-free silicon single crystals. Dr. Talanin was an assistant professor (2002–2003) and an associate professor (2003–2006) and is a full professor (since 2006). Since 2012, he has been working as the full professor and vice chief of Computer Science and Software Engineering Department at the Zaporozhye Institute of Economics and Information Technologies. Dr. Talanin was awarded a scholarship at the Ukraine Ministry Cabinet (2002–2003) and President Grant for Scientific Researches (2007, 2008). He was awarded the title Doctor of Science, honoris causa (2012), and Full Professor (2013) at the Russian Academy of Natural History. Dr. Talanin has published extensively in material science and semiconductor physics and also in Russian history and genealogy (as hobbies). He has written 10 monographs and ca. 200 scientific papers. Dr. Talanin is a member of editorial board of the *Journal of Crystallization Process and Technology* (2011–present) and *Journal of Science and Technology* (2011–present). Dr. Talanin was a member of a program committee at the I (2013), II (2014), III (2015), and IV (2016) and is a member of a program committee at the V (2017) World Conference on Information Systems and Technologies. Dr. Talanin was a member of a Program Committee at the I (2015) and II (2016) International Conference on Semiconductor Physics and Devices.

Contents

Preface

As well known, silicon has a large impact on the modern world economy. Most free silicon is used in the steel refining, aluminum casting, and fine chemical industries. Silicon is the basis of the widely used synthetic polymers called silicones. The pure silicon such as hydrogenated amorphous silicon and upgraded metallurgical-grade silicon are used in the production of such applications as liquid crystal displays and thin-film solar cells. The very highly purified monocrystalline silicon is used to produce silicon wafers used in the semiconductor industry, in electronics, and in photovoltaic applications. In silicon photonics, silicon can be used as a continuous wave Raman laser medium to produce coherent light. Silicon has become the most popular material for both high-power semiconductors and integrated circuits because it can withstand the high temperatures and greatest electrical activity. Monocrystalline silicon is expensive to produce and is usually used in the production of integrated circuits. In common integrated circuits, a wafer of monocrystalline silicon serves as a mechanical support for the circuits, which are created by doping and insulated from each other by thin layers of silicon oxide. In such systems, the tiny crystal imperfections (defects) can interfere with circuit paths. The quality of very highly purified silicon, which is used in semiconductor electronics, depends on such crystal imperfections (defects). Therefore, the modern technology—most computers, cell phones, and others—depends on it.

During the last 60 years, many aspects of silicon, especially the monocrystalline silicon, were seriously studied. However, some questions, such as structural perfection and some new applications of silicon, need further investigation and clarification. Such wide range of aspects, from fundamental problems to technical and technological applications, conditioned the theme of this book. The overall purpose of this book is to provide timely information on selected topics in silicon technology. The articles for this book have been contributed by the much respected researchers in this area and cover the most recent developments and applications of silicon technology and some fundamental questions. I hope that the scientists and all readers will find this book useful.

I thank all contributors for this book and give my special thanks of Ms. Mirena Čalmić for her assistance during the write-up of this book and its final preparation.

My greatest acknowledement goes to my family who have helped me all my life, to my Dad Prof. Dr. Igor Evgenievich Talanin and to my Mom Lubov' Ivanovna Talania for their patience and support.

Prof. Dr. Vitalyi Igorevich Talanin
Department of Computer Science & Software Engineering
Institute of Economics and Information Technologies
Zaporozhye, Ukraine

Chapter 1

Introductory Chapter: Description of the Real Monocrystalline Structure on the Basis of the Vlasov Model for Solids

Vitalyi Igorevich Talanin,
Igor Evgenievich Talanin and
Vladislav Igorevich Lashko

Additional information is available at the end of the chapter

http://dx.doi.org/10.5772/intechopen.68684

1. Introduction

As well known, all characteristics of initial materials for following applications depend on the crystal structural perfection. In the last few decades, our knowledge of the nature of the solid state has sharply increased. At the same time, the areas of application of crystalline solids in various fields of technology are increased. The needs of technology stimulate the rapid development of a relatively young area of modern natural science—solid state chemistry. The main tasks of solid state chemistry are (i) the synthesis of solids, (ii) the research of their physicochemical properties and reactions with their participation, and ultimately (iii) the creation of materials with predetermined properties.

All real solids contain structural defects. Structural defects are violations of the periodicity of the spatial arrangement of atoms. The impact of defects on the physical properties of crystals is extremely diverse. The introduction or change in the concentration of defects in a solid can change the physical properties of the matter. Such operations lead to the creation of new optical, electronic, mechanical, and other characteristics of the solids. Creating a fairly complete picture of the nature and behavior of various defects is a prerequisite for a scientific approach to control structurally sensitive properties and processes in solids. Therefore, in the research of solids, the most informative is the way of a joint examination of their structure, formation, and properties.

Of all the variety of solids, semiconductor silicon occupies a completely unique position. Semiconductor (or electronic) silicon is the basis of the electronic industry at present and in the foreseeable future. On the basis of silicon, discrete device and integrated circuits are manufactured. For the requirements of microelectronics, it is necessary to grow single crystals (i) in which there are no dislocations completely, (ii) with a uniform distribution of doping and background impurities, and (iii) with a controlled and limited content of intrinsic point defects. Stringent requirements stimulate intensive research into the nature of defect formation and improvement of methods for obtaining modern dislocation-free single crystals of silicon.

Requirements for the quality of modern microelectronic devices have led to the fact that silicon is the most pure material in the world. To obtain single crystals of silicon, the Czochralski method and the floating zone method are used. Both methods ensure the production of the initial silicon with the total content of residual impurities 10^{11} to 10^{12} cm^{-3}. Semiconductor silicon (a highly pure material with the perfect structure) can be considered as the initial model for the construction of theoretical models of defect formation in other semiconductor materials and metals. The obtained mathematical models and methods proposed for their solution in silicon make it possible to formulate and solve many problems on the kinetics of diffusion processes in solids.

The growth of dislocation-free silicon single crystals by the Czochralski method and floating zone melting is accompanied by the formation of structural imperfections known as grown-in microdefects. The formation of grown-in microdefects is caused by complex processes of interactions between intrinsic point defects (vacancies and interstitial silicon atoms) and impurity atoms of oxygen and carbon in cooling silicon ingots. Depending on the thermal conditions of crystal growth, grown-in microdefects are precipitates of impurities, microvoids, or interstitial dislocation loops [1]. The presence of grown-in microdefects in dislocation-free silicon single crystals significantly affects their mechanical, electrical, optical, and other properties. The problem of controlling the nature, content, size, and nature of the distribution of defects present in the dislocation-free silicon single crystal is primarily related to the development of effective methods for influencing the state of the ensemble of interacting point defects in the grown ingot. Advances in the control of the condition of the ensemble of point defects determine, ultimately, the success of the quality management of electronic devices. The problem of the formation of grown-in microdefects has both the technological aspect (control of the defect structure of the crystal during the growth for the purpose to increase its quality) and the fundamental scientific importance because its solution makes it possible to describe the physics of defect formation in ultra-pure dislocation-free silicon single crystals. The construction of a theoretical model of the formation of grown-in microdefects in dislocation-free silicon single crystals is an important scientific and practical problem from the viewpoint of the development of physical principles of the purposeful creation of new materials with a desired set of physical properties.

In 2010, we created a model of high-temperature precipitation of impurities [2]. In this model, the inclusion of thermal growth conditions of silicon single crystals in the form of the dependence $T(t)$ (where T is temperature, t is time) allows us to theoretically describe the conditions for nucleation, growth, and coalescence of the precipitates in the range from the crystallization temperature to room temperature. The model of high-temperature precipitation is not only adequate to the experimental results of the investigation of grown-in microdefects but also does not contradict the classical theory of nucleation of second-phase particles. This model extends the field of application of the mathematical apparatus of the classical nucleation theory to the formation of nuclei in a solid during its cooling after the growth. The introduction and consideration of thermal conditions of crystal growth in the initial equations of the classical nucleation theory make it possible to demonstrate that the formation of critical nuclei of small sizes, with a higher probability, occurs in the vicinity of the crystallization front. Cooling of the crystal is accompanied by the growth and coalescence of precipitates. The interaction of impurities with intrinsic point defects has a fundamental character and determines the defect structure of the crystal from the onset of the crystallization to the fabrication of the finished device [3].

In general, in the last century to describe the physical properties of solids with a point of view of their atomic structure, two approaches were suggested. The classical approach posits the notion of the crystal lattice [4]. The classic approach assumes (i) the location of each atom in the vicinity of the fixed lattice site, (ii) the consistency of introducing the concept of probability and of the mechanical description of the behavior of particles, and (iii) the premise that the totality of atoms in a crystal is an integer. Statistical and quantum mechanical methods are used in the classical theory of the crystal. This theory has resulted in important results. In particular, one of such results is the creation of the classical theory of nucleation and growth of second-phase particles in solids [5]. With the help of this theory, we can describe the formation and development (transformation) of the defect structure of a solid, both in the process of its growth and as a result of various technological impacts [3].

In the second approach, a periodic structure of the crystal is not a consequence restriction on the freedom of movement of the atoms in the crystal. Periodic structure is caused by the specificity of statistical laws of motion of particles, where a periodic structure agrees with the freedom of movement of atoms. As a result, the probability of finding an atom in interstitials of the crystal lattice is always different from zero [6, 7]. The second approach is based on the solution of the Vlasov kinetic equation. Vlasov equation represents a system of equations that describe the evolution of the particle continuum with potential of the pair interaction [8]. Vlasov created the physical and mathematical apparatus of system of many particles. He has developed and entered into the reference the concept of collective variables. Vlasov assumed that the method used by him to solve the problem of the plasma has a universal character. This method can be applied to the description of plasma, gas, liquids, and solids. The interaction in those systems can be a short range or long range, weak or strong [9, 10].

For the first time, a solution for solid state was conducted by Vlasov. Vlasov showed that the spatial periodic distribution is one of the particular conditions of particle motion [6, 7]. Vlasov in particular investigated using the linearized Vlasov equation and found the criteria for the origin of the periodic structure in terms of the temperature, density, and microscopic interaction of particles of the medium.

It is noted that in his 1946 paper [11], Lev Landau wrote a very sharp remark: "*these equations were used by A.A. Vlasov for investigation of the vibrations of plasma. However most of his results are incorrect.*" In Ginzburg et al. [12], which is not published in the English version, Lev Landau, Vitalyi Ginzburg, and others were even tougher: "*Recently (in 1944-1945) in print appeared a number of works by AA Vlasov, dedicated to generalization of the concept of the electron plasma and solid state theory. In these works, the author concludes that in the self-consistent field method accounting of the interaction forces at large distances reveals new dynamic properties of polyatomic systems and leads to a change in our perceptions of gas, liquid, solid in the direction of combining them with the concept of plasma etc. Consideration of these works Vlasov led us, however, to the conclusion about their complete failure and about the absence of the results of scientific value.*"

In the field of plasma physics, history has judged critics and Vlasov [13]. However, in his 2000 paper [14] Vitalyi Ginzburg wrote: "*I do not know about any Vlasov achievements in solid state theory and the general theory of many particles (outside of plasma physics).*" Indeed, the Vlasov model for a solid was used only for ideal crystals. Hitherto, this model for solid state for the description of the real structure of the crystal has not been used.

Only in 2016 in the journal "Physics of the Solid State," we have considered the complex formation in the semiconductor silicon in accordance with the Vlasov model for solid state [15]. It was demonstrated that the Vlasov model for solid state can be applied not only for the research of the hypothetical ideal crystals but also for the description of the formation of structure of real crystal. In this article, the formation of silicon-carbon and silicon-oxygen complexes during cooling after the growth of dislocation-free silicon single crystals has been calculated using the Vlasov model of crystal formation. It has been confirmed that the complex formation begins in the vicinity of the crystallization front. The characteristic numbers of complexes (silicon-oxygen and silicon-carbon) were determined.

2. Mathematical model

Vlasov created a physical and mathematical apparatus of many-particle system, developed and introduced the concept of collective variables into circulation. The primary concept of Vlasov physics is the distribution function. The particle is characterized by spectrum of geometric and kinematic properties. Thus, the category of motion is included on the same level with the primacy of categories of space and time. Therefore, in Vlasov model for solid state, the periodic distribution of probability density of the particles is the state (motion) of the particles of the system, not the design.

The Vlasov model for solid state is based on the following fundamental physical positions [6, 10]: (1) rejection of the principle of spatial and speed localization of the particles (in terms of classical mechanics), which takes place regardless of the force interactions; (2) introduction of force interactions in analogy with classical mechanics, but taking into account the new principle of non-localization of particles; (3) the behavior of each particle of the system is described by means of an extended phase space f-function. In this approach, the ideas of continuity and of corpuscular are combined. The method of describing the motion of particle associated with the extended function, and the particle in the form of a point occurs only in the particular case [6].

In general, the equation of Vlasov describes the evolution of the function of distribution $f(x, v, t)$ of the continuum of interacting particles in Euclidean space for speed v and for coordinate x at time point t. The equation has the form

$$\frac{\partial f}{\partial t} + \left(\frac{\partial f}{\partial x}, v\right) + \left(\frac{\partial f}{\partial v}, F\right) = 0 \tag{1}$$

$$F = -\frac{\partial}{\partial x}\int K(x, y)f(y, v, t)dvdy$$

where K is the pair interaction potential, which is in real problem depends on the distance $|x-y|$; F is the total force, with which all the particles act on one of them, which is located at time t at point x [8]. In order for distinguishing between the types of interactions, the systems of equations of Vlasov (equations of the Vlasov-Poisson, Vlasov-Maxwell, Vlasov-Einstein, and Vlasov-Yang-Mills [16]) are usually discussed about.

For a description of stationary properties of the crystal, the concept of distribution density of particles $\rho(r)=\int f(r,v)dv$ is used. The molecular field is determined only by the probable locations of atoms that are described by the potential function. The potential function contains the density of probability of the particles considering the temperature distribution of the particles [6]. The choice of potential of the pair interaction depends on the problem under consideration. Nonlocal model of a crystal is described by nonlinear equations. These equations make it possible to calculate the molecular potential and density of location of particles in the conditions of temperature equilibrium [6]:

$$V(r)=\lambda kT\int_{-\infty}^{\infty}K_{1,2}(r)\exp\left(-\frac{K_{1,2}(r)}{kT}\right)dr \tag{2}$$

$$\rho(r)=\lambda kT\exp\left(-\frac{K_{1,2}(r)}{kT}\right)$$

where k is the Boltzmann constant, $K_{1,2}$ is the potential of the pair interaction, λ is some characteristic number, and T is the temperature. The initial equations are equations for the two particles in a stationary condition $\left(\frac{\partial}{\partial t}=0\right)$ [6]. Under the characteristic number where the values of a parameter λ are to be understood, Eq. (2) has solutions different from the trivial [6]. If the position of one of the particles is taken as the origin of coordinates, it is possible to determine $\rho(0)=\lambda kT$ [6]. The important task of Vlasov model for the solid state is the determination of characteristic numbers.

The characteristic number λ is determined from the basic criterion of the existence of the crystal state. The conditions of crystallization can be written as follows:

$$\frac{4\pi N}{kT_m}\int_{0}^{\infty}K_{1,2}^{*}(\rho)\rho^2d\rho=1 \tag{3}$$

where N is the number of particles, T_m is the temperature of melting (crystallization) of crystal, $K_{1,2}^{*}=-K_{1,2}$ [6].

Eq. (2) is written for the conditions of thermal equilibrium of system. The minima of interatomic potential correspond to the stable equilibrium position of atoms in the complexes (silicon-oxygen and silicon-carbon). Then, we can determine the density of the distribution of complexes as a function of the cooling temperature of the crystal

$$\rho(T) = \lambda kTexp\left(-\frac{V_{1min,\,2min}}{kT}\right) \tag{4}$$

The formation of new phase nuclei takes place near the crystallization front of the crystal [2]. In article [2], the model of high-temperature precipitation of impurities is described. At the second stage of the precipitation process, clusters grow without a change in their number. Assuming that the precipitates have a spherical shape, it is possible to calculate the average radius of the precipitate at the growth stage:

$$R(t) = \sqrt[3]{\frac{3bi(t)}{4\pi}} \tag{5}$$

where b is the quantity of order of the distance between the particles in the cluster; $i(t) = [N(0) - N(t)]/N_c$ is the average number of particles at the nucleation centers; $N(0)$ is the monomer concentration at the initial instant of time; $N(t)$ is the changes in the concentration of monomers over time; N_c is the concentration of nucleation centers; and t is the time [17].

At the third stage, when the new phase particles are large and supersaturation is low, new particles are not formed and the main role is played by coalescence, which is accompanied by the dissolution of small-sized particles and the growth of large-sized particles. Under condition $R(t)/R_{cr}(t) \approx 1$ (where $R_{cr}(t)$ is the critical radius of the precipitate), the precipitate is in equilibrium with the solution ($dR/dt = 0$). The precipitate grows at $R(t) > R_{cr}(t)$ and dissolves at $R(t) < R_{cr}(t)$ [17, 18].

It is possible to determine the critical size of the precipitates in accordance with [19]

$$\begin{aligned} R_{cr}^{O}(t) &= \frac{2\sigma uV_p}{kT(t)\ln(S_0 S_i^{-\gamma_i} S_v^{\gamma_v}) - 6\mu\delta\varepsilon uV_p} \\ R_{cr}^{C}(t) &= \frac{2\sigma uV_p}{kT(t)\ln(S_c S_i^{\gamma_i} S_v^{-\gamma_v}) - 6\mu\delta\varepsilon uV_p} \end{aligned} \tag{6}$$

where $S_o = \frac{C_o}{C_o^{eg}}$, $S_c = \frac{C_c}{C_c^{eg}}$, $S_i = \frac{C_i}{C_i^{eg}}$, $S_v = \frac{C_v}{C_v^{eg}}$ are the supersaturations of the oxygen atoms, carbon atoms, intrinsic interstitial silicon atoms, and vacancies, respectively; σ is the density of the surface energy of the interface between the precipitate and the matrix; μ is the shear modulus of silicon; δ and ε are the linear and volume misfit strains of the precipitate and the matrix, respectively; γ_i and γ_v are the fractions of intrinsic interstitial silicon atoms and vacancies per impurity atom attached to the precipitate, respectively; V_p is the molecular volume of the precipitate; and $u = (1 + \gamma_i x + \gamma_v x)^{-1} \cdot (\frac{1+\varepsilon}{1+\delta})^3$.

The average size of precipitates at the stage of the coalescence is proportional to the cube root of time [17]:

$$R_{av}(t) = \sqrt[3]{R_{cr}^3(t_0) + \frac{4D\beta t}{9}} \quad (7)$$

where D is the diffusion coefficient of impurity atoms; $\beta = \left(\frac{\sigma\Omega}{kT}\right)N(0)$; $R_{cr}(t_0)$ is the initial critical radius; σ is the surface tension at the precipitate-solid solution interface; and Ω is the atomic volume.

As is known, the formation and development of the structure of grown-in microdefects in silicon are determined by thermal conditions of growth and cooling of the crystal [1, 3]. The temperature distribution along the length of the ingot during its cooling varies depending on the thermal parameters of the growth according to the expression $1/T = 1/T_m + Gz/T_m^2$, where z is the distance from the crystallization front; G is the axial temperature gradient at the crystallization front; and T_m is the crystallization temperature [3]. It should be noted that, in the general case, it is necessary to take into account the radial inhomogeneity of the temperature field. We introduce the crystal growth rate (V) into this formula and obtain

$$T(t) = \frac{T_m^2}{T_m + VGt} \quad (8)$$

3. Results and discussion

In the paper [15], the interactions between the atoms of the substances and, consequently, their properties are determined on the basis of information about the potential of interatomic interaction. The exact form of the interaction potential of two atoms is determined from the quantum-mechanical calculations. The obtained potentials are usually described by the functions with a large number of parameters, which complicates their analytical treatment. Therefore, it is usual practice to operate model potentials that contain a small number of parameters.

An unambiguous determination of the interatomic potential for silicon crystals is complicated. Therefore, theoretical predictions within their formalism have used different fitting parameters whose behavior under various temperature conditions is difficult to predict or justify [15]. In the paper [15], characteristic numbers were determined from the basic criterion of the crystalline state (3) with the help of a model potential Mee-Lennard-Jones.

In this paper, the calculations were performed using the following parameters: $U_{1min} = 2.84$ eV (SiO_2), $U_{2min} = 2.71$ eV (SiC), $\lambda_1 = 4.482 \times 10^8 eV^{-1}$($SiO_2$), $\lambda_2 = 1.099 \times 10^9 eV^{-1}$(SiC); а также $V_p = 4.302\times10^{-2}$ nm^3 (SiO_2), $V_p = 2.04\times10^{-2}$ nm^3 (SiC), $\sigma = 310$ erg/cm^2 (SiO_2), $\sigma = 1000$ erg/cm^2 (SiC), $\mu = 6.41\times10^{10}$ Pa, $\delta = 0.3$, $\varepsilon = 0.15$, $D_O = 0.17 \exp(-2.54 eV/kT)$, $D_c = 1.9 \exp(-3.1 eV/kT)$, $b = 0.25$ nm, $k = 8.6153 \times 10^{-5}$eV/K, $N_c = 10^{12}$ cm^{-3}.

We performed three separate groups of calculations that simulated the processes of precipitation during the growth of crystals of large and small diameters with the use of the Czochralski method (CZ-Si) and crucibleless floating zone melting (FZ-Si). Calculations of the first group (I) were performed using the following parameters: the crystal growth rate was $V = 0.6$ mm/min, the axial temperature gradient was $G = 25$ K/cm, the oxygen concentration was $N(0) = 10^{18}$ cm^{-3}, and the carbon concentration was $N(0) = 10^{18}$cm^{-3}. The corresponding parameters used in calculations of the second group (II) were as follows: the crystal growth rate was $V = 0.3$ mm/min, the axial temperature gradient was $G = 25$ K/cm, the oxygen concentration was $N(0) = 10^{18}$ cm^{-3}, and the carbon concentration was $N(0) = 10^{18}$cm^{-3}. These conditions correspond to the growth of large-sized silicon single crystals with the use of the Czochralski method. For calculations of the third group (III), we used the following parameters: the crystal growth rate was $V = 6$ mm/min, the axial temperature gradient was $G = 130$ K/cm, the oxygen concentration was $N(0) = 10^{18}$ cm^{-3}, and the carbon concentration was $N(0) = 10^{18}$cm^{-3}. Group III corresponds to the conditions for the growth of small-sized silicon single crystals with the use of crucibleless floating zone melting.

In this, the computing experiment is assumed that the concentration of nucleation centers for complexes constitutes $\sim 10^{12}$ cm^{-3}. This value corresponds to the experimental data obtained by transmission electron microscopy [1]. The densities of the distribution of complexes of silicon-carbon for crystals I and II groups are shown in **Figure 1**. The densities of the distribution of complexes of silicon-carbon for crystals III are shown in **Figure 2**.

In accordance with the Vlasov model for solid state, the process of precipitation begins in the vicinity of the crystallization front due to the disappearance of excess intrinsic point defects on sinks whose role is played by oxygen and carbon impurities. The results of the calculations performed in groups I, II, and III allow us to compare the processes of precipitation in the CZ-Si and FZ-Si crystals and to analyze them for almost maximum contents of the oxygen and carbon impurities. Changing growth conditions of crystal (in particular, an increase in the growth rate and the axial temperature gradient) results in reduced time of the nucleation stage. Besides, the calculations in group III showed that the change in thermal conditions of the growth leads to a decrease in the average radius of precipitates in the FZ-Si crystals as compared to the CZ-Si crystals at the stage of the growth of precipitates, as well as to the corresponding decrease in the precipitate sizes at the stage of the coalescence.

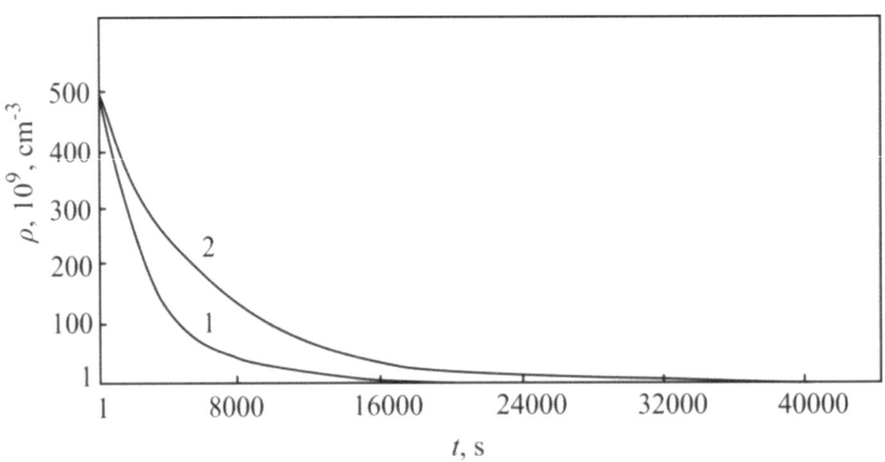

Figure 1. Changing the density of distribution of complexes silicon-carbon for crystals I (1) and II (2) groups.

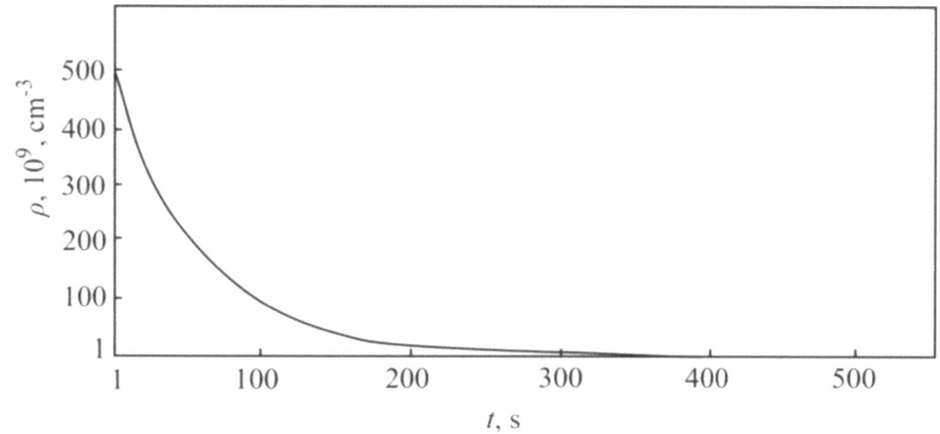

Figure 2. Changing the density of distribution of complexes silicon-carbon for crystals III group.

The condition providing changeover to the stage of the coalescence is written in the form $R(t) \approx R_{cr}(t)$, which is satisfied at the temperature $T \approx 1423$K. The stage of the coalescence in large-sized silicon single crystals begins at temperatures close to the temperatures of the formation of clusters of intrinsic point defects [17]. **Figure 3** shows the variation of the average size of oxygen and carbon precipitates at the stage of the coalescence in the temperature range of cooling from 1423 to 300 K.

Changing of growth conditions for small-sized FZ-Si single crystals (high growth rates and axial temperature gradients) gives what the stage of the coalescence begins far in advance (at $T \approx T_m - 20$K).

In the classical theory of the formation of second-phase particles during heat treatments of single crystals, the formation of critical nuclei of small sizes occurs at low temperatures. In this case, an increase in the temperature leads to the growth and coalescence of precipitates. In real silicon single crystals, there are grown-in microdefects that, during heat treatments, serve as sinks for intrinsic point defects and impurity atoms. Therefore, the growth of precipitates formed during the growth of single crystals can occur simultaneously with the formation of new precipitates.

However, the application of the Vlasov model for solid state to the processes of formation complexes during heat treatment of silicon crystals gives different result (**Figure 4**).

In accordance with the Vlasov model for solid state, the formation of complexes at temperature $T \approx 723$K is the unlikely. This temperature corresponds to the average temperature of the formation of thermal donors. The formation of complexes in silicon during heat treatment is possible only at high temperatures. Excess intrinsic point defects and impurity atoms disappear at sinks whose role in this process is played by initial structural defects. This process leads to the growth and transformation of the initial defect structure of crystal.

At the same time, it is known that the thermal treatments of silicon lead to the formation of electrically active centers. If the formation of complexes in the temperature range $300\ \text{K} \leq T \leq 600\ \text{K}$

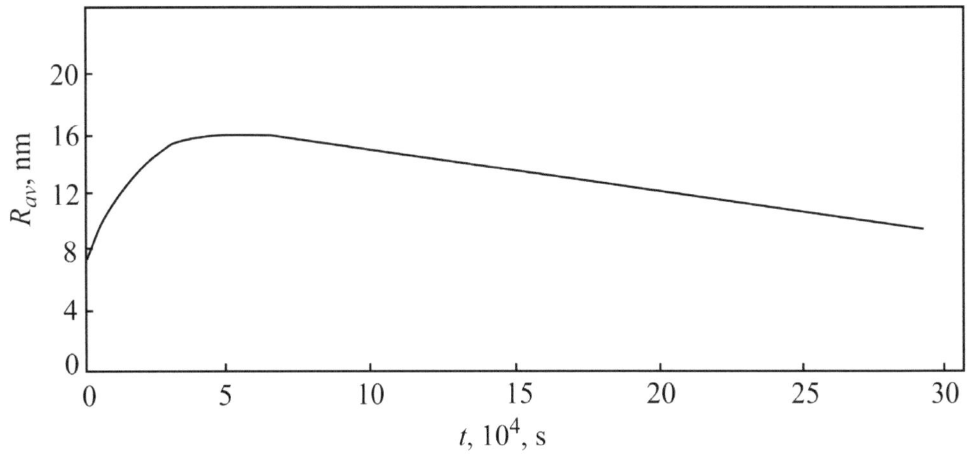

Figure 3. Variation of the average size R_av of (1) oxygen precipitates at the stage of the coalescence during cooling of the CZ-Si crystals in the temperature range from 1423 to 300 K.

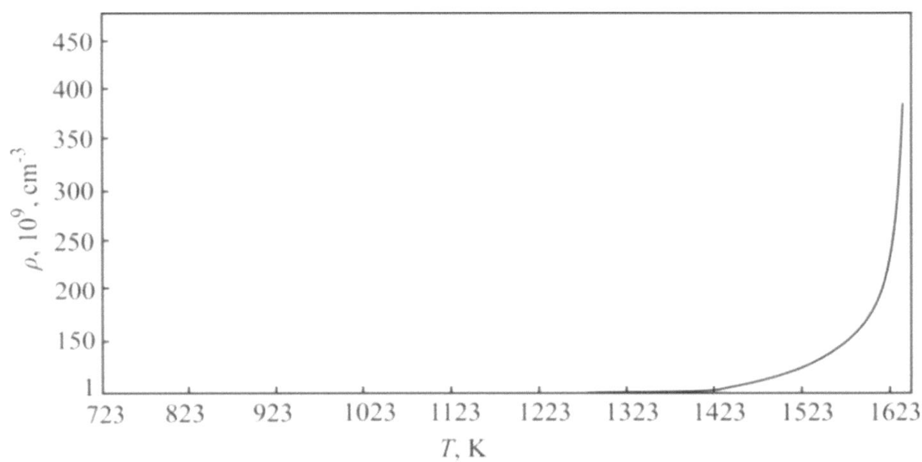

Figure 4. Dependence of the density of distribution for the silicon-oxygen complexes from the temperature during the annealing of the crystal.

is unlikely, then the processes of coalescence occur in crystals of any diameter. Then, in accordance with the Vlasov model for solid state, it can be assumed that the grown-in precipitates, which dissolve, lead to the formation of electrically active centers.

In conclusion, we note that two theories of nucleation second-phase particles, which are based on different gnoseological approaches (classical theory of nucleation and growth of second-phase particles in crystals and the Vlasov model for solid state), lead to identical results. Both approaches describe the processes of high-temperature precipitation of impurities. In turn, the high-temperature precipitation of impurities is the basis of the process of defect formation in crystals.

4. Conclusions

The main task of the solid body is a description of the physical properties of solids in terms of their atomic structure. Despite the successful development of the classical approach in the last 100 years, the theoretical description of the process of defect formation in crystals has not received a satisfactory solution.

Until now, the probabilistic approach (Vlasov model for the solid state) for real crystals has not been applied. At the same time, joint use of both approaches would allow for a fresh look at the already-known facts and discover new phenomena and laws in the study of real solids. This work represents the first attempts in the last 70 years to obtain the solution of Vlasov model for solid state. Questions of the application of the Vlasov model for solid state for the description of the nucleation of defect structure, taking into account the thermal conditions of the crystal growth, were discussed. There are the following main results:

1. The concepts and principles of Vlasov physics are fully applicable to solid state.
2. The Vlasov model for solid state describes the processes of complexation during the growth of real crystals adequately classical theory of nucleation and growth of second-phase particles in solids.
3. The method of calculation of the initial defect structure of crystals was proposed. This method includes the Vlasov model for solid state and the classical theory of nucleation and growth of second-phase particles in solids.
4. With the help of the Vlasov model for solid state, it was shown that during low-temperature treatments of crystals, the complexation is unlikely. In this case, we have the processes of coalescence of grown-in defects.

Author details

Vitalyi Igorevich Talanin*, Igor Evgenievich Talanin and Vladislav Igorevich Lashko

*Address all correspondence to: v.i.talanin@mail.ru

Department of Computer Science & Software Engineering, Institute of Economics & Information Technologies, Zaporozhye, Ukraine

References

[1] Talanin VI, Talanin IE. Formation of grown-in microdefects in dislocation-free silicon monocrystals. In: Elliot TB, editor. New Research on Semiconductors. New York, USA: Nova Science; 2006. pp. 31–67.

[2] Talanin VI, Talanin IE. Kinetics of high-temperature precipitation in dislocation-free silicon single crystals. Physics of the Solid State. 2010;**52**:2063–2069.

[3] Talanin VI, Talanin IE. Diffusion model of the formation of growth microdefects: A new approach to defect formation in crystals (review). Physics of the Solid State. 2016;**58**:427–437.

[4] Born M, Huang K. Dynamical Theory of Crystal Lattices. Oxford, UK: Clarendon Press; 1954.

[5] Cristian JW. The Theory of Transformations in Metals and Alloys. Oxford, UK: Pergamon Press; 1965.

[6] Vlasov AA. On the theory of the solid state. Journal of Physics. 1945;**9:**130–139 (in USSR)

[7] Vlasov AA. Thread-like and plate-like structures in crystals and liquids. Theoretical and Mathematical Physics. 1970;**5:**1228–1241.

[8] Kozlov VV. The generalized Vlasov kinetic equation. Russian Mathematics Surveys. 2008;**63:**691–726.

[9] Vlasov AA. Many-Particle Theory and its Application to Plasma. New York, USA: Gordon & Breach; 1961.

[10] Vlasov AA. Nonlocal Statistical Mechanics. Moscow, Russia: Nauka; 1978.

[11] Landau LD. On the vibration of the electronic plasma. Journal of Physics-USSR. 1946;**10:**25–34.

[12] Ginzburg VL, Landau LD, Leontovich MA, Fok VA. About inconsistency of works by A. A.Vlasov on general theory of plasma and physics of solid body. Journal of Experimental Theoretical Physics (Russia). 1946;**16:**246–252.

[13] Pegoraro F, Califano F, Manfredi G, Morrison PJ. Theory and applications of the Vlasov equation. European Physics Journal D. 2015;**69,68**:1–3.

[14] Ginzburg VL. About some grief-historians of physics. Questions of History of Science and Technology (Russia). 2000;**4**:5–14.

[15] Talanin VI, Talanin IE. Complex formation in semiconductor silicon within the framework of the Vlasov model of a solid state. Physics of the Solid State. 2016;**58:**2050–2054.

[16] Vedenyapin VV. Boltzmann and Vlasov Kinetic Equations. Moscow, Russia: Fizmatlit; 2001.

[17] Talanin VI, Talanin IE. Kinetic model of growth and coalescence of oxygen and carbon precipitates during cooling of as-grown silicon crystals. Physics of the Solid State. 2011;**53:** 119–126.

[18] Slezov VV, Sagalovich VV. Diffusive decomposition of solid solutions. Uspekhi Fizicheskikh Nauk. 1987;**151**:67–104.

[19] Vanhellemont J. Diffusion limited oxygen precipitation in silicon: Precipitate growth kinetics and phase formation. Journal of Applied Physics. 1995;**78**:4297–4299.

Chapter 2

Theoretical and Experimental Characterization of Silicon Nanoclusters Embedded in Silicon-Rich Oxide Films

Néstor David Espinosa Torres,
José Álvaro David Hernández de la Luz and
Javier Martínez Juárez

Additional information is available at the end of the chapter

http://dx.doi.org/10.5772/67614

Abstract

We present theoretical calculations using DFT method and the Global Reaction Model (GRM) for molecular structures and photoluminescence (PL) and Fourier Transform Infrared (FTIR) spectroscopy for silicon nanoclusters (Si-NCs) embedded in silicon rich oxide (SRO) films. Correlations between theoretical predictions and experimental results are made taking as reference experimental results obtained from measurements performed on SRO thin films obtained by the Hot Filament Chemical Vapor Deposition (HFCVD) technique. Our theoretical predictions are general since they do not depend on the particular technique used to obtain such films but rather the suggested SinOn structures. A good correlation exists for Eg values for films grown at 1300°C corresponding to Si_8O_8 and $Si_{16}O_{16}$ molecular structures suggested and for films grown at 1150°C with Si_9O_9. Regards PL correlation, a film grown at 900°C gives a spectrum peaked at 440nm and 548nm while theoretical one shows peaks at 471nm and 549.8 nm for a structure $Si_{16}O_{16}$. Such sample with a further annealing displays peaks at 405nm, 749nm and 820nm with theoretical predictions at 415nm using Si_6O_6. As for FTIR, theoretical calculations predict vibrational mode frequencies of bonds Si-O and Si-H whose values are well located in the experimental frequency range corresponding to the structure $Si_{16}O_{16}$.

Keywords: silicon-rich oxide, luminescence, DFT, GRM

1. Introduction

It is well known that the crystalline silicon has no photoluminescence due to multiple phenomena of non-radiative recombination between electrons and holes, and that the SiO_2 presents photoluminescence both by its amorphous nature and by their large gap. The latter has been taken as a starting point to consider silicon-confined systems of great interest because they offer the possibility of light emission from silicon-based materials. Following the initial report of the light emission from porous silicon, announced by Canham, this has been a novel subject of intense scientific activity currently along with other confined systems. Particularly, from the latter, it results in emphasizing the importance of the silicon rich oxides thin films. The physical microscopic structure of SiO_x is still a subject of discussion nowadays, where its structural arrangement is a key knowledge to get a deep understanding of radiative emission mechanisms in this kind of nanostructured materials.

The silicon rich oxide (SRO) is a silicon oxide with silicon excess or at amount of oxygen less than that of a silicon dioxide (SiO_2), so that the best appropriate notation would be SiO_x, with $x = 0,\ldots,2$. This material, after heat treatment, has a phase separation being composed of silicon nanocrystals (Si-NCs) surrounded by silicon oxide (Si-NCs/(SiO_x), regardless of the technique employed to obtain them (for instance, LPCVD, PECVD, etc.) [1]. The density and size of the Si-NCs have a strong dependence on the annealing temperature, the concentration of silicon, the characteristics of the silicon substrate, the partial pressure ratios between the reactant gases, the used production technique and the surrounding atmosphere during the annealing. It is remarkable to point out that the presence of phase separation (Si-NCs/(SiO_x), with a fixed value of x, can produce Si-NCs embedded in SiO_x with different values of x respect to SiO_2 and, consequently, multiple defects appeared at the borders of the Si-NCs and also within them and probably porous silicon may be present there. The defects generated during the production of the material disappear with the annealing process, particularly, at temperatures greater than 1000°C. It has been reported that for annealing temperatures less than 1100°C, the generation of amorphous silicon clusters is favoured [2], while for higher temperatures, the number of interfacial states is increased between Si-NCs and SiO_x, therefore the number of defects is also increased. A great field of active investigation has been opened in order to understand the underlying physical mechanisms of the SRO, which give rise to its crystalline structure and optical properties because it is a material, which promises important applications mostly in optoelectronic devices.

In the study of the SRO, one branch of research is focused on investigating the main mechanisms, which generate the luminescence phenomena in this material. In the case of structures such as SRO thin films, we found in the literature different approaches. So, it has been proposed several mechanisms involved in the luminescent emission observed, these include among others: quantum confinement of excitons, luminescence due to chemical species (such as siloxanes, silicon oxides and sub-oxides), interfacial states, defects and strain-related luminescence. Although there is a growing consensus that the quantum confinement effects (QCE) may explain some of the features of the two luminescence spectra [3], it is clear that in some cases, at least, other mechanisms are present. Up to now, there are no models that include several of these emission mechanisms.

In the proposal of quantum confinement as the dominant mechanism in luminescent effect, Qin and Li [4] proposed a limit on the size of the Si-NCs, considered as a critical size, above which dominates the quantum confinement effect while below it the effect of interfacial states prevails above all. With respect to photoluminescence in SRO including Si-NCs, it has been reported that it is dominated by strong quantum confinement above from a critical size of Si-NCs which may be about 1.1 nm in diameter, it corresponds to 10 times the Bohr radius of silicon, below this critical size, the photoluminescence may be dominated by the weak quantum well confinement as the interfacial defects.

Moreover, luminescent phenomenon in SRO structures can be excited by different forms, so we find that experimentally, luminescence spectra are generated by photoluminescence (PL), electroluminescence (EL) and cathode luminescence (CL) mechanisms. The fundamental physical mechanisms, which explain correctly the origin of luminescent phenomena in each one of these different forms of excitation, are still an active field of research.

Today, a few models are frequently used to describe a SRO network, namely the Mixture Model (MM) by Bell and Ley [5], the Random Bonding Model (RBM) by Philipp [6], and the Intermediate Model (IM) introduced in 2011 by Novikov and Gritsenko [7]. In 2012, Davor et al. [8], in an extensive review, considered that the actual structure seems to be greatly determined by the deposition procedure. In some works, the SiO_x structures films obtained by radiofrequency sputtering and physical evaporation were claimed to correspond to RBM, whereas the SiO_x films obtained by magneto-sputtering, Plasma-Enhanced Chemical Vapor Deposition (PECVD), have been assigned to MM. IM was used to describe SiO_x layers prepared by low pressure chemical vapour deposition (LPCVD) technique using SiH_4 and N_2O as a reactant precursors at 750°C. Finally, we find a new model, the Global Reaction Model (GRM) suggested lately by Espinosa-Torres et al. [9]. This model describes, first, the global and partial reaction(s) compulsory to produce the oxide matrices (SiO_2, Si_2O_3, SiO and Si_2O), second, the annealing reactions for elucidating the compositional changes after and before the thermal treatment, and subsequently, the variations in the intensity of luminescence spectra, and also it describes a set of secondary reactions of the oxide matrices with the hydrogen produced in the reaction chamber to obtain the charged species that could be associated to the emission in SRO thin films with explicit defects. This work is focused on studying the theoretical predictions about PL luminescence of SRO thin films based on DFT method and the GRM and make correlations with experimental PL results provided by measurements from SRO thin films obtained experimentally by the hot filament chemical vapour deposition (HFCVD) technique.

2. Relevant characteristics of silicon-rich oxide (SRO)

2.1. Some techniques of production of SRO

Among the different techniques used for depositing thin films, we can mention those which are commonly employed to grow SRO as a nanostructured material. One such technique is called hot filament chemical vapour deposition (HFCVD). This technique is known by different names like initiated-CVD (I-CVD), catalytic-CVD (Cat-CVD) and hot wire-CVD

(HWCVD). Amorphous and micro-crystalline silicon could also be obtained with this method [10]. The HFCVD technique gives some important advantages, it does not need a sophisticate temperature control and vacuum system, in addition, the deposition is free of plasma and we can obtain the precursors by a solid source or mixtures of gases. HFCVD system utilizes quartz-like solid source and makes use of the collision theory in order to describe the SRO deposition process [11]. The solid source to obtain SRO is made of Quartz, and it is etched by H^0.

Another technique is that known as low pressure chemical vapour deposition (LPCVD), it allows us to get silicon-rich oxide layers using oxide species like nitrous oxide (N_2O) and silicon compounds (silane, SiH_4) as reactants gasses. Silicon excess is easily controlled by changing the partial pressure ratio R_o between N_2O and SiH_4 defined by $R_o = P(N_2O)/P(SiH_4)$ or simply $R_o = N_2O/SiH_4$.

When we make a SRO layer, first, the partial pressures must be calculated by choosing a suitable R_o and P_T value. The suitable pressure for each reactant gas is calculated by means of relations,

$$P_{SiH_4} = \frac{P_T}{\frac{R_o}{20}+1} \ ; \ P_{N_2O} = \frac{R_o P_{SiH_4}}{20} \tag{1}$$

where P_{SiH_4} is the partial pressure of diluted silane at 5% (by this reason, term 20 appears) in nitrogen, P_{N_2O} is the nitrous oxide partial pressure and P is the total pressure. Si excess in the deposited films is controlled by the partial pressure ratio R_o of the reactants gases.

The optical characteristics of SRO, obtained by this technique, can be varied with the excess silicon in the films, making SRO attractive to fabricate optoelectronic devices. Therefore, Si-NCs embedded in a SiO_2 matrix are currently attracting much interest to be a good candidate for building efficient light-emitter devices [12–14]. Such devices are highly desirable for the integration of optical signal and electronic data processing circuits on the same chip. Furthermore, its fabrication process is compatible with the present large-scale integration technologies [15, 16]. The native band gap of Si-NCs is enlarged with respect to one of the bulk material; this fact brings the possibility to observe an intense visible PL spectrum at room temperature. The PL spectrum consists generally of intense emission peaks in the near-infrared (NIR) and visible (VIS) regions. It was established that blue and green PLs are caused by various emitting centres in the silicon oxide, while the nature of the intense PL in orange-red region is still discussed [17–19].

Experimentalists have reported in the literature that emission spectra of the SRO films vary in the interval of wavelengths from 400 to 850 nm for layers deposited by LPCVD. It has not been observed experimental emission outside of this range for samples obtained by this technique. Nevertheless, by preliminary theoretical studies, using the density functional theory (DFT), it has been found as isolated molecules, presumably found in SRO thin films that could emit in UV region.

In addition, for SRO thin films deposited by LPCVD, PL is only observed in annealed samples. In fact, only thin films annealed at 1100°C produce high emission, and the highest photoemission is obtained for SRO films with $R_o = 30$, which corresponds to a silicon excess about 3%, and R_o = 10 corresponds with a silicon excess about 12%. The PL is only obtained from the visible (VIS) to NIR range, and its intensity reduces as R_o decreases. In contrast, we find that powders and thin films SRO deposited by other technique such as HFCVD technique, display a reduction in photoluminescence intensity in most cases after being annealed, and only in very particular conditions (deposited parameters) the PL is increased.

By using the LPCVD technique, the SRO refractive index can be changed, it has been reported that the index varies from 1.45 to 1.94 when the silicon excess is changed. It is observed that the refractive index of the SRO on silicon substrate increases with the increment of Si excess, it means that as R_o augments, the index tends towards the refractive one of SiO_2, contrarily, it tends towards that of Si. Then, by measuring the refractive index, the inclusion of Si in these SRO films is easy to detect. For example, it has been found that for R_o = 10, after thermal annealing, the refractive index reaches values greater than 2. In contrast, when the reactant ratios are R_o = 20, 30 and 40, the refractive indexes diminish with a clear tendency to reach values less than 1.5 [20].

Relating to transmittance of SRO obtained by LPCVD (an important optical parameter measured in experiments), we find that the transmittance spectra for R_o = 10, 20, 30 films deposited on sapphire substrates are reported in Ref. [2]. According to this reference, the observed transmittance of these films deposited on sapphire is high (>80%) and vanishes to zero when wavelength reduces. The annealing time parameter does not produce a clear variation on transmittance; however, when R_o parameter undergoes increments it causes a clear shift of the curves towards lower wavelength. The absorption coefficients are determined from transmission spectra, and it can be observed that the position of the absorption edge moves towards higher energy when R_o increases.

Other important technique is the sputtering one, where sputtering means an ionic bombardment process in vapour phase at room temperature. By this technique, plasma is formed by the ionized process gas, usually Argon, due to the presence of a strong electric field. The high voltage between cathode and anode causes that ions of the process gas hit the target with enough energy to pull out atoms from the surface of the target (cathode) by a momentum transfer process. The multiple collisions enable some atoms of the material acquire enough energy to leave the surface, then reach the substrate and stick to it. Sputtering technique can be divided into reactive and non-reactive. It is denominated no reactive sputtering to one in which the process gas does not react with the deposited material. Typically, Argon is used as inert gas due to the high yield obtained and its minor cost. When sputtering is made in the presence of a reactive gas as oxygen, the sputtering is called reactive. In this case, the presence of ionized oxygen causes oxidation of the material, thus obtaining films whose properties depend on the concentration of reactive gas in the chamber during the process.

In the sputtering technique, the growth model of SRO films is based in the work proposed by Nyberg and Berg [21], where they describe the reactive sputtering processing behaviour in many different ways to carry out. This model has been specially adapted to the growth of SiO_x films where the growth kinetics is dependent on the flux of oxygen onto the substrate surface relative to the flux of Si incorporated into the film.

As to the optical properties of the Si/SO_x, they differ due to the synthesis technique, indicating structural differences of the Si-NCs (as silicon nanocrystals, amorphous silicon and defects) and their environment [22]. Raman scattering measurements are frequently used to prove the presence of Si-NCs. With this fast and non-destructive method, it is possible to determine whether silicon particles are amorphous or crystalline. Moreover, information about mean size (MS) and sizes distribution (SD) of nanocrystallites can be obtained from the peak position and shape of the Raman band. For samples using Si wafers as substrates, Raman measurements must be undertaken with care because the strong Raman peak from the crystalline Si (c-Si) substrate masks the signal of the Si-NCs. To avoid this problem, sapphire substrates should be used. Then, the obtained Raman data can clearly be related to the deposited thin film, not to the substrate.

2.2. Morphological aspects of SRO films

Figure 1(a) shows SEM micrograph of SRO film deposited by HFCVD at 409°C. The characteristics of the films depend on the deposition conditions such as substrate temperature, deposition time, solid sources and kind of substrate. So there are several factors that can change the morphology of the films. In this figure, we can see agglomerate structures, which form isomers Si_7 with sphere shape; this geometric shape is a consequence of the elevated temperature. Also, we can appreciate that there are interspaces among agglomerates, so it leads to the formation of a porous film.

On the other hand, silicon-rich oxide powders (SROP) agglomerates and frost can be seen more clearly in SEM Micrographs shown in **Figure 1(b)** and **(c)**, where we can see that the big silicon rich oxide powders (SROPs) are formed by the agglomeration of small nanoparticles. The excess of precursors in gas phase is called supersaturation, and it is present in all SROP samples. In HFCVD technique, the filament-source distance controls the supersaturation. Therefore, the morphology is controlled by the saturation and growth temperature. This process can be described in the following manner: the precursor reacts in gas phase, and it creates clusters in the environment. Some clusters could be composed by silicon, silicon-hydrogen, silicon monoxide and silica as powder. The temperature between filament and source is very high, but it decreases between source and substrate due to a temperature gradient within the filament. When the distance of source-substrate increases, the growth temperature decreases, and with this conditions, the environment is supersaturated and it is partially condensed.

Relevant differences in SEM images shown in **Figure 1(a)–(c)**, are originated owing to the deposition temperatures, which were 409, 368 and 320°C, respectively. Grains formed at the highest temperature are bigger due to the matrix SiO_2 that embedded silicon agglomerates or nanocrystals. Presence of silicon agglomerates was corroborated using micro Raman. Images of **Figure 1(a)** and **(b)** were taken with the secondary electron detector for high-resolution images. This detector uses a lower voltage, which favours the superficial resolution of images. In **Figure 1(b)** is observed a porous surface, probably formed by smaller isomers than those formed at 409°C, where their size is considerably greater than 8 μm.

Finally, **Figure 2** displays results of band gap values obtained by applying the Kubelka-Munk method to SiO_x powders deposited by HFCVD at the temperatures of 300, 320 and 539°C and filament-source distances of 2, 9 and 10 mm whose values are 3.42, 3.63 and 3.72 eV, respectively [23].

http://dx.doi.org/10.5772/67614

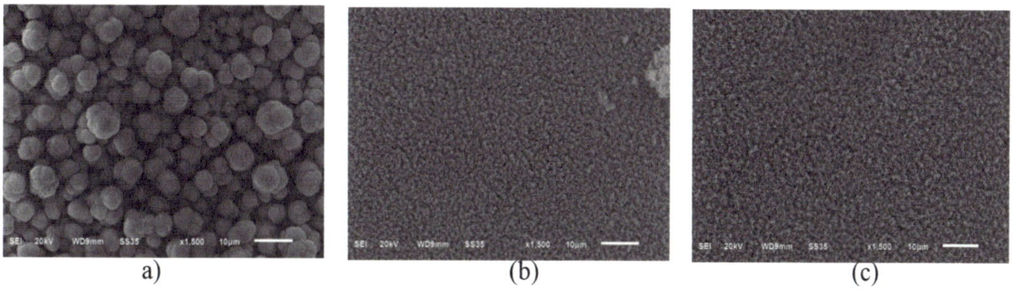

Figure 1. SEM micrograph of a SRO film deposited by HFCVD (a) at 409°C, (b) at 368°C and (c) at 320°C.

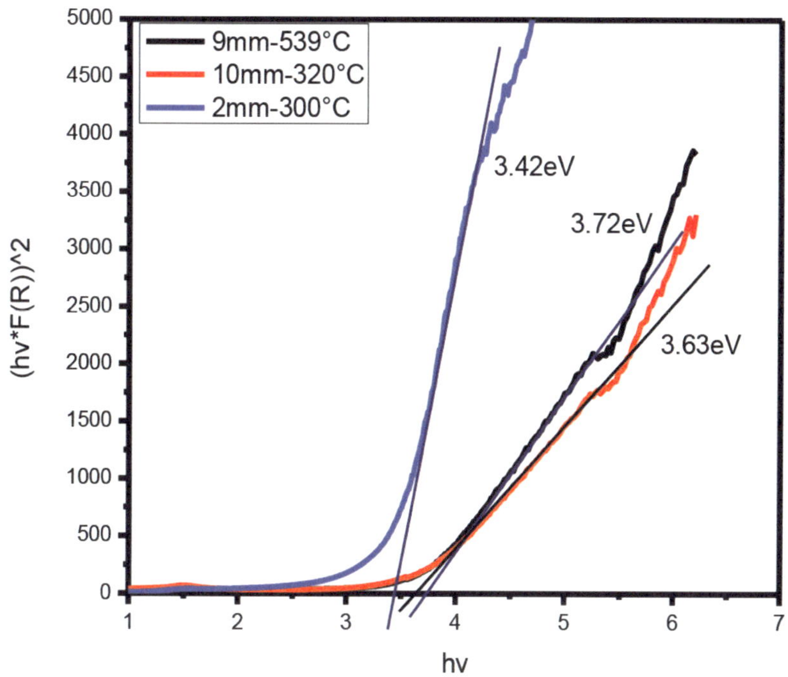

Figure 2. Band Gap values of SRO films measured by Kubelka–Munk's method.

3. Correlation between experimental results and theoretical predictions of the Global Reaction Model (GRM)

3.1. Global Reaction Model

In this section, we present for first time a new model, which considers the global and partial reaction(s) necessary to generate the oxide matrices (SiO_2, Si_2O_3, SiO and Si_2O), the annealing reactions for explaining the compositional changes after and before the thermal treatment and consequently the changes in luminescence spectra intensity and a set of secondary reactions of the oxide matrices with the hydrogen produced to obtain the ions that could be associated to the emission in SRO thin films with specific defects. When SRO is prepared by LPCVD, a gas mixture of N_2O and silane is habitually used [24, 25] and the excess Si content can be modified by the gas flow ratio $R_o = \left[\frac{N_2O}{SiH_4}\right]$. The silicon excess can be as high as 17% for $R_o = 3$; and experimentally stoichiometric SiO_2 (a non-free silicon film) can be obtained for $R_o \geq 50$ [26]. Theoretically, $R_o = 40$ corresponds to the stoichiometric silica, when a mixture silane-nitrogen at 5% is used. Experimentally, there is enough evidence that SRO thin films are constituted by a silicon oxides mixture and not only by one of them, independently of the value of R_o. We can establish the global reactions of the chemical combination of N_2O, N_2 and SiH_4, which generates the theoretical composition of SRO given by

$$\mathrm{SiO_2} = \frac{(1-x)}{(4+z+v)}\ ;\ \mathrm{Si_2O_3} = \frac{(1-v)}{4+z+v}\ ;\ \mathrm{SiO} = \frac{(1-y)}{(4+z+v)} \tag{2}$$

$$\mathrm{Si_2O} = \frac{(1-z)}{(4+z+v)}\ ;\ \mathrm{Si} = \frac{(x+y+2z+2v)}{(4+z+v)} \tag{3}$$

where the parameters x, y, z, v satisfy the conditions $0 \leq x, y, z, v \leq 1.0$ and $0 \leq x+y+2x+v \leq 1.0$.

When SRO thin films are annealed, some oxides are degraded. The plausible 'annealing reactions' proposed are as follows:

$$\mathrm{SiO_{2(solid)}} + \mathrm{Heat} \rightleftarrows (1-\xi)\mathrm{SiO_{(solid)}} + (0.5+0.5\xi)O_{2(gas)} + \xi\,\mathrm{Si_{solid}} \tag{4}$$

$$\mathrm{Si_2O_{3(solid)}} + \mathrm{Heat} \rightleftarrows (1-\xi')\mathrm{Si_2O_{(solid)}} + (2+\xi')\mathrm{O_{2(gas)}} + 2\xi'\,\mathrm{Si_{solid}} \tag{5}$$

Double arrow stands for denoting equilibrium condition, ξ *and* ξ' are the progress of the annealing reactions. The extent of progress of reaction is defined as the ratio between the total change in the number of moles of a species and their stoichiometric coefficients.

As is known, a mole of any substance contains the number of atoms equal to the Avogadro's number, $N_A = 6.023 \times 10^{23}$. In order to perform first-principles calculations, the problem must be modelled with a maximum of 100 atoms, consequently, the oxides must be rewritten in terms of the number of atoms, that is, Si_nO_{2n}, $Si_{2n}O_{3n}$, Si_nO_n and $Si_{2n}O_n$. The hydrogen gases can react accordingly as $H_2 = [H]^+ + [H]^-$.

The hydrogen ion formed, in turn, reacts with the silicon oxides to form ions such as $[Si_nO_{2n}H]^+$, $[Si_nO_{2n}H]^{2+}$ etc. During heat treatment the reactions produce a dehydration of these cations, resulting in the formation of new ones, for example, $[Si_nO_{2n}H_2]^{2+} \Leftrightarrow [Si_nO_{2n-1}]^{2+} + H_2O\uparrow$, and so on. Finally, the moieties obtained from these reactions result as follows:

$$[\mathrm{Si}_n\mathrm{O}_{2n-1}]^{2+}\ ;\ [\mathrm{Si}_{2n}\mathrm{O}_{3n-1}]^{2+}\ ;\ [\mathrm{Si}_n\mathrm{O}_{n-1}]^{2+} \text{ and } [\mathrm{Si}_{2n}\mathrm{O}_{n-1}]^{2+}. \tag{6}$$

These anions formed are different oxides matrices containing vacancies or defects of silicon which may or not be present in the films of SRO.

3.2. Comparison between experimental results and theoretical predictions of GRM.

In this section, we will display and discuss the results employing DFT to evaluate theoretically structures type Si_nO_n, where $5 \leq n \leq 26$ is the number of silicon atoms. The GRM already discussed is employed to explain the physical microscopic structure of SRO thin films regardless of the technique used to obtain the SRO structure since we are only interested in its molecular composition and atomic arrangement. Also, we will show the results of the evaluation of structural and optical properties for a wide set of moieties, seemingly found in SRO. DFT results predict emission in visible region for molecules with a number of atoms less than 14 silicon ones. Results obtained in this research predict luminescence in visible region for just about half calculated structures, especially for $n \leq 16$ silicon atoms, while large structures with $n \geq 17$ display luminescence in ultraviolet region.

3.3. Measurements of band gap E_g for SiO_x thin films

We begin by taking SiO_x films obtained by HFCVD technique from our experimental group (results not even published). From these samples, we take into account the band gap parameter as the first one to be analysed. The approximate experimental values of the energy band gap E_g are obtained by the relationship known as Tauc plot [27] as shown in **Figure 3(a)**. The methodology for obtaining the E_g has been described in a previous work [28]. It is found that E_g decreases when the substrate temperature is raised. The band gap of the SiO_x films lies in the range of 1.8–2.5 eV. When x decreases out from 2.0 in a $-SiO_x$ (amorphous oxide), the valence band edge moves up, as the increased Si—Si bond states are gradually overlapped with the oxygen non-bonding states (ONS), and finally spread out into the Si valence band. Simultaneously, the conduction band edge also moves down, giving rise to the band gap decreasing nonlinearly when the Si concentration is increased continually.

In **Figure 3(a)**, we observe that by making an extrapolation through the straight line according to Tauc procedure, we obtain the approximated bang gaps which have values of 1.95eV and 2.15 eV for **Figure 3(b)**. At first glance, the variations of E_g are originated by the difference in temperatures then it is observed that E_g reduces as temperature increases. These two experimental values of E_g fit well with those obtained theoretically using DFT as shown in **Table 1** further below that are results for structures type Si_nO_n.

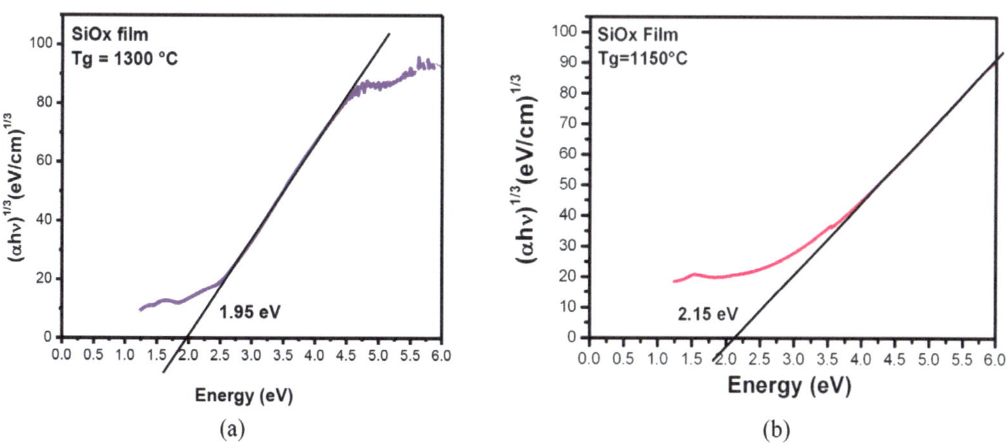

Figure 3. Determination of E_g for a SiO_x film by making use of the relationship known as Tauc plot $[\alpha(h\nu)]^{1/3}$ versus energy ($h\nu$) for two temperatures (a) T_g=1300°C and (b) T_g=1150°C.

Number of silicon atoms	HOMO eV	LUMO eV	GAP eV
5	–6.19	–3.14	3.05
6	–6.01	–3.46	2.55
8	–5.51	–3.58	1.93
9	–5.34	–3.28	2.06
11	–5.93	–3.22	2.71
12	–5.97	–2.96	3.01
13	–5.61	–3.30	2.31
14	–6.04	–2.79	3.25
15	–6.11	–3.39	2.72
16	–6.16	–4.20	1.96

Table 1. Highest Occupied Molecular Orbital (HOMO), *Lowest Unoccupied Molecular Orbital* (LUMO) and HOMO–LUMO gap calculated using DFT for $[Si_nO_n]$ structures with $5 \leq n \leq 16$.

We have seen that SiO_x thin films, deposited at T_g = 1300°C, resulted with a bang gap of 1.95 eV, calculated using Tauc technique.

Now, by observing **Table 1**, we can correlate two possible theoretical values either E_g = 1.93 eV corresponding to Si_8O_8 molecule or E_g = 1.96 eV corresponding to $Si_{16}O_{16}$ one, existing the possibility of a mixture of them. On the other hand, SiO_x film deposited at 1150°C corresponds to an approximated experimental band gap of 2.15 eV, which correlates well with a theoretical value of E_g = 2.06 eV associated to Si_9O_9 molecule in accordance with **Table 1**.

In regard to the origin of the PL emission in Si-NCs, it is still a subject of debate; however, we can find some proposals of models suggested explaining this phenomenon. One of this model relates the PL to quantum confinement effects (QCEs) [29, 30]. The other model relates the PL to defects in the oxide matrix or at the interface SiO_2/Si-NCs [28, 31].

Broadly, both common accepted proposals make use of approximated quantum methods in order to solve the Schrödinger equation associated with quantum confinement of the electron restricted to move in so small spatial dimensions, thence their predictions about luminescent phenomenon are limited. Considering this important fact, we hope that an analysis of this phenomenon made from the view point of composition and molecular structures can significantly contribute to a better knowledge of luminesce in SRO considered as an arrangement of Si-NCs embedded in oxide matrices.

3.4. Thermal effects in PL of the SiO_x thin films

Now, we proceed to our analysis in relationship to PL phenomenon. Inspecting, in **Figure 4(a)**, the experimental PL spectrum measured from the SiO_x film as grown at T_g = 900°C, we locate two emission bands both covering a wide spectral range from 380 to 750 nm (from violet to red).

The left band peaked around 440 nm while the right one peaked around 548 nm as can be confirmed by de-convolution curves. After making a further annealing to the same SiO_x film, the PL spectrum **Figure 5(a)** displays two main bands, one is the A band, which lies in the violet-blue range (380–495 nm) peaked at 405 nm with a relatively weak PL intensity. The other band is the B one, which lies in the orange-near infrared range (590–875 nm), this band shows a strong intensity and two main peaks, one located at 749 nm and other at 820 nm as confirmed again by de-convolution curves.

Looking at **Figures 4(a)** and **5(a)** and making comparisons between them, we infer that the PL intensity that decreases after thermal annealing [32] due to the PL intensity is lower after a further annealing. According to this result, we conclude that annealing process stimulates the formation of crystalline silicon (c-Si) as well as the formation of defects both contributing to the PL emission. For this reason the PL spectrum shifts to the red region.

With regard to the theoretical PL spectrum, **Figure 4(b)** exhibits this one, for the SiO_x film as grown at T_g = 900°C, that was calculated considering the arrangement of a $Si_{16}O_{16}$ molecule. With this molecular structure, we obtained the PL spectrum which has two remarkable peaks, one of them located at 471 nm with the highest intensity and the other has two located peaks closer at 549.8 and 556 nm. In addition, an excited state at 683 nm was also found, but its intensity was too negligible or non-detectable.

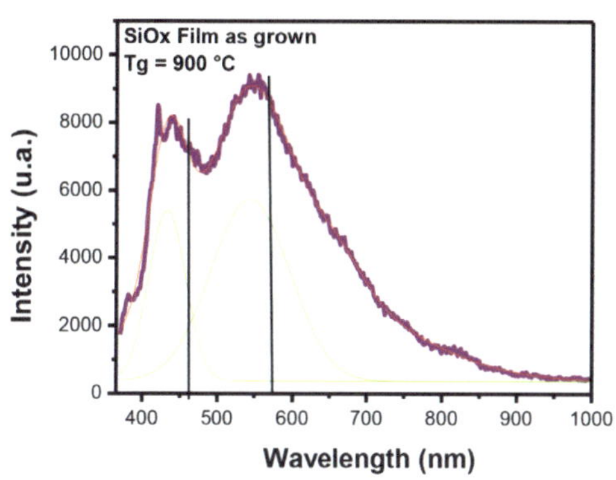

a

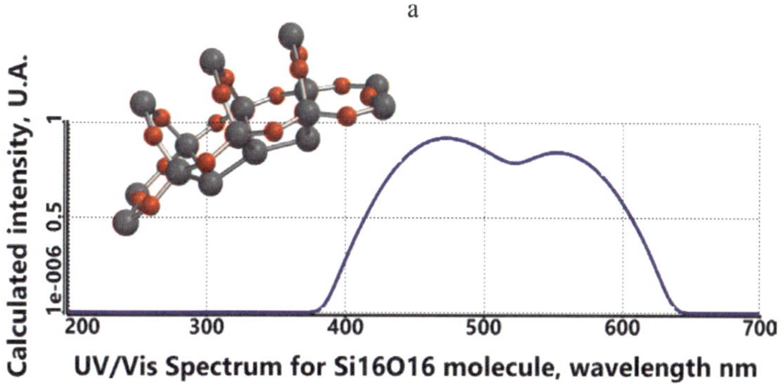

b

Figure 4. (a) Experimental PL spectrum from a SiO_x film grown at 900°C. (b) Theoretical PL spectrum calculated for $Si_{16}O_{16}$ structure.

To complement the study of the PL spectrum including annealing effects, in **Figure 5(b)** we display results for the theoretical PL spectrum calculated by considering a small molecule type Si_6O_6. With this molecular structure we predicted excited states emitting in the region from 393.78 to 415.59 nm in addition to a region peaked at 772.8 nm, which corresponds approximately to the left side of the wide shoulder observed in the experimental spectrum displayed in **Figure 8** where the de-convolution curve points out a peak at 749 nm. We emphasize that in this theoretical spectrum the presence of the peak at 820 nm found in the experimental PL spectrum is not clear.

In relation to the study of the configuration of the molecular structure, we present in **Figure 6(a)** the two structures corresponding to the case of the SiO_x film as grown at T_g = 900°C and after a further annealing **Figure 6(b)**, both structures correspond to

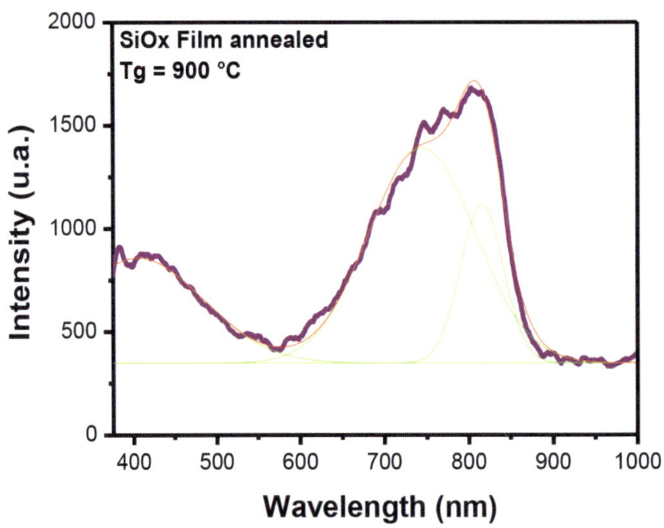

a

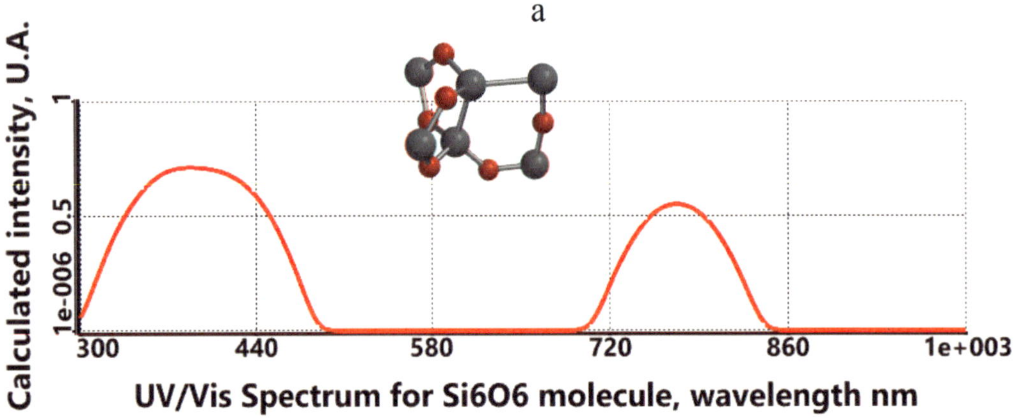

b

Figure 5. (a) Experimental PL spectrum from a SiO_x film after a further annealing. (b) Theoretical PL spectrum correlated to the experimental one of (a), including the annealing effects. This spectrum is reproduced by means of considering the small molecule type Si_6O_6.

$Si_{16}O_{16}$ molecule. The structure at the top of the figure consists of 16 hydrogen atoms linked to eight silicon atoms with tetra valences forming the $Si_{16}O_{16}:H_{16}$ arrangement. On the other hand, in the bottom we display the $Si_{16}O_{16}$ molecule after annealing. It could be described as a set of seven rings constituted of three member's mini rings each one with two bonds Si—O and a third link Si—Si. There are four mini rings with a semi-circular arrangement and the rest is in orthogonal position.

Regarding the molecular structure proposed for molecule $Si_6O_6:H_8$ (as grown), it is represented at the top of **Figure 6(c)**, whereas at **Figure 6(d)** we localize the corresponding molecule Si_6O_6 after a further annealing. This molecule was proposed for modelling the SiO_x film after being annealed at 900°C. The structure of this molecule consists only of three mini-rings, two of which are three member rings with two Si—O bonds and one Si—Si bond. The third ring has four silicon atoms with two Si—O bonds and a Si—Si—Si chain. Within the molecule as grown, that is $Si_6O_6:H_8$, the six silicon atoms have tetra valence.

We now deal with the situation where the growth temperature of H_8SiO_x film is increased. This implies that the molecular structure should be modified giving rise to another new one with different properties. In **Figure 7(a)**, we display the experimental PL spectrum measured from a SiO_x film as grown at temperature of 1150°C.With the naked eye a wide band is identified at right hand and approximately half wide band at left side. Through de-convolution curves of this spectrum, we can locate the positions of maxima of both bands. The half wide band is peaked around 385 nm and the right band around 690 nm.

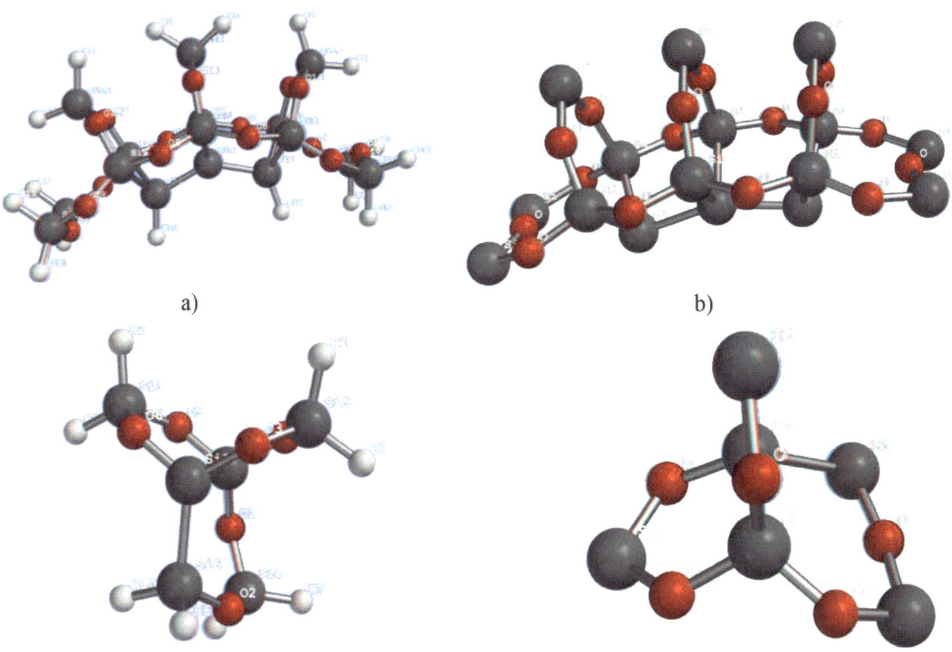

Figure 6. Molecular structure suggested for the SiO_x film: (a) $Si_{16}O_{16}:H_{16}$, as grown at T_g = 900°C (b) $Si_{16}O_{16}$ after a further annealing. Molecular structure proposed for modelling the SiO_x film after being annealed at 900°C, (c) as grown is type $Si_6O_6:H_8$ and (d) after annealed is type Si_6O_6.

Comparatively by **Figure 7(b)** we can observe the correlated theoretical PL spectrum calculated for $Si_{11}O_{11}$ molecule. With this molecular structure, we have predicted a PL spectrum having only two bands where the half left band possesses the highest intensity being peaked at 380 nm, and the other one with lower intensity and its maxima corresponds to an excited state of emission at 699 nm. In this case, we did not look for triplets for Si_{11} O_{11} molecule, taking for granted that if they exist they would have a negligible intensity.

Furthermore, the molecular structure suggested for $Si_{11}O_{11}$ molecule is represented in **Figure 8(a)** that corresponds to as grown molecule $Si_{11}O_{11}$:H_{12}. In the arrangement of this structure we figure out that there are 12 hydrogen atoms joined to six silicon ones with tetra valence. The 'backbone of this molecule' is a silicon atom joined to four silicon atoms shaping a tetrahedral arrangement, with six mini rings constituted where five of them have four silicon atoms and two Si—O bonds each one and a Si—Si—Si chain. The sixth mini ring has only one Si—O bond and a large silicon chain Si—Si—Si—Si. **Figure 8(b)** represents the after-annealed condition for the molecular arrangement of $Si_{11}O_{11}$.

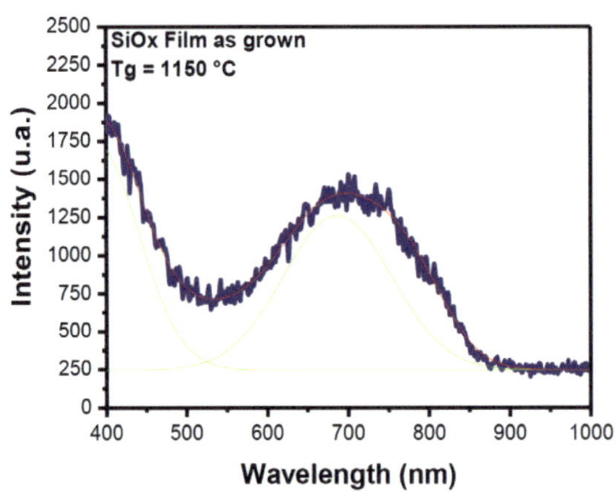

a

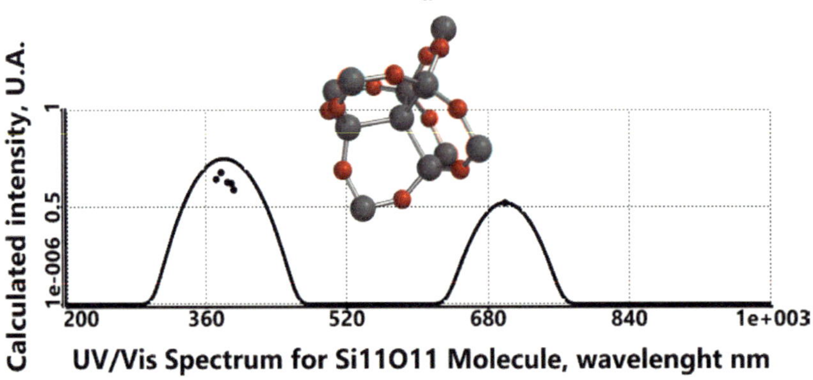

b

Figure 7. (a) Experimental PL spectrum of a SiO_x film as grown at 1150°C. (b) Correlated theoretical PL spectrum of (a). In this case the SiO_x film is modelled with a molecular structure of type $Si_{11}O_{11}$ molecule.

http://dx.doi.org/10.5772/67614

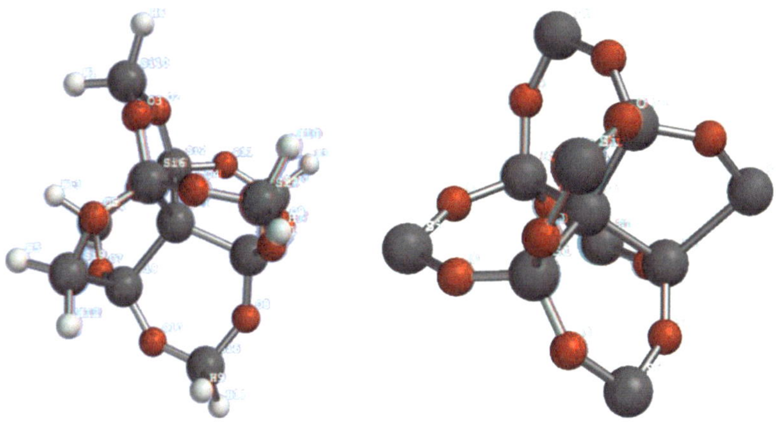

a) b)

Figure 8. Molecular structure suggested. (a) As grown it corresponds to $Si_{11}O_{11}$:H_{12} arrangement, (b) after annealed, the molecular arrangement is type $Si_{11}O_{11}$.

3.5. FTIR spectra of the SiO_x thin films

In order to get a more complete evaluation of the optical and structural properties of SRO structures, we now proceed to make a theoretical analysis of Fourier Transform Infrared (FTIR) spectroscopy. In practice, FTIR is a technique used for structural characterization of materials with which it is possible to study atomic bonds between elements that are present in a given film. The various bonds are manifested as different absorption bands which lie in different wavelength ranges. The position and shape of these bands are related to the density, stoichiometry and the nature of the bond primarily. The infrared energy causes vibrational motion of atoms in a molecule identified as rocking, stretching, wagging and bending when they interact with such energy. A fraction of the incident radiation is absorbed at specific wavelengths. A molecule must vibrate so that there is a displacement from the electrical centre and absorbed radiation in the infrared region, that is, there must be a change in the dipole moment.

In **Figure 9(a)** we display the experimental FTIR spectrum of a SiO_x film as grown at 900°C. According to Ay and Aydinly [33] they have associated these vibrational frequencies as indicated in **Table 2**. In such **Table 2**, we make comparisons between vibrational frequencies as found experimentally in **Figure 9(a)** and theoretical frequencies as obtained in this report when considering a $Si_{16}O_{16}$ molecule in addition to being identified with the different vibrational modes reported in literature. Particularly, the experimental frequencies at 654 and 875 cm^{-1}, in **Figure 9(a)**, have been associated as reported in the literature to Si—H vibrational frequencies. Comparing these frequencies with those found by our theoretical calculations, we find discrepancies since we observed such frequencies at 652 and 885 cm^{-1} which are due to Si—Si and Si—O bonds.

Apart from discrepancies in their values between experimental and theoretical frequencies, we also find differences between their associated intensities as is evident from **Figure 9(a)** when is compared with **Figure 9(b)**. Two very remarkable differences in their intensities are found in peaks located at 875 and 1051 cm^{-1} in the experimental FTIR spectrum and their correlated theoretical values located in **Figure 9(b)** approximately at 885 and 1058 cm^{-1}, in fact, if we observe carefully theoretical and experimental intensities are inverted in magnitude.

On the other hand, we stress that plot in **Figure 9(b)** corresponds to a Gaussian curve fitted to FWHH = 40, whereas plot in **Figure 9(c)** corresponds to a Gaussian curve fitted to FWHH = 80. We can observe clearly that the difference in Full Width at Half Maximum (FWHH) values gives rise to the effect that the peak at 1058 cm^{-1} in **Figure 9(b)** is transformed into a shoulder of the peak at 885 cm^{-1} as is shown in **Figure 9(c)**. The effect of varying FWHH to higher values causes the FTIR spectrum to become more intense in **Figure 9(c)**, it means physically the SiO_x film were thicker; in addition to this, FTIR spectrum resembles more to the experimental one of **Figure 9(a)** although discrepancies in wavenumbers continue to persist.

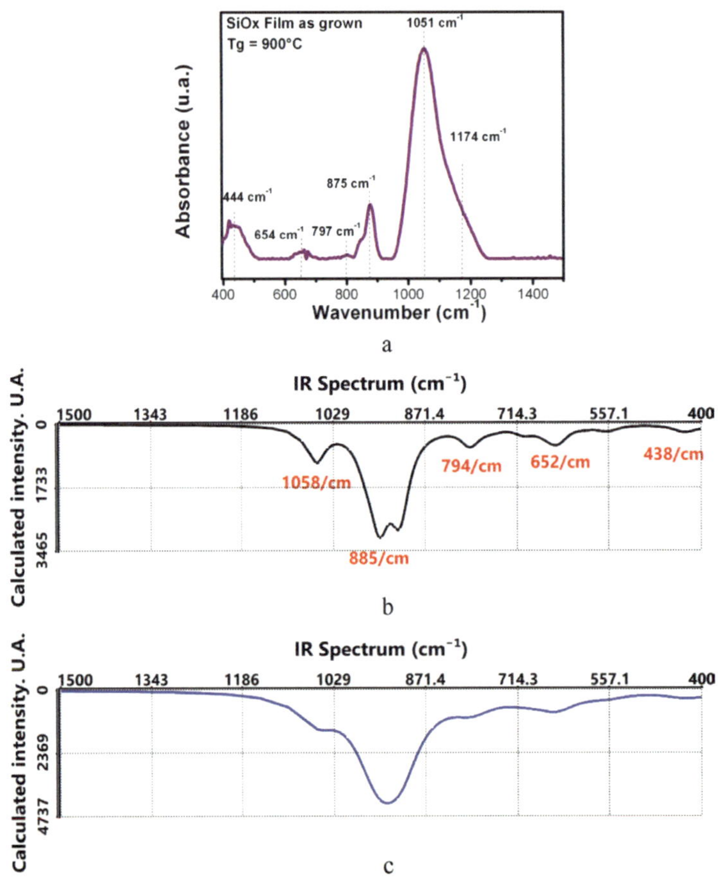

Figure 9. (a) Experimental FTIR spectrum from a SiO_x film as grown at 900°C, (b) theoretical FTIR spectra calculated using DFT for a $Si_{16}O_{16}$ molecule with FWHH = 40. Vibrational frequencies found in this spectrum when are correlated to those of (a) show discrepancies. This plot corresponds to a Gaussian curved fitted to FWHH = 40 and (c) same as (b) but FWHH = 80.

Now, we consider important to study the influence of annealing process on FTIR spectra. For this, we focus on a SiO_x thin film deposited at 900°C with a further annealing, normally carried out in the (1100–1150)°C range. **Figure 10** displays the experimental FTIR spectrum for this sample. It is worthwhile to mention that after the films have been heat-treated (annealed), the FTIR spectrum peaks corresponding or associated to Si–H vibrational frequencies disappear. Also, the uppermost peak which corresponds to Si–O stretching mode (1051 cm^{-1}) is considerably shifted to higher frequencies (1082 cm^{-1}) indicating a phase separation.

It is possible then, for example, that a molecule like $Si_{16}O_{16}$ can be transformed in a smaller molecule like $Si_{11}O_{11}$ or Si_6O_6 (it depends mainly on time and annealing temperature). On the other hand, it has been reported that amorphous silicon (a-Si) as well as crystalline one (c-Si)

	Wavenumber, cm^{-1}, SiO_x film as grown from 900 to 1150°C				
Associated in literature to	Si-O rocking	Si-H wagging	Si-O bending	Si-H bending	Si-O stretching
Experimental	429–444	645–654	797–810	875–885	1048–1064
Calculated using DFT		Si-Si bending in $Si_{16}O_{16}$ molecule		Si-O in 'external rings'	
Calculated	438	652	794	885	1058

Table 2. Comparison of vibrational frequencies observed experimentally in a SiO_x film as grown at 900°C versus calculated FTIR spectrum for a $Si_{16}O_{16}$ molecule.

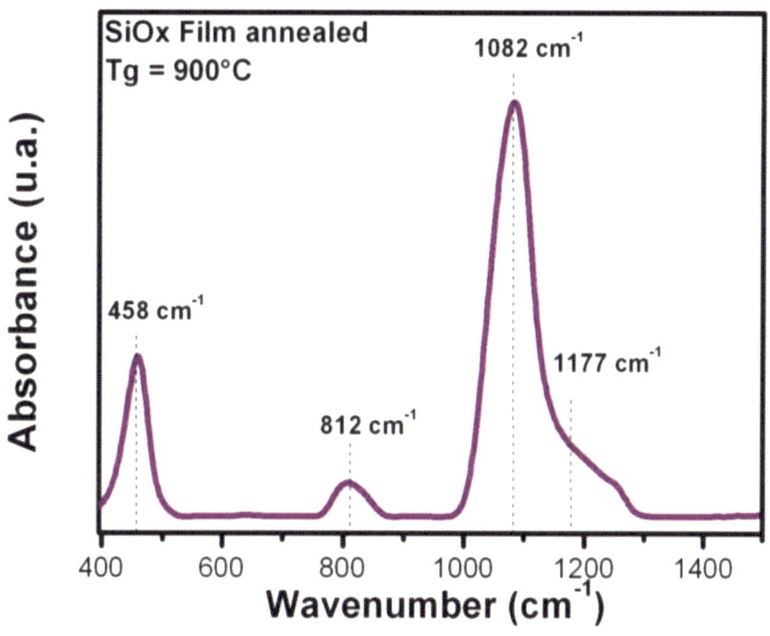

Figure 10. Experimental FTIR spectrum from a SiO_x thin film after further annealing.

could be identified by Raman spectroscopy [34]. A broadband around ~480 cm^{-1} is typically associated to a-Si, while bulk silicon has a sharp intensity peak around 521 cm^{-1}. As to Si-NCs, it is found that intensity peaks are shifted to smaller wavenumbers where this change is a function of decreasing Si-NCs size, this event has been extensively attributed to quantum confinement effects.

For example, in **Figure 11** which corresponds to a theoretical Raman spectrum that we have calculated for a $Si_{11}O_{11}$ molecule suggested, we can clearly identify frequencies at 460 and 508 cm^{-1}, respectively.

We have explained previously that the molecule $Si_{11}O_{11}$ has two chains, one is type Si—Si—Si and the other is a larger chain type Si—Si—Si—Si (tetragonal arrangement shape).

3.6. PL spectra of porous silicon in SiO_x thin films

Keeping in mind that in SRO thin films the generation of porous silicon as a consequence of phase separation of Si-NCs/SiO_x are present, as just mentioned in Section 1, its results are attractive as we apply our theoretical methods, used along the optical and structural analyses on SiO_x structures, to get a major comprehension of the radiative mechanisms linked with the atomic arrangement of molecular structures which contribute to the PL in this type of films. So, hereafter we focus specifically on the study of porous silicon which was obtained by HFCVD technique by our experimental group (unpublished results). **Figure 12** shows photoluminescence spectra between 40 and 296 K. PL measurements were carried out in vacuum and the sample was excited with a 407 nm excitation light source and an incident power of about 60 mW/mm^{-2}.

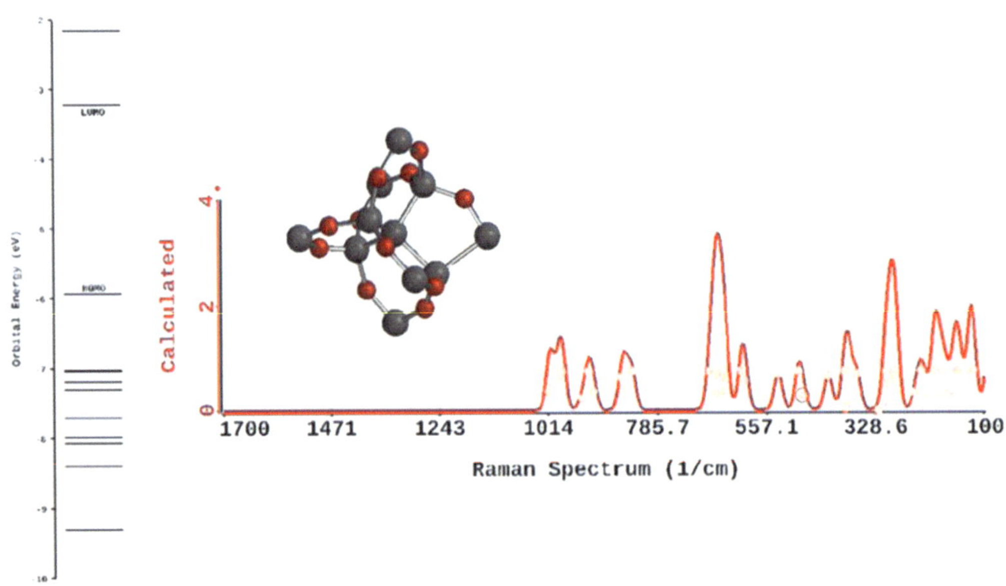

Figure 11. Theoretical Raman spectrum calculated for [$Si_{11}O_{11}$] configuration. We identify in the centre of figure the atomic structure and the left side inset displays the orbital energy in units of (eV).

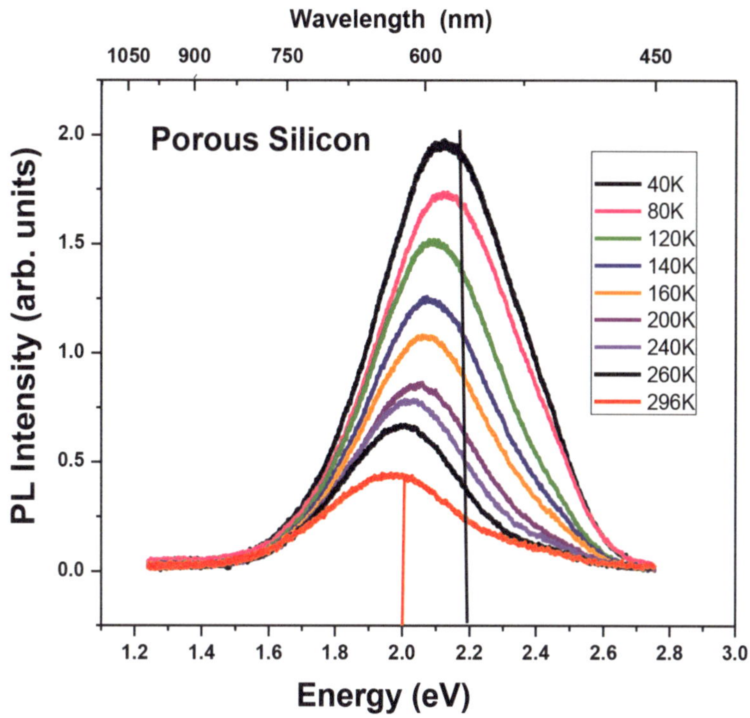

Figure 12. Experimental porous silicon PL spectra at several temperatures.

It is easily identified that the PL intensity increases as temperature decrease and at the same time the energy peak is shifted towards higher energies when decreasing the temperature of the sample. It can be seen that each peak is symmetric which means a good surface morphology and a material without defects. The shift towards higher energies is given at a rate of 5.85 E-4 eV/K. Because in our theoretical analysis neither the technique used to obtain the material nor the experimental process are relevant, since we make only use of atoms in a specific arrangement, more details about experimental conditions to grow the sample are not necessary.

In accordance with information given by PL spectra in **Figure 12**, we have that the energies of peaks of the spectra are contained approximately between 1.65 and 2.6 eV (475–750 nm). The highest intensity for each peak fit well with temperature by the following equation: $E(\text{eV}) = 2.1636 - 0.000585\,T(\text{K})$ for $40\,\text{K} \leq T \leq 300\,\text{K}$. Likewise, we can calculate the wavelength of the peak, if its associated energy is known, by using the simple equation $\lambda(\text{nm})$ = 1240 eV. In order to make the correlation between experimental PL results and theoretical ones, we evaluated four isomers for Si_{18} agglomerates. These correspond to isomers identified as 18A, 18B, 18C and 18D from which isomer 18D did not achieve convergence, consequently, information about it is not available. **Table 3** contains information about isomer 18A, 18B and 18C. We can appreciate that isomer 18C is not the most stable because it has the highest energy. Contrary, the isomer Si_{18A} (18A) possesses the lowest energy (a difference of 0.96 eV as indicated in third column) by this reason is the most stable of all. Other parameters of interest were calculated such as, the band gap, the dipole moment, the polarizability and the ovality.

Isomer	Energy (au)	Rel E (eV)	Band gap eV	Dipole debye	Polarizability	Wavelength nm, corresponding to highest emission intensity	Ovality
18A	–5210.83526	0.00	2.2550	2.84	75.20232	552.447567	1.25050
18B	–5210.80158	0.92	1.3883	0.04	75.20910	692.813713	1.25806
18C	–5210.80010	0.96	1.3232	0.01	75.21392	1732.74488	1.25816

Table 3. Parameters calculated for isomers Si_{18}.

With regard to Si_{18} cluster, four low-lying isomers were considered. The elongated isomer 18A has the lowest-energy at the HF/6–31G* level of theory. Moreover, the Isomer18A has its structure similar to the ground-state structure predicted by Rata et al. [35]. It contains a magic-number-cluster Si_6 unit and a hexagonal chair unit. A slight structural perturbation to this C_{3V} isomer followed by a geometry relaxation gives isomer 18B with C_S symmetry. Both 18B and 18C with C_{2V} symmetry contain tri-capped-trigonal-prism unit and are also very viable in stability compared to 18A because of the calculated difference in energy as 0.92 and 0.96 eV, respectively. Isomer 18D is a new isomer with high symmetry but relatively high energy. It is composed of two capped tetragonal anti-prisms, and we did not obtain convergence at HF/6-31G* level of theory for this isomer. We also observe from **Table 3** the predicted wavelength of the highest emission intensity corresponding to each isomer under the following order: 552.447567 nm for 18A isomer, 692.813713 nm for isomer 18B and 1732.74488 nm for isomer 18C. With this latter information we will try to make the correlation with experimental results regards PL curves shown in **Figure 12** as far as possible.

Prior to making the correlation, firstly we find out the information given in **Figure 13**, there with the naked eye, we locate at the top of the figure the FTIR spectrum and at the bottom of the one the PL spectrum both correspond to silicon agglomerate Si_{18} calculated using DFT. In the PL spectrum the appearance of the uppermost intensity peak predicted at 679.9 nm is outstanding according to our theoretical approach. Taking into account this wavelength value and making use of the simple equation to get its corresponding energy value, as just mentioned above, we obtain 1.82379 eV. This energy value when located in the experimental PL spectrum in **Figure 12** has a position as indicated by the black vertical line. As can be seen this energy may be correlated in a good approximation to the lowest PL spectrum, which corresponds to a temperature of 296 K, since it is closer to its maximum.

Evidently this theoretical energy value is shifted-left from the energy associated to the peak of more intensity of the experimental PL spectrum which is identified by the red vertical line. Although it has been possible to suggest a molecular structure in order to reproduce experimental measurements of PL spectra of porous silicon, it is obvious that our proposal may be substantially modified because of the significant discrepancies found between experimental results and theoretical predicted ones. It implies that we must reformulate our theoretical model which permits us better theoretical predictions in order to elucidate the correct molecular structure along with its composition which is responsible for radiative processes in the porous silicon; so it is a task that remains to be done. However, we may consider these results as preliminary ones which can guide us along correct direction to be followed. Finally, we can see in the middle of **Figure 13** the suggested molecular structure corresponding to the isomer Si_{18}. Furthermore, the top graph in **Figure 13** represents the FTIR spectrum of the isomer 18C; it presents the two most intense peaks at the vibrational frequencies of 179 and 262 cm^{-1} where we point out that two remaining calculated wavenumbers were imaginary. The three smallest peaks appear at frequencies of 305, 390 and 434 cm^{-1}.

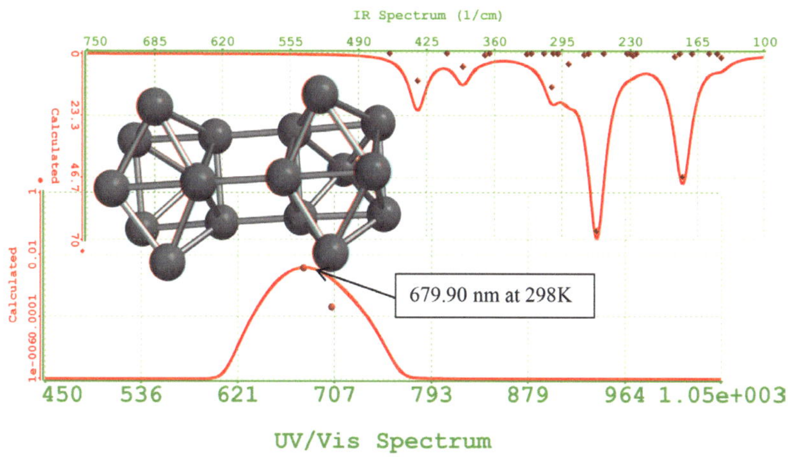

Figure 13. Top spectrum is FTIR and bottom is PL spectrum for Si_{18} C. Between graphs is represented the molecular structure.

4. Conclusions

Theoretical calculations about optical and structural properties of SiO_x films obtained experimentally by HFCVD technique have been performed using the DFT method and the GRM including temperature effects. We made correlations between theoretical results and experimental ones with respect to band gaps values, Photoluminescence and FTIR spectra. The predicted theoretical band gaps for the case of SiO_X films grown at 1300 and 1150°C are correlated with experimental values finding a good correlation since for the case of samples grown at 1300°C; it is found that $E_g = 1.95$ eV and the corresponding theoretical value is either $E_g = 1.93$ eV for a molecular structure Si_8O_8 or $E_g = 1.96$ eV for a molecular structure $Si_{16}O_{16}$. Likewise, for samples grown at 1150°C, $E_g = 2.15$ eV while theoretical results give us a value of $E_g = 2.06$ eV with an associated molecule type Si_9O_9. Moreover, when we correlated PL spectra with and without annealing effects, we find that a SiO_x film grown at 900°C, without annealing effects, gives a PL spectrum with two main peaks at 440 and 548 nm while theoretical spectrum shows peaks at 471and around 549.8 nm and the corresponding molecular structure is type $Si_{16}O_{16}$. In addition, the same sample with a further annealing displays luminescence peaked at 405, 749 and 820 nm and in this case theoretical results predict only one correlated peak at 415 nm using a molecule type Si_6O_6. Also, in relation to the FTIR correlation, theoretical calculations predict frequencies of vibrational modes of bonds Si—O (rocking, bending and stretching) and Si—H (wagging, bending) whose values are well located in the experimental frequency range and the atomic structure is type $Si_{16}O_{16}$. Except for the case of the PL spectrum with annealing effects where the prediction of one peak was only possible, all theoretical predictions are well correlated with experimental measurements within experimental accuracy. Noticeable limitations in theoretical predictions are found in the case of PL porous silicon since an important difference between energies corresponding to peaks of theoretical and experimental spectra were observed, which indicates that better models must be used for this type of material.

Acknowledgements

Néstor Espinosa wants to acknowledge the support given by CONACYT (Consejo Nacional de Ciencia y Tecnología) under postdoctoral grant (reference CVU 229741).

Authors' information

Néstor David Espinosa Torres has obtained his Ph.D. at Researching Center for Semiconductors Devices (IC-CIDS) in the Science Institute from Autonomous University of Puebla, México. He worked on a theoretical study of the luminescent phenomena in silicon and silicon-rich oxide. His researching interests include modelling using molecular mechanics, semi-empirical methods, Hartree Fock and density functional theory, material science including methods for deposition and characterization techniques of semiconductors, superconductors and energy storage devices. He also has explored other topics including phtalocyanines, graphene, ZnO, TiO_2, SnO_2, MoS_2, polyoxomethalates and single and multi-wall carbon nanotubes. Currently he is a post-doctoral fellow at Instituto de Energías Renovables, UNAM campusTemixco.

José Álvaro David Hernández de la Luz is currently a researcher and professor in the Science Institute—Center of Investigation for Semiconductors Devices (IC-CIDS) from Autonomous University of Puebla, México. He has been working on optical properties of semiconductors in the framework of local and nonlocal theory, Casimir forces with dispersive spatial effects and luminescent effects in compound semiconductor. Recently, his research interest is focused on spintronics in semiconductors materials and luminescence in graphene and nanotubes of carbon.

Javier Martínez Juárez is currently a researcher and professor in the Science Institute—Center of Investigation for Semiconductors Devices (IC-CIDS) from Autonomous University of Puebla, Mexico. He has been working on families compound semiconductor III–V (GaAs, GaSb, GaAlAs, GaAlSb, GaAlSbAs, InGaSbAs), II-VI (ZnO, ZnS, CdO, CdS, CdSe, CdSSe) and IV-VI (PbS), WO_3, Cu_2O, epitaxial growth by LPE, synthesis by spray pyrolysis and chemical bath, semiconductor devices (LD and APDs), structural and photoluminescence characterization of materials and devices.

http://dx.doi.org/10.5772/67614

Author details

Néstor David Espinosa Torres[1]*, José Álvaro David Hernández de la Luz[2] and Javier Martínez Juárez[2]

*Address all correspondence to: ndet@ier.unam.mx

1 Institute of Renewable Energies (IER), National Autonomous University of Mexico (UNAM), Temixco, Morelos, Mexico

2 Researching Center for Semiconductors Devices (CIDS), Sciences Institute (ICUAP), Meritorious University Autonomous of Puebla (BUAP), Puebla, Mexico

References

[1] E. Werwa, A. A. Seraphin, L. A. Chin, C. Zhou, and K. D. Kolenbrander. Appl. Phys. Lett. 64, (1994) 1821.

[2] X. X. Wang, J. G. Zhang, L. Ding, B. W. Cheng, W. K. Ge, J. Z. Yu, and Q. M. Wang. Phys. Rev. B, 72, (2005) 195313

[3] L. T. Canham. Phys. Status Solidi B, 190, (1995) 9.

[4] G. G. Qin, and Y. J. Li. Phys. Rev. B, 68, (2003) 85309.

[5] F. G. Bell, and L. Ley. Phys. Rev. B, 37, (1988) 8383.

[6] H. R. Philipp. J. Phys. Chem. Solids, 32(8), (1971) 1935–1945.

[7] Y. N. Novikov, and V. A. Gritsenko. J. Appl. Phys., 110, (2011) 014107.

[8] D. Ristić, M. Ivanda, G. Speranza, Z. Siketić, I. Bogdanović-Radović, M. Marciuš, M. Ristić, O. Gamulin, S. Musić, K. Furić, G. C. Righini, and M. Ferrari. J. Phys. Chem. C., 116, (2012) 10039–10047.

[9] N. D. Espinosa-Torres, J. F. J Flores-Gracia, J. A. Luna-López, D. Hernández de la Luz, and J. Martí-nez-Juárez. Inte. J. Sci. Res, 3(7), (2014) 82–92 ISSN No 2277–8179.

[10] H. Wiesmann, A. K. Ghosh, T. McMahon, and M. Strongin. J. Appl. Phys., 50(5), (1979) 3752.

[11] X. QI, Z. Chen, and G. Wang. J. Mat Sci. Tech, 19(3), (1993) 235–239.

[12] M. Nayfeh, S. Rao, N. Barry, J. Therrien, G. Belomoin, A. Smith, S. Chaieb. Appl. Phys. Lett., 79, (2002) 1249–1251.

[13] L. Pavesi, L. Dal Negro, C. Mazzoleni, G. Franzo, and F. Priolo. Nature., 480, (2000) 440–444.

[14] K. Luterova, I. Pelant, I. Mikulskas, R. Tomaisiunas, D. Mueller, J.-J. Grob, J.-L. Rehspringer, and B. J. Honerlage. Appl. Phys., 91, (2002) 2896–2900.

[15] G. G. Sui, X. L. Wu, Y. Gu, and X. M. Boa. Appl. Phys. Lett., 74, (1999) 1812–1814.

[16] H. Tamura, M. Ruckschloss, T. Wirschem, and S. Veprek. Appl. Phys. Lett., 65, (1994) 1537–1539.

[17] A. J. Kenyon, P. F. Trwoga, C. W. Pitt, and G. Rehm. J. Appl. Phys., 79, (1996) 9291–9300.

[18] X. Y. Chen,Y. F. Lu,Y. H. Wu, B. J. Cho, M. H. Liu, D. Y. Dai, and W. D. Song. J. Appl. Phys., 93, (2003) 6311–6319.

[19] J. A. Luna-López, J. Carrillo-López, M. Aceves-Mijares, A. Morales-Sánchez, C. Falcony. Superficie y Vació, 21(4), (2009) 11–14.

[20] A. Barranco, F. Yubero, J. P. Espinos, F. Groening, and A. R. Gonzales-Felipe. J. Appl. Phys., 97, (2005) 113714.

[21] S. Berg, and T. Nyberg. Thin Solid Films, 476, (2005) 215–230.

[22] F. F. Gracia, M. Aceves, J. Carrillo, C. Domínguez, and C. Falcony. Superfices y Vacio, 18, (2005) 7–13.

[23] A. Benítez-Lara, G. García-Salgado, D. E. Vázquez-Valerdi, A. Morales-Sánchez, N. D. Espinosa-Torres, and J. A. Luna-López. Adv. Powder Technol., 26(1), (2015) 163–168 doi:10.1016/j.apt.2014.09.005

[24] C. Banerjee, J. Sritharathikhum, A. Yamada, and M. Konagai. J. Phys. D: Appl. Phys., 41, (2008) 185107.

[25] V. A. Gritsenko, J. B. Xu, R. W. M. Kwok, Y. H. Ng, and I. H. Wilson. Phys. Rev. Lett., 81, (1998) 1054.

[26] H. R. Philipp. J. Non-Cryst Solids., 8(10), (1972) 627–632 © North-Holland Publishing Co.

[27] J. I. Pankove. Optical Process in Semiconductors; Prentice Hall; Englewood; Cliffs; NJ; (1971) 34.

[28] J. A. Luna López, J. Carrillo López, D. E. Vázquez Valerdi, G. García Salgado, T. Díaz Becerril, A. Ponce Pedraza, and F. J. Flores Gracias. Nanoscale Res. Lett., 7, (2012) 604.

[29] Y. C. Fang, W. Q. Li, L. J. Qi, L. Y. Li, Y. Y. Zhao, Z. J. Zhang, and M. Lu. Nanotechnology., 15, (2004) 495–500.

[30] M. Aceves, A. Malik, and R. Murphy, In: Sensors and Chemometrics, eds: Maria Teresa Ramirez-Silva, (Research Signpost, ISBN: 81-7736-067-1, 2001) p.1–25.

[31] Y. Matsumoto, S. Godavarthi, M. Ortega, V. Sánchez, S. Velumani, and P. S., Mallick. Thin Solid Films., 519, (2011) 4498–4501.

[32] D. E. Vázquez-Valerdi, J. A. Luna-López, J. Carrillo-López, G. García-Salgado, A.-F. Benítez-Lara, and N. D. Espinosa-Torres, Nanoscale Res. Lett., 9 (2014) 422.

[33] F. Ay, and A. Aydinly,: Opt. Mater., 26, (2004) 33–46.

[34] K. Ragnar, J. A. Luna-López, O. D. Guilherme, A.-M. Mariano, and J.-B. S. Willibrordus. J. Mex. Chem. Soc., 52(3), (2008) 212–218. Sociedad Química de México. ISSN 1870-249X.

[35] I. Rata, A. A. Shvartsburg, M. Horoi, Th. Frauenheim, K. W. M. Siu, and K. A. Jackson. Phys. Rev. Lett., 85, (2000) 546.

Chapter 3

The Formation and Migration of Molten Inclusions in Silicon

Arkady Skvortsov

Additional information is available at the end of the chapter

http://dx.doi.org/10.5772/67787

Abstract

The mechanisms of formation and migration of molten inclusions in silicon are described. Consideration of them is necessary for development of technological approaches to formation of conducting regions (characteristic sizes of 10 μm–30 nm) and doped conducting channels in silicon. The problem of formation of local conducting regions is topical, e.g., for single electronics. It is known that sizes of a "Coulomb island" (about 10 nm) present difficulties to modern technologies. That is why the development of alternative procedures to form a "Coulomb island" (in particular, those based on electromigration of molten regions in semiconductor bulk) is a topical problem. Besides, high heat loads on the "classical" silicon structures, owing to specific character of their work (operating current densities may be as high as $j \sim 10^9$–10^{10} A/m^2), promote their active degradation. The appearing molten regions coagulate into drops and are displaced by a current along the electric field lines. Therefore, a detailed analysis of the mechanisms of inclusions formation and migration in silicon will make it possible to predict more accurately the "critical" modes of silicon structures operation (e.g., in power electronics). The main mechanisms of molten inclusions formation (contact melting in a metal-semiconductor system, decomposition of supersaturated solid solutions in heavily doped semiconductors) are considered, as well as the procedures to form and control mobility of such objects: electric current, temperature gradient, structural inhomogeneities, etc. The problems of molten inclusions migration in silicon single crystals occurring in electric field with contact melting electric transport of melt components and thermoelectric effects at interphase boundaries are considered in detail. The molten region is moving along the electric field lines, thus enabling to exert control over the trajectory of its motion. The results of author's original investigations on formation and migration of molten silver- and aluminum-based inclusions in silicon are presented, as well as the experimental data by other authors. The main migration mechanisms related to melting-crystallization processes at interphase boundaries and components diffusion through the melt region are discussed in detail. The problems to defect formation in silicon after crystallization of inclusions are also considered.

Keywords: silicon single crystals, contact melting, crystallization, melt inclusion, formation of molten regions in a metal-semiconductor system, electromigration of inclusions by the melting-crystallization mechanism, diffusion in the melt, gradients of concentration and temperature, formation of structural defects in silicon single crystals with inclusions

1. Introduction

Formation and dynamics of impurities in semiconductors as melt regions occur, as a rule, when the critical operating conditions for power semiconductor devices (transistors, Peltier elements, etc.) are reached [1–3]. In this case, as is known [4, 5], the near-contact regions and metal-semiconductor contacts may heat up to temperatures over the melting points for eutectics of relatively fusible systems (Al-Si, Ag-Si, Au-Si; the melting point for Au-Si eutectic $Te = 370°C$).

To illustrate, during operation of a thermoelectric generator for automobile internal-combustion engines, certain modules or their parts may be heated for a short time to temperatures over 330°C [6]. The result can be fusion of structure fragments and their failure. And this happens despite equalizing of temperature field (provided with the construction of heat-exchange apparatus) at the hot side of thermoelectric modules [6]. Obviously, failure of electric contact in a thermoelectric module because of overheating will lead to failure of the whole system, thus considerably reducing the efficiency of thermoelectric generator.

During operation of such devices, a direct electric current consumed by the resistive load is flowing through each module. At overheating and beginning of fusion, the current flows through the liquid phase-crystal contact. This makes it necessary to study the electromigration processes at the semiconductor surface and in the bulk involving melt regions that are proceeding under the action of direct and pulse currents. In those cases, migration of the produced liquid phase in electric field is possible in the system studied. The field of crystal structural nonuniformity (gradient of dislocation density) also can promote migration of the melt regions [7, 8]. These are just the above factors that cause the interest of researchers in the electromigration processes at the semiconductor surface and in the bulk involving melt regions [9].

As a rule, migration of liquid inclusions in semiconductors is related to transfer of matrix atoms from one inclusion boundary to another through the inclusion itself. An inclusion migrates because the matrix atoms are dissolving at the forward inclusion wall, then they are diffusing through the inclusion and at last become deposited onto the matrix at the back wall, with formation of new atomic layers. In the case of liquid inclusions, solubility of matrix atoms in the inclusion and the coefficient of diffusion of these atoms in the inclusion bulk are sufficiently high, so such a migration mechanism is crucial [10, 11].

With such a mechanism, flowing of electric current through the liquid phase-crystal contact is accompanied by Peltier heat release at the one interphase boundary and absorption at the other interphase boundary: $\pm Q_p = Pjt$. (Here P is the Peltier coefficient at the melt-crystal interface, j is the current density, t is the time of current action.) This leads to variation of concentrations at the inclusion walls and requires accounting for thermoelectric effects in the course of mass transfer. Besides, one should also take into account the effect of electromigration that creates concentration nonuniformity near the interphase boundaries, depending on the value and sign of the effective charge Z^* of semiconductor atoms in the melt [10, 11].

When determining the predominant role of the above mechanisms, one should note that for inclusions of sizes

$$\ell \gg \sqrt{\frac{D\tau}{\pi}}, \tag{1}$$

(τ is the current pulse duration), only the contribution from Peltier heat is significant at migration.

A qualitative experimental confirmation of this statement was made for elementary semiconductors of p- and n-type. At pulse duration $\tau = 150$ μs and pulse height $I = 6 \times 10^3$ A, migration of drops of the Al, Ga, Ag melt occurred towards anode and was determined by the thermoelectric effects at the interphase boundaries. In the stationary regime, inversion of drop velocity was observed (the drops moved towards the negative electrode). This indicated the predominant role of electromigration. Unfortunately, application of nonstationary heating of semiconductor and surface migration only prevented obtaining dimensional dependence of migration velocity.

To determine the nature of mass transfer, let us assess the path velocity in accordance with the known mechanisms [12].

In the case of a vacancion mechanism of inclusion transfer, a vacancy current in the matrix creates a counter current and an equal current of matrix atoms. In this case, the speed of inclusion migration, w, can be expressed by the following ratio:

$$w = -\frac{2(\rho_s - \rho_\ell)}{2\rho_s + \rho_\ell}\frac{D_s e Z^*}{fkT}E_\infty. \tag{2}$$

Here, D_s is the self-diffusion coefficient of semiconductor atoms, E_∞ is the density of the electric field away from the inclusion, f is a correlation factor, and ρ_s and ρ_ℓ are the electrical conductivities of the matrix and inclusion, respectively.

The effective charge of the atoms in a semiconductor matrix is known to be negative. This makes the activated atoms of the semiconductors move toward the anode, while vacancies move toward the cathode. By occupying vacancies at the leading edge, atoms of inclusions ensure that the edge is pushed toward the positive electrode. Therefore, inclusions should be displaced against the electric field lines, a prediction that is inconsistent with the experimental results. In addition, the assessed velocity, w, for aluminum inclusions in silicon is $w = 5 \times 10^{-15}$ m/s in the case where the silicon self-diffusion coefficient is $D_s = 3 \times 10^{-18}$ m²/s (temperature of T~1043 K) [13]. Here, $E_\infty = 10$ V/m, $\rho_\ell = 2 \times 10^{-6}$ Ω·m, $\rho_{ol} = 10^{-5}$ Ω·m, $Z^* = -10$, $f = 0.8$, and $R = 8.31$ J/(K·mol). Both the operator of the migration velocity w and its seventh-order (10^7)

lower value are inconsistent with the experiment. Therefore, the contribution of this mechanism can be considered insignificant.

As the molten zones are displaced, surface diffusion [12] is visible only for low-conductivity inclusions. Its importance obviously increases at low temperatures, when the surface/bulk diffusion ratio is high. The velocity, w, determined by this mechanism can be assessed using the following equation:

$$w = -3\frac{D_s' ae Z^*}{HfkT} E_\infty, \tag{3}$$

where $D_s' = 10^{-12}$ m²/s, $a = 5 \times 10^{-5}$ m, $H = 10^{-4}$ m, and $T = 1043$ K. The expected velocity value determined by this mechanism is three times lower than the experimental value.

The movement of inclusions in the electric field associated with the current of matrix atoms in bulk inclusions can be rather fast if the melted material is a conductor. A theoretical estimate of w for this case can also be provided using the known ratios [12], assuming targeted transport of matrix atoms in the melt under the electric transport effect.

When $T = 1043$ K and $j = 10^6$ A/m², the displacement velocity for the diffusion and kinetic modes equals

$$w = -\frac{V_s}{V_\ell}\left[\frac{3\rho_s}{2\rho_\ell + \rho_s}\right]\frac{DeZ^*C}{fkT\ell} E_\infty = 2 \cdot 10^8 \text{ m/s}, \tag{4}$$

$$w = -V_s\left[\frac{3\rho_s}{2\rho_\ell + \rho_s}\right]\frac{\beta e Z^* C\ell}{fkT} E_\infty = 4 \cdot 10^8 \text{ m/s}. \tag{5}$$

From this point onward, V_s and V_ℓ represent the atomic volumes of the matrix and inclusion, respectively.

The theoretical estimates, w, for the diffusion and kinetic modes are consistent with the experiment in terms of quality. However, the ratios used in this case do not show the complexity of the phenomena that occur when current passes through the inclusion. One can easily notice that Eq. (4) results in a size dependence of approximately $w = 1/\ell$, while Eq. (5) yields $\frac{w}{j} = \text{const} \cdot \ell$, which does not fully describe the experiment. The observed discrepancy in the data obtained through calculation and experiment requires that more than just electric transport should be taken into account in the case of targeted migration.

Therefore, by analyzing the movement of secondary phases in the bulk semiconductor (**Figure 1**), we take into account the fact that movement is performed not only by means of electric transport but also under the influence of thermoelectric phenomena (the Peltier effect) at the interface between the liquid inclusion and solid phase [14–16].

If targeted migration is determined via dissolution-crystallization at the phase interfaces (kinetic control $\beta\,\delta \ll D$), then the equation for w takes the following form [17, 18]:

$$\frac{w}{j} = -\frac{V\beta\bar{C}}{N_A}\left(\frac{\delta\rho e Z^*}{kT} + \frac{V P_{LS} L}{N_A 2\lambda k T^2} \cdot \ell\right). \tag{6}$$

In the case of diffusion control ($\beta\,\delta \ll D$), the equation becomes

$$\frac{w}{j} = -\frac{VD\bar{C}}{N_A}\left(\frac{\rho e Z^*}{kT} + \frac{V P_{LS} L}{\delta N_A 2\lambda k T^2} \cdot \ell\right). \tag{7}$$

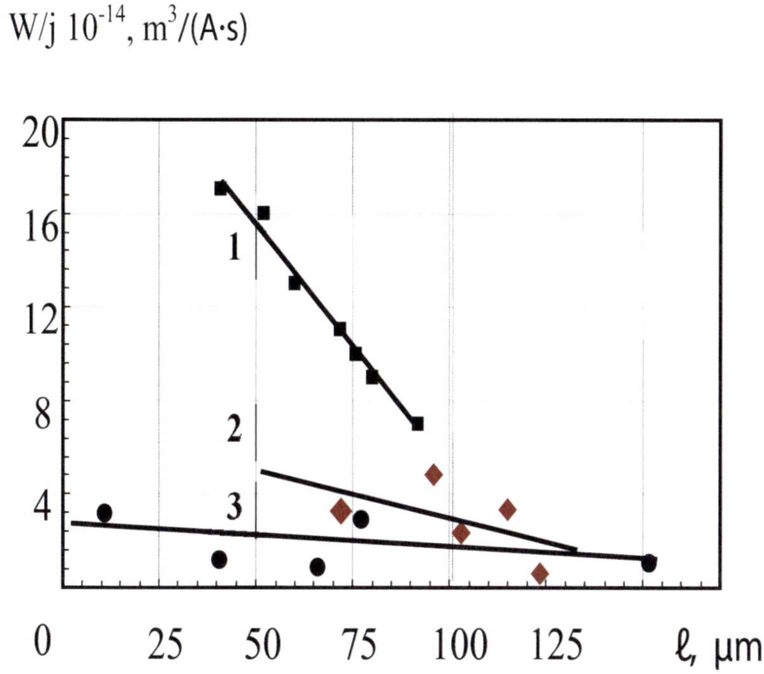

Figure 1. Dimensional dependence of migration of Al-Si inclusions in the bulk of single crystalline silicon at T = 1073 K at flowing: 1—direct current of 12 A; 2—rectangular current pulses of height I_{max} = 24 A and duration τ = 300 μs; 3—I_{max} = 24 A and τ = 900 μs.

where P_{LS} is the Peltier coefficient of the melt-crystal interface, V is the specific volume of the melt, $\overline{c}$ and D are the equilibrium density and diffusion coefficient of the conductor molecules in the melt, respectively, N_A is Avogadro's number, δ is the thickness of the diffusion layer at the phase interface, L is the temperature of the transition of one volumetric unit of the solid phase into the melt, and λ_e is the thermal conductivity coefficient usually calculated according to the Wiedemann-Franz law.

2. Materials and methods

The experiments were performed in a vacuum chamber for silicon (n-type, resistivity of 10 Ω·cm) specimens that were previously cut into 4 × 4 × 15 mm blocks (**Figure 2**). After lapping with M-15 micropowder and with ACM-5 diamond paste, the specimens were treated with ethanol and then were clamped between two tantalum electrodes by the procedure described in Refs. [10, 11] in a quartz electric resistance furnace. A 5 mm thick graphite spacers were put between the specimen and electrodes to prevent their contact interaction. The electrodes were connected to the dc source through an ammeter using copper wires. When performing experiments with current pulses, we used a rectangular pulse generator. The pulse duration and height were τ = 100–1000 μs and I = 5–30 A, respectively.

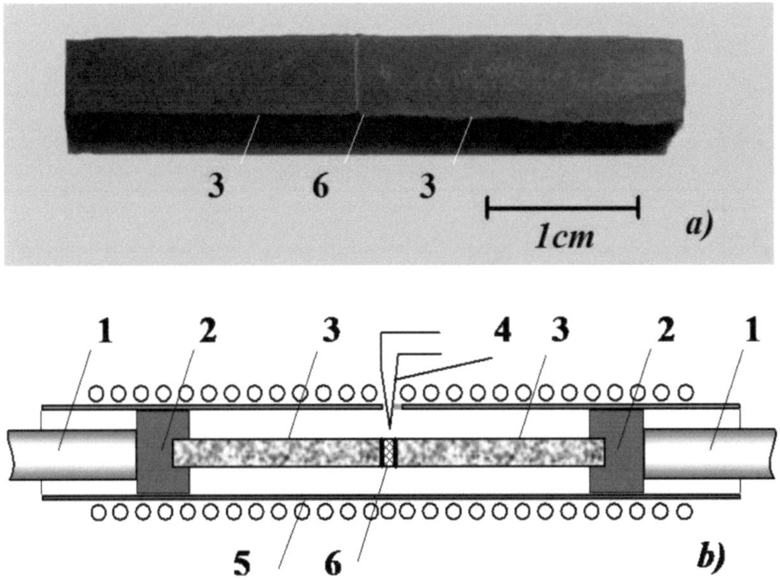

Figure 2. (a) Visual appearance of samples and (b) annealing configuration: 1—electrodes; 2—graphite interlayers; 3—samples; 4—thermocouple; 5—quartz-based resistive furnace; 6—metal film.

At migration in the bulk, an aluminum plate (mass of 0.6–0.9 mg) was put between the {111} planes of two specimens. Then they were heated up to a temperature close to the experiment one ($T = 1000–1123$ K). The experiment temperature was determined by the net effect from the space heater and electric current flowing through the specimen.

In the course of experiment performance, the temperature of the specimen was determined either using the thermocouple method or from the specimen luminosity. In the latter case, the temperature regimes in the experiments were registered with a pregraduated photocell located on one of the eyepieces of a high temperature microscope. The temperature measurement error was not more than ±10%. The distribution of temperature T and concentration C of semiconductor atoms (dissolved in inclusion) during migration of the melt region are presented in **Figure 3**.

An occurrence of melt regions in the silicon matrix bulk at stationary electric annealing is related to formation of an alloy between an Al film and Si single crystal. This mechanism obeys the reaction-diffusion law: a new phase is formed that was not found before interaction between components. Its growing is given by the following expression [15]:

$$x = 2\beta\sqrt{D\tau_k}. \tag{8}$$

Here D is the reaction-diffusion coefficient; β is an argument of the Kramp function that is determined from the diagram of phase equilibrium. Thereby, a melt film is formed at the metal-semiconductor interface as the proper temperature is reached. Thermodynamic

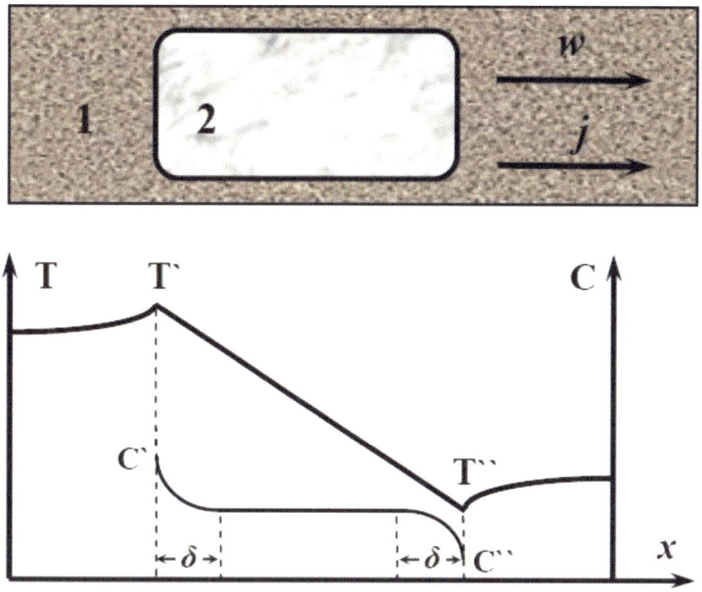

Figure 3. Schematic of migration of melt region: 1—semiconductor matrix; 2—melt inclusion. Distribution of temperature *T* and concentration *C* of semiconductor atoms (dissolved in inclusion) during migration of the melt region.

instability of the film promotes its dispersion into drops. The inclusions generated in such a way drift in an electric field.

An analysis of numerous experimental data obtained for the Al-Si system showed that, in the course of migration, melt inclusions are faceted by specific crystallographic planes that shape inclusions into a distinguishing feature (**Figure 4**). In this case, the electric current flowing through the inclusion does not distort the form of drifting regions.

To determine crystallographic indices of the faceting planes at migration of the Al-Si system, the equation of structural crystallography for a cubic lattice [12] was used:

$$\cos\varphi = \frac{h_1 h_2 + k_1 k_2 + l_1 l_2}{\sqrt{h_1^2 + k_1^2 + l_1^2} + \sqrt{h_2^2 + k_2^2 + l_2^2}} \tag{9}$$

Here φ is the angle between the planes; h_i , k_i, and l_i are the indices of crystallographic planes. According to the experimental data, an angle φ between the {111} plane and those with unknown indices $\{h_2,k_2,l_2\}$ is 35° or 145° for which cosφ = +0.819. A good agreement between the experimental and theoretical values of cosφ can be obtained if (110) and $(\bar{1}\bar{1}0)$ will correspond to the indices h_i , k_i, and l_i: substitution of the above values in Eq. (4) gives cosφ = +0.821.

The inclusion size and depth of inclusion penetration in the matrix from the starting position were determined by sequential removal of *N* layers (at every 5–10 μm) [10, 11, 13] with subsequent identification of inclusions using an optical microscope.

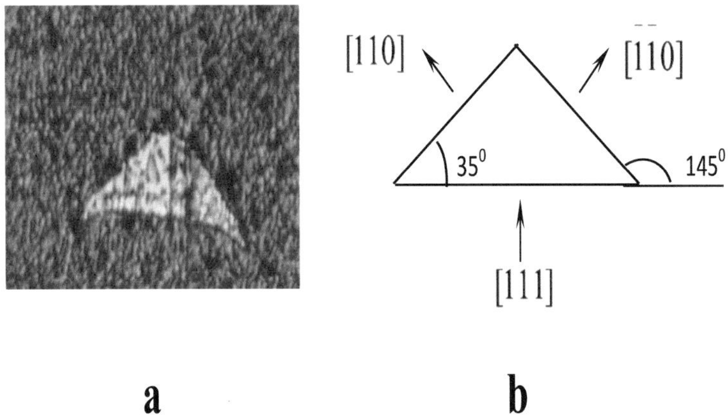

Figure 4. A pattern of Al-Si inclusion in single crystalline silicon. The section plane: ⊥(111). Magnification: ×200.

3. Analysis of experimental results

The typical results of investigations are presented in **Figure 4**. As before [10, 11, 14], the dimensional dependence of migration velocity is linear. To identify the origin of driving forces that determine inclusion migration in electric field in the case of predominance of transfer processes in inclusion bulk, let us use Eq. (7).

For a stationary electric annealing (1, **Figure 4**), Eq. (7) agrees fully with experiment if P is considered negative and Z^* is considered positive. At pulse current action, the character of dimensional dependence is changed—see Eqs. (3) and (4), **Figure 4**. Reduction of pulse duration leads to increase of velocity for inclusions of the same size. We relate the main responsibility for the observed variations to reduction of the contribution from electromigration.

Indeed, at $P < 0$ the concentration nonuniformity at interphase boundaries is provided by heat release (absorption) Q_p at the cathode (anode) interphase boundary. This promotes inclusion migration because of melting-crystallization at the interphase boundaries along the electric field lines (just as observed in the experiment). Electromigration ($Z^* > 0$) will produce the opposite gradient, thus reducing $\partial C/\partial x$ of the thermoelectric origin.

So, migration of inclusions at stationary electric annealing of Si crystals is determined by heat release at the interphase boundaries that ensures migration of inclusions toward the negative electrode. Electromigration produces the opposite gradient, which decreases the resulting inclusion velocity. The smaller the length of electric current pulses is, the weaker is their effect on reduction of the contribution from electromigration to the resulting inclusion velocity. This shows itself most clearly at pulse duration $\tau = 300$ μs.

The crystallization of molten inclusions in a semiconductor is accompanied by active formation of defects, namely dislocations. The results of independent research (**Figure 5**) and our studies (**Figure 6**) both indicate the formation of dislocations during the crystallization of the molten zone near the crystallization front.

During the crystallization of single silicon crystals, the area occupied by the drop will contain approximately 10% of the extrinsic stacking (interstitial) atoms because of the higher (approximately ~10%) density of liquid silicon compared with a single crystal [19]. At a certain drop size, extrinsic stacking atoms may transform into a dislocation loop (**Figure 6**).

The shape of second-phase inclusions depends not only on the crystallographic properties of the matrix but also on the reactivity of the melt. If there is an impurity near the drop, it will diffuse into the liquid and dissolve within it because of the higher solubility of impurities in a liquid phase. In this case, the infrared light-absorption coefficient will increase and the molten zone will be stabilized.

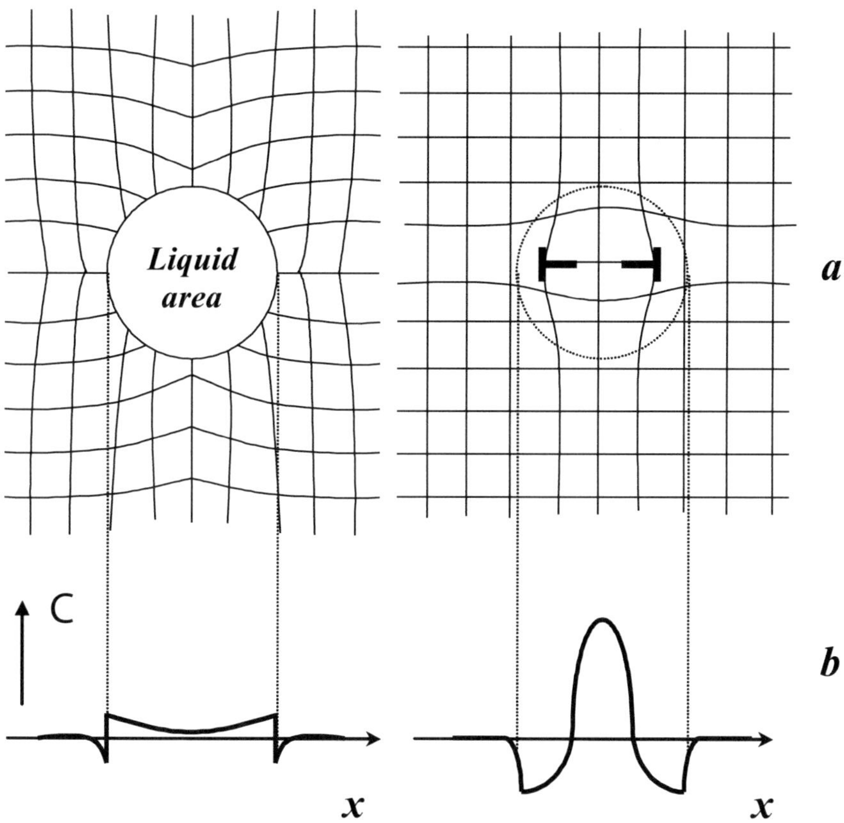

Figure 5. Diagram showing liquid-drop transformation into a dislocation [19]. (a) Lattice distortion and (b) impurity-concentration profile.

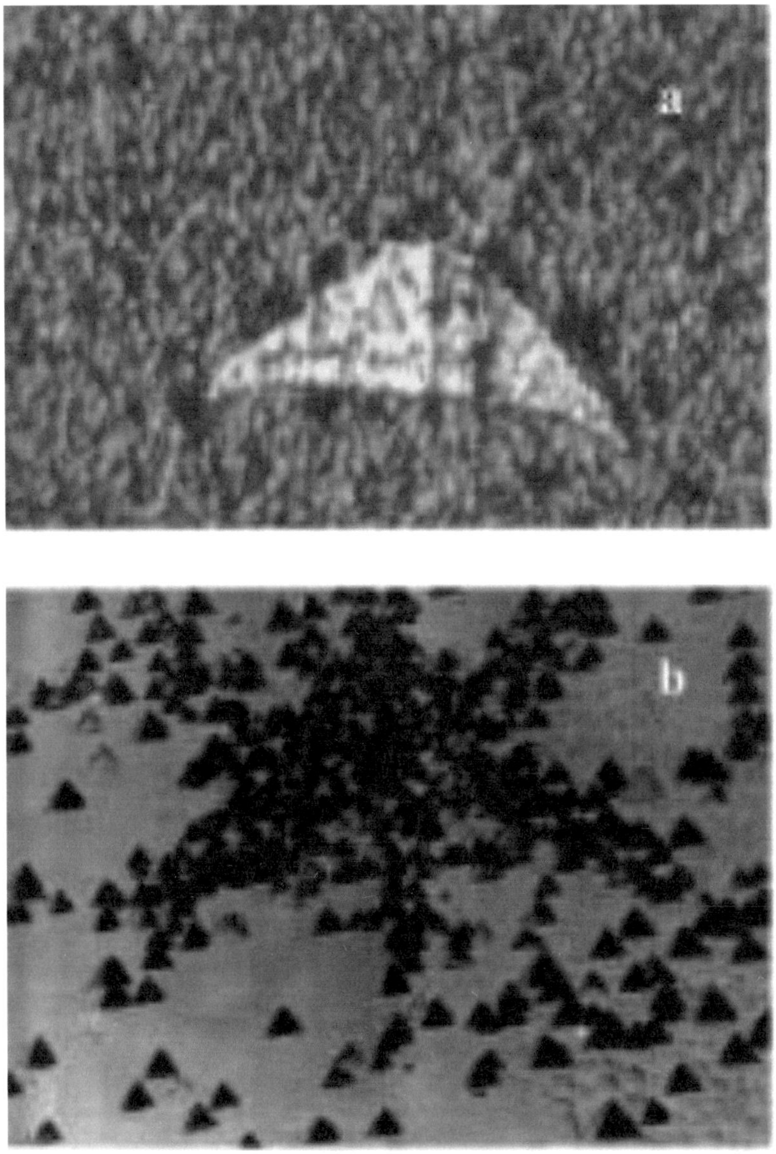

Figure 6. (a) photo of the surface of the polished section containing the Al-Si inclusion in silicon and (b) distribution of dislocation etch pits around the inclusion after four hours of annealing at $j = 100$ A/cm^2 and a temperature of 800°C. Polished section plane (111). 200× zoom.

4. Comparative analysis of the formation and electromigration of molten inclusions (AL-SI) in silicon and aluminum

In this context, the present investigation was aimed at an experimental study difference of the electromigration of melted Al-Si inclusions in silicon and aluminum crystals.

The experiments were performed in a setup that is schematically depicted in **Figure 7**. The samples had the form of silicon bars (4 × 4 × 15 mm; *n*-type, 20 Ω·cm, growth dislocation density not exceeding 4×10^3 1/m^2) and aluminum cylinders (length, 15 mm; diameter, 4 mm; average grain size, 500 μm). For investigations of the electromigration in silicon, a metal (Al) plate weighing 0.6–0.9 mg was placed between the {111} edge planes of silicon samples and the system was heated to a temperature above the eutectic (T = 850 K). The sample temperature was determined by the joint action of the external heater and the current passing through the sample. The temperature was measured by a thermocouple situated in the immediate vicinity of the sample.

For the formation of melted Al-Si inclusions in aluminum, particles of highohmic silicon with diameters of ~300–500 μm were placed into preliminarily made depressions on the edge surface of one aluminum cylinder. Then, the aluminum cylinders were matched with their edges, placed inside the quartz furnace tube (with a diameter of 5 mm), and pressed between the current carrying electrodes with graphite spacers. At temperatures above the eutectic, the components exhibited contact melting and formed melted film zones, which rapidly separated into separate drops (inclusions) with dimensions of 20–350 μm. These inclusions were always driven by the current (j =1.4 × 10^5 A/m^2) along the lines of electric field, but the velocity of this migration in both silicon and aluminum was dependent on the inclusion size l.

It was established that the formation of inclusions from a liquid film in silicon is accomplished during a 5–10 μm penetration of the melt into the crystal matrix volume, while in aluminum this process was accomplished within a 30–50μm thick layer. **Figure 8** shows micrographs of drifting Al-Si inclusions as observed after the electroannealing and crystallization.

Figure 8 shows the typical results as represented by the empirical plot of

$$W/j \;=\; a + b\ell, \tag{10}$$

where w/j is the specific migration velocity defined as the ratio of the velocity w of a melted inclusion to the current density j in the sample; a and b are empirical coefficients, the values of which are given in **Table 1**.

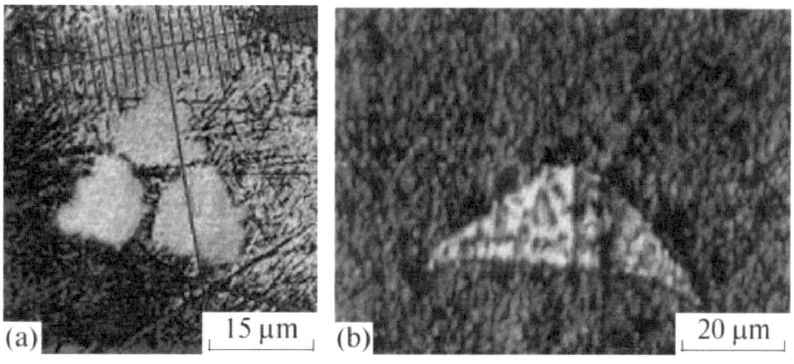

Figure 7. Micrographs showing Al-Si inclusions observed after electroannealing (j = 1.4 × 10^5 A/m^2) in (a) polycrystalline aluminum (T = 893 K) and (b) single crystalline silicon (T = 1043 K; section ⊥(111); magnification, ×200.

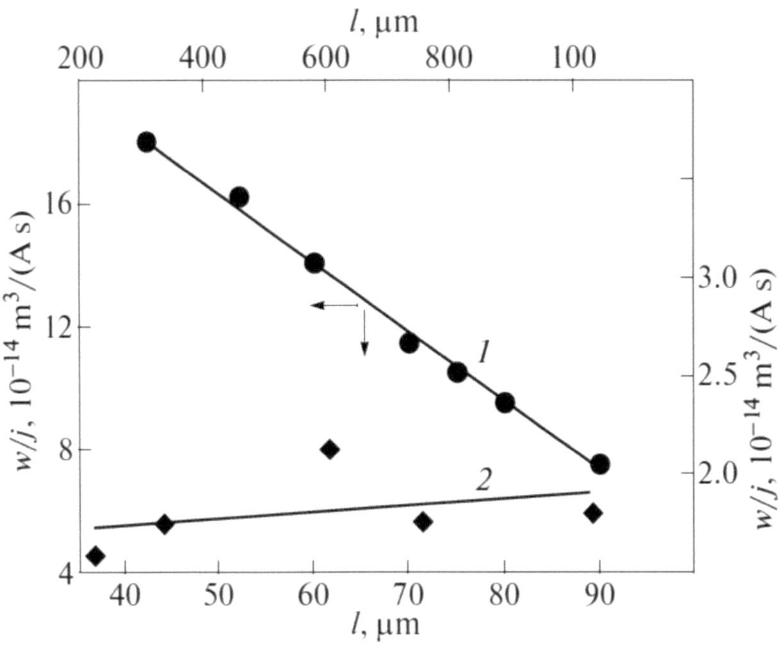

Figure 8. Plots of the specific migration velocity w/j versus size l of Al-Si inclusions in (1) silicon T = 1043 K and (2) aluminum at T = 893 K.

No	Parameter	Units	Temperature, K	
			1043	893
1.	Matrix material	–	Silicon	Aluminum
2.	Melt composition	$C_{Si}\backslash C_{Al}$, at%	76\24	10\90
3.	a, 10^{-14}	$m^3/(A{\cdot}s)$	27.4	1.7
4.	b	$m^2/(A{\cdot}s)$	-2.4×10^{-9}	2.2×10^{-12}
5.	P_{ℓ}, 10^{-7}	Ω·m	36	4.5
6.	M, 10^{-3}	кg/mol	27	27
7.	V, 10^{-6}	m^3/mol	10.9	11.2
8.	L, 10^{9}	J/m^3	3.9	1.0
9.	λ	W/(m·K)	7.1	48
10.	δ	μm	200	200
11.	C, 10^{28}	$1/m^3$	5.5	5.4
12.	D, 10^{-9}	m^2/s	6.1	2.8
13.	Z^*	–	–1.1	–1.0
14.	P, 10^{-3}	V	230	–8.7

Table 1. Calculated and experimental data on the migration of melted Al-Si inclusions in polycrystalline aluminum and single crystalline silicon.

It was previously demonstrated that the transport of liquid inclusions in these systems is driven by two competitive mechanisms: thermoelectric phenomena (Peltier heat evolution) at interphase boundaries and electric field induced redistribution of components dependent on their effective atomic charges in the melt. The specific migration velocity w/j is proportional to the Peltier coefficient P of the crystal-melt interface and the effective charges Z^* of semiconductor and metal atoms in the eutectic melt.

We have described the migration of melted Al-Si inclusions in silicon and aluminum taking into account the condition $\beta\delta > D$ and data from the table, using Eqs. (7) and (10), and evaluating coefficients a and b from our experiments (**Figure 8**). The P and Z^* values were calculated using published data [11, 12] on the composition dependence of the resistivity of Al-Si melts; the heat conductivity was evaluated using the Wiedemann-Franz law. Since no data were available on the temperature dependence of the coefficient of mutual diffusion in Al-Si melts, the D values at e temperatures studied were taken from monographs [11, 13, 14]. The results of P and Z^* calculations for the electromigration of melted Al-Si inclusions are presented in **Table 1**.

An analysis of the obtained results allows us to ascertain that the electromigration of melted Al-Si zones in silicon, similar to the analogous processes in other semiconductors [8–11, 17, 18], is determined by the electromigration of melt components in the volume of inclusion and by the thermoelectric phenomena at interphase boundaries. The negative value of the effective charge on semiconductor atoms in the melt is indicative of their directed transport to the anode. This migration results in silicon depletion of the melt in the diffusion layer of the front part of the inclusion and silicon enrichment of the boundary layer in the trailing part of the melted zone. Accordingly, the electromigration favors the displacement of inclusions toward the negative electrode. An important feature of the obtained results (**Figure 8**, curve 1) is a clear manifestation of the mutually compensating contributions due to the P and Z^* values. A positive P value ensures evolution of the Peltier heat at the rear side of the inclusion and the absorption of this heat at the frontal boundary during the current passage. Therefore, under the action of thermoelectric factors, the inclusion must diffuse in the counter direction relative to the electric field lines. Moreover, this mechanism would predominate with increasing size of inclusions, which accounts for the fact that the electromigration and thermoelectric effects in experiments with silicon crystals appear as competitive (counteracting) factors.

As for aluminum, the size dependence of the specific migration velocity in this system is much less pronounced as compared to silicon, since the $P = P_S – P_L$ value (where P_S and P_L are the Peltier coefficients of the solid phase and melt, respectively) on the interphase boundary is also significantly smaller than in the semiconductor matrix and has the opposite sign ($P < 0$). For this reason, the motion of melted Al-Si zones in polycrystalline aluminum is also determined by the electromigration of Al atoms in the liquid phase and by thermoelectric effects, the role of which increases with the size of inclusions, but in this case the two mechanisms act complementarily to each other.

5. Conclusions

Thus, the electromigration of extrinsic zones in Si has been established to be determined by the dissolution-crystallization process at the interface between the solid and liquid phases. This happens under the influence of Peltier heat localized at the phase interfaces and the forces of electric transport in the bulk inclusion. The size dependence of the inclusion displacement velocity, w, is shown to be governed by a linear law.

The contribution of electric transport and Peltier heat to the resulting velocity of zone displacement by the current is determined. It varies for inclusions of different sizes; all other conditions being constant, the contribution of electric transport decreases with the increase of inclusion thickness.

Acknowledgements

The author is grateful to Professor Anatoliy M. Orlov for the interest and support of this research area. This study was partially supported by the Ministry of education and science of the Russian Federation (project 8.5171.2017).

Author details

Arkady Skvortsov

Address all correspondence to: skvortsovaa@outlook.com

Moscow Polytechnic University, Moscow, Russia

References

[1] Perepezko J.H, Wilde G. Melt undercooling and nucleation kinetics. Current Opinion in Solid State and Materials Science. 2016; **20(1):** 3–12. DOI: 10.1016/j.cossms.2015.07.001

[2] Brassart L., Liu Q., Suo, Z. Shear, dilation, and swap: mixing in the limit of fast diffusion. Journal of the Mechanics and Physics of Solids. 2016; **96:** 48–64. DOI: 10.1016/j.jmps.2016.06.013

[3] Perepezko JH, Hoffmeyer MK, De Cicco MP. Analysis of melt undercooling and crystallization kinetics. Metallurgical and Materials Transactions A: Physical Metallurgy and Materials Science. 2015; **46(11):** 4898–4907. DOI: 10.1007/s11661-015-2970-9

[4] Tian Z, Lee S, Chen G. Heat transfer in thermoelectric materials and devices. Journal of Heat Transfer. 2013; **135(6):** 061605. DOI: 10.1115/1.4023585

[5] Sharmin A, Rashid M, Gaddipati V, Sadeque A, Ahmed S. Multiscale design of nanostructured thermoelectric coolers: effects of contact resistances. Journal of Electronic Materials. 2015; **44(6):** 1697–1703. DOI: 10.1007/s11664-014-3520-8

[6] Khripach NA, Papkin BA, Korotkov VS, Zaletov DV. Study of the influence of heat exchanger body design parameters on the performance of a thermoelectric generator for automotive internal combustion engine. Biosciences Biotechnology Research Asia. 2015; **12(Spl. Edn. 2):** 677–689. http://dx.doi.org/10.13005/bbra/2084.

[7] Liu D-W, Li J-F. Electrocrystallization process during deposition of Bi-Te films. Journal of the Electrochemical Society. 2008; **155(7):** D493–D498. DOI: 10.1149/1.2907398

[8] Rodbell KP, Shotynsky SR. Electromigration in sputtered Al-Si thin films. Thin Solid Films. 1983; **108:** 95–102. DOI: 10.1016/0040-6090(83)90045-7

[9] Skvortsov AA, Zuev SM, Koryachko MV, Glinskiy V. Thermal shock and degradation of metallization systems on silicon. Microelectronics International. 2016; **33(2):** 102–106. DOI: 10.1108/MI-05-2015-0049

[10] Orlov AM, Skvortsov AA. Temperature gradient effect on the electromigration of Ag-Te melt inclusions in tellurium. Inorganic Materials. 2007; **43(5):** 471–474. DOI: 10.1134/S0020168507050056

[11] Orlov AM, Skvortsov AA, Salanov AA. Electric transport in gallium antimonide single crystals involving molten GaSb-Sn inclusions. Semiconductors. 2004; **38(4):** 376–379. DOI: 10.1134/1.1734661

[12] Gershanov VYu, Garmashov SI. Non-stationary nonlinear effects at mass transfer in small volumes of solution in melt enclosed in anisotropic crystal. Journal of Crystal Growth. 2009; **311(9):** 2722–2730. DOI: 10.1016/j.jcrysgro.2009.01.080

[13] Magomedov MN. On self-diffusion and surface energy under compression of diamond, silicon, and germanium. Technical Physics. 2013; **58(12)**: 1789–1799. DOI: 10.1134/S1063784213120153

[14] Zhang J.-S., Wu Y.-P., Wang Y.-G., Tao Y. Thermomigration in micro interconnects in integrated circuits. Journal of Materials Science-Materials in Electronics. 2010; **21(10):** 1090–1098.

[15] Gottstein G. Physical Foundations of Material Science. Berlin-Heidelberg: Springer-Verlag. 2004. 347 p.

[16] Geguzin YaE, Kaganovskiĭ YS. Diffusive mass transfer in island films. Soviet Physics – Uspekhi. 1978. **21(7)**:611–629. DOI: 10.1070/PU1978v021n07ABEH005667

[17] Belashchenko DK, Orlov AM, Orlova SP. Crystallization of semiconductors in a constant electric field. High Temperature. 1976; **14(3):** 516–520.

[18] Orlov AM, Belashchenko DK, Derikova SA, Parkhomenko VI. Mechanism and kinetics of the mass transfer of impurities in polar A3B5 compounds. Soviet Physics Journal. 1976. **19(5):** 576–580. DOI: 10.1007/BF00952553

[19] Ravi KV. Imperfections and Impurities in Semiconductor Silicon. A. Wiley-Interscience Publication. 1981. 348 p.

Chapter 4

Study on the Preparation of Solar Grade Silicon by Metallurgical Method

Dawei Luo

Additional information is available at the end of the chapter

http://dx.doi.org/10.5772/67643

Abstract

The global PV industry has rapidly developed over the past decade, which has led to a large demand for silicon materials. Solar cells are currently fabricated from a variety of silicon-based materials. However, current market is difficult to ensure a steady supply for this material. Development of a new process to produce silicon at low cost is definitely necessary. Metallurgical grade silicon (MG-Si) with the purity of 98%, which is produced by carbothermic reduction in electric arc furnaces, has been considered as a cheap starting material for conversion to purity of 99.99%. Many alternative methods for purifying MG-Si to Solar grade silicon (SoG-Si) have been explored, for example, (1) pyro metallurgical processes, (2) hydrometallurgical processes and (3) electrochemical methods. Metallurgical route is recognized as a promising process to meet market demand for solar energy silicon material, which is different from the traditional Siemens process. This chapter focuses on the introduction of three kinds of typical impurity removal methods in metallurgical process, and the impurity removal effect of different processes was analysed and discussed.

Keywords: metallurgical grade silicon, solar grade silicon, acid leaching, electron beam melting, electromagnetic induction slag melting

1. Introduction

The rapid development of photovoltaic industry results in lack of silicon feedstock which meets technical and economic requirements of the photovoltaic cells. More and more attentions have been paid to the research of producing solar grade silicon (SOG-Si) by refining metallurgical grade silicon. Metallurgical route is recognized as a promising process to meet market demand for solar energy silicon material, which is different from the traditional Siemens process. There are many ways to remove impurities in metallurgical process, this chapter focuses on the three typical kinds of impurity removal process in detail, which is the process of acid leaching process, electron beam process and electromagnetic induction slagging process.

2. Purification of metallurgical grade silicon (MG-Si) by acid leaching

2.1. Characterization of MG-Si

Pre-treatment of metallurgical grade silicon (MG-Si) is an important part in metallurgical purification process because it can remove the metallic impurities such as aluminium, iron and calcium effectively under acid leaching process. Among above methods, hydrometallurgical method is recognized as the best pre-treatment process for silicon refining because metallic impurities can be removed from MG-Si by this process without large amounts of energy consumption and capital investment of equipment. A series of comparative experiments were designed to study the effect of acid leaching parameters on impurities removal efficiency in detail. The effects of acid agent, acid concentration, time, temperature and silicon particle size on impurities removal efficiency were investigated, respectively.

Impurities contents and their segregation coefficients in MG-Si are listed in **Table 1** [1]. The major metallic impurities are aluminium (0.271%), iron (0.234%) and calcium (0.058%).

Microstructures of MG-Si are shown in **Figure 1**. The bright phases at the grain boundaries are the segregation of metal impurities. The qualitative composition of bright phases and their relative content in MG-Si are listed in **Table 2**. It has been studied that the sensitivity of phases in the samples to hydrochloric acid and hydrofluoric acid, all of the phases in the samples can be removed by the hydrofluoric acid and hydrochloric acid but the Fe-Si and Fe-Ti-Si phases cannot be attacked by hydrochloric acid. According to the sensitivity of these phases to the

Element	Concentration ($mg \cdot kg^{-1}$)	Segregation coefficient
Al	2709	2.0×10^{-3}
Fe	2339	8×10^{-6}
Ca	575	8×10^{-3}
Ti	198	2×10^{-6}
Cu	27	4×10^{-4}
Mn	124	1.3×10^{-5}
Mg	21	3.2×10^{-6}
Zn	<2	1×10^{-5}
Cr	<2	1.45×10^{-5}
B	15	0.8
P	20	0.35

Table 1. Impurity contents and their segregation coefficients in MG-Si.

acids in **Table 2**, hydrochloric acid and hydrofluoric acid are selected as acid agents in this chapter. The removal efficiency of aluminium, iron and calcium after the acid leaching by hydrochloric acid and hydrofluoric acid are checked.

2.2. Effect of various acid leaching parameters on the impurity content

MG-Si was crushed by the crushing mill and grinded with the attritor which were composed of adamantine spar container and Al_2O_3-sintered hard ceramic balls before the leaching experiments. And then silicon particles were sieved to different particle sizes for leaching trials. Different particle sizes silicon were disposed by acetone solution under ultrasonic wave and washed by the deionized water. Leaching trials were done with aqueous thermostat after the silicon powders were dried. The mass of silicon powders employed at a time was 10 g and the liquid-solid weight ratio of 10:1. After each experiment, silicon powder was filtered and washed with deionized water. The contents of aluminium, iron and calcium under different acid leaching conditions were analysed by inductively coupled plasma atomic emission spectrometer (ICP-AES), respectively. A larger number of leaching trials were performed in a systematic way in order to study the effect of the acid agent, acid concentration, time, temperature

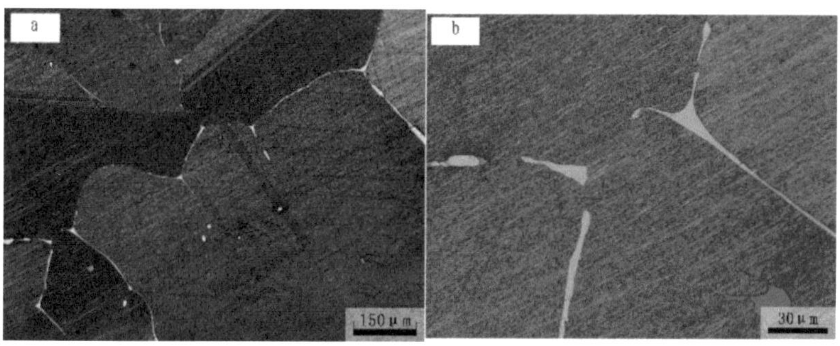

Figure 1. The metallographs of MG-Si: (a) the low power microstructure and (b) local amplification bright phases at the grain boundaries.

Composition of phases	Relative abundance	Sensitivity	
		Hydrochloric acid	Hydrofluoric acid
Si-Fe	***		*
Si-Fe -A1	***	*	*
Si-Fe-Ti	**		*
Si-Fe-A1-Ti	***	*	*
Si-Fe-A1-Ca	**	*	*
Si-Fe-A1-Ca-Ti	**	*	*
Si-Ca	*	*	*

Table 2. The main precipitated phases and their sensitivity for hydrochloric acid and hydrofluoric acid in MG-Si [2].

and silicon particle size on the removal efficiencies of impurities element. Specific results were described in the following sections in more detail.

2.2.1. Effect of acid agent and acid concentration on the impurity content

The change of impurity content as increasing the concentration of the hydrofluoric and hydrochloric acids was plotted in **Figure 2**. The content of aluminium and iron gradually decreased with the increasing concentration of hydrochloric acid but the change regularity of calcium content was in equable. Calcium content reduced first and then grew little with the increasing concentration of hydrochloric acid. The minimum value achieved at 15%. This was because the silicon powders generated passivation when the concentration of hydrochloric acid exceeded 15%. The change tendency for the content of aluminium, iron and calcium after the hydrofluoric acid treatment were very different. Aluminium content increased and iron content gradually decreased with the increasing concentration of hydrofluoric acid. Calcium content reduced first and then grew little, the minimum value was achieved at 2%. The reason for the unusual change regularity of aluminium and calcium was most probably due to the formation of insoluble fluorides.

2.2.2. Effect of acid temperature on the impurity content

The effect of temperatures on the impurity contents was demonstrated in **Figure 3**. The content change of aluminium, iron and calcium with temperatures was coincident after leaching with hydrochloric acid. They all gradually decreased with temperatures. But the results were diverse after leaching with hydrofluoric acid. As the increasing temperatures, the content of iron gradually decreased, but the contents of aluminium and calcium reduced first and then grew little. The minimum value was achieved at 40 and 60°C, respectively. The reason for above phenomenon was most probably due to the formation of insoluble fluorides.

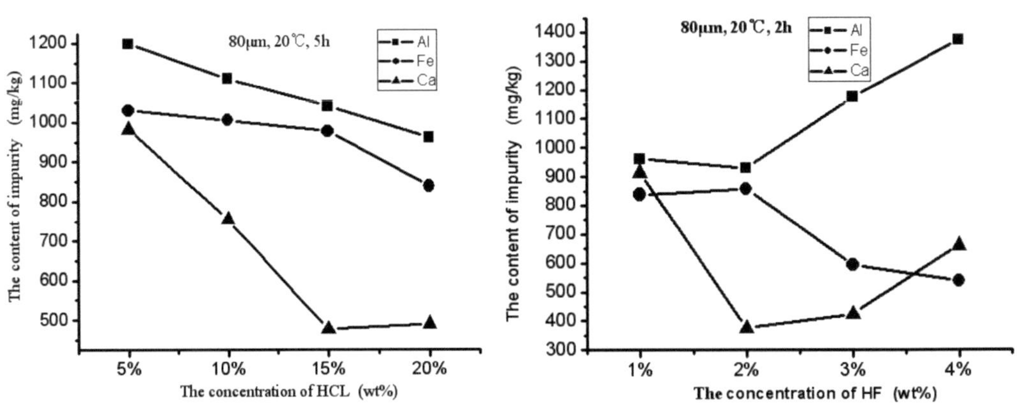

Figure 2. Final impurity content after leaching with hydrochloric acid and hydrofluoric acid at different concentrations.

2.2.3. *Effect of acid time on the impurity content*

The effect of the leaching times on the impurity contents is shown in **Figure 4**. The content change of aluminium, iron and calcium with the increasing leaching times was the same after leaching with hydrochloric acid. They all reduced first and then grew little; the minimum value was achieved at 10 h. The content change of iron and calcium with times after hydrofluoric acid leaching was the same as that after hydrochloric acid leaching. The minimum value achieved at 2 h. But the aluminium content gradually decreased with the increasing times.

2.2.4. *Effect of particle sizes on the impurity content*

The effect of the particle sizes on the impurity contents is demonstrated in **Figure 5**. The content change of aluminium, iron and calcium with particle sizes was alike after leaching with hydrochloric and hydrofluoric acids. The contents of aluminium, iron and calcium gradually depressed with the increasing particle sizes.

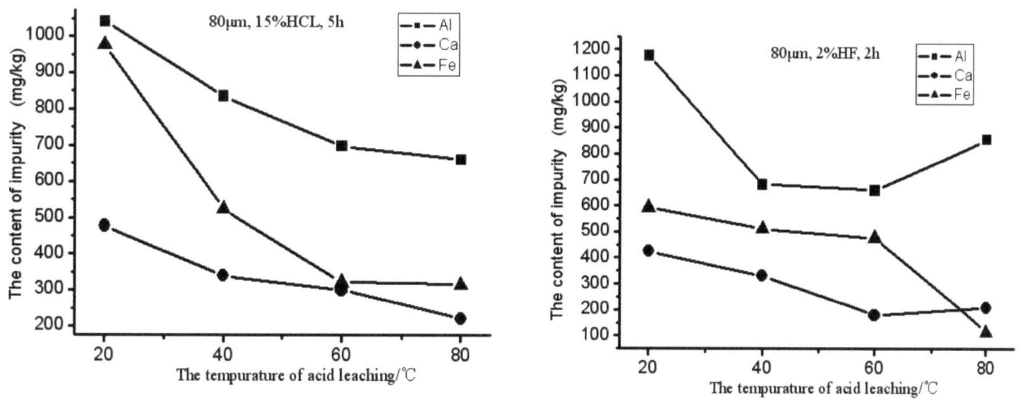

Figure 3. Final impurity content after leaching with hydrochloric acid land hydrofluoric acid at different temperatures.

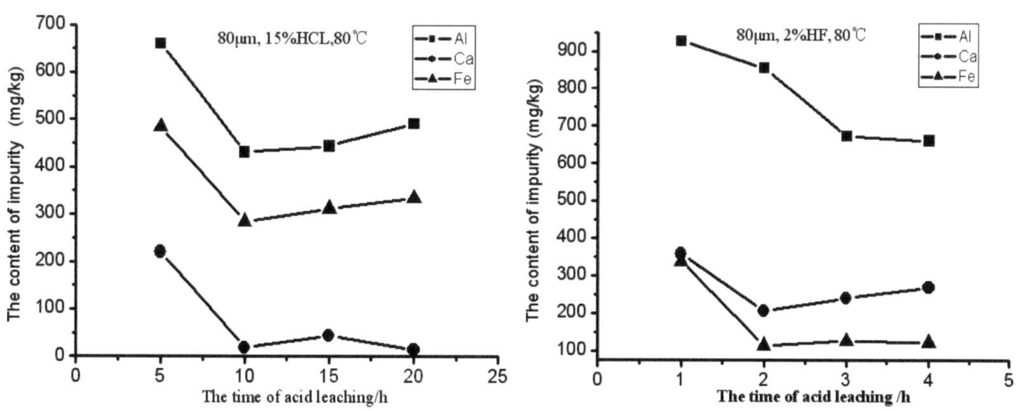

Figure 4. Final impurity content after leaching with hydrochloric acid and hydrofluoric acid at different leaching times.

The total content of aluminium, iron and calcium impurities reached minimum at 100 μm. This result was opposite to some previous research. In order to explain this consequence, the content of major impurities in MG-Si at different particle sizes was analysed by ICP-AES before acid leaching. The measurement result is illustrated in **Figure 6**. The content of major impurities in silicon particles with 30 μm was higher than that with 80 μm. It suggested that more friable phases that included more impurities were aggregated to the finer fraction during sieving. A similar grain size effect was found for other minor impurities. Impurities such as the doping elements boron and phosphorus with unfavourable segregation coefficients were not greatly affected by acid leaching. However, they must be removed by pyro metallurgical treatment or a similar method [3].

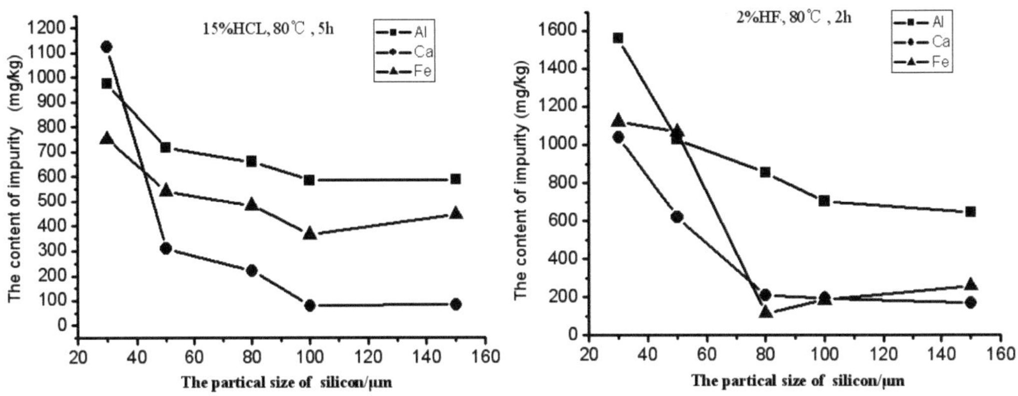

Figure 5. Final impurity content after leaching with hydrochloric acid and hydrofluoric acid at different particle sizes.

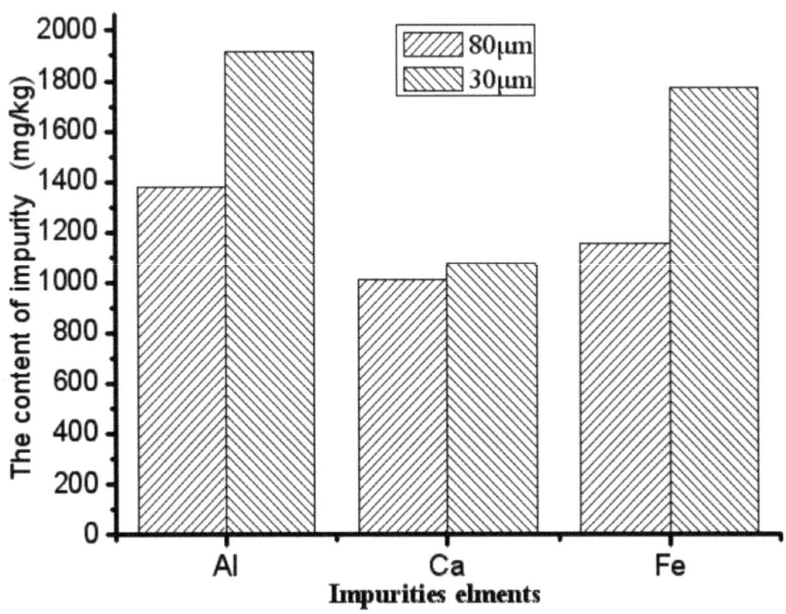

Figure 6. The concentration of major impurities at different particle sizes.

2.2.5. The final purity of silicon at the optimum processing parameters

From the above analysis, the optimum process parameters for acid leaching are as follows: hydrochloric acid, 15%, 80°C, 10 h and 100 μm. The final content of major impurities after leaching with hydrochloric acid and hydrofluoric acid solutions at the optimum processing parameters is represented in **Figure 7**. Although the acid leaching parameters are the optimum processing parameters, final total content of major impurities treated with hydrochloric acid is lower than that treated with hydrofluoric acid. The final purity of silicon powders after leaching with hydrochloric acid and hydrofluoric acid solutions at the optimum processing parameter is denoted in **Figure 8**. The final purity of silicon powders after leaching with hydrochloric acid solution is better than that after leaching with hydrofluoric acid solution, and the final purity of silicon powders after leaching with hydrochloric acid and hydrofluoric acid mixture solutions is a little better than that after leaching with hydrochloric acid alone. Meanwhile, there is no significant differences between silicon treated in one step with mixture acid and treated in two steps, first with hydrochloric acid and then with hydrofluoric acid. When hydrochloric and hydrofluoric acids are used for leaching simultaneously, care must be taken because it might cause explosions under this condition. Hydrogen is generated during acid leaching process and, in addition, because of the presence of silicide (especially calcium and magnesium silicide) in MG-Si, self-igniting silicon hydrides are formed. So adequate safety precaution should be adopted in order to avoid the danger mentioned above and it is better to select hydrochloric acid as the acid leaching solution.

2.3. The acid leaching critical and discussion

From the results obtained till now, the following can be stated: (1) The optimum processing parameters for acid leaching are as follows: hydrochloric acid, 15%, 80°C, 10 h and 100 μm.

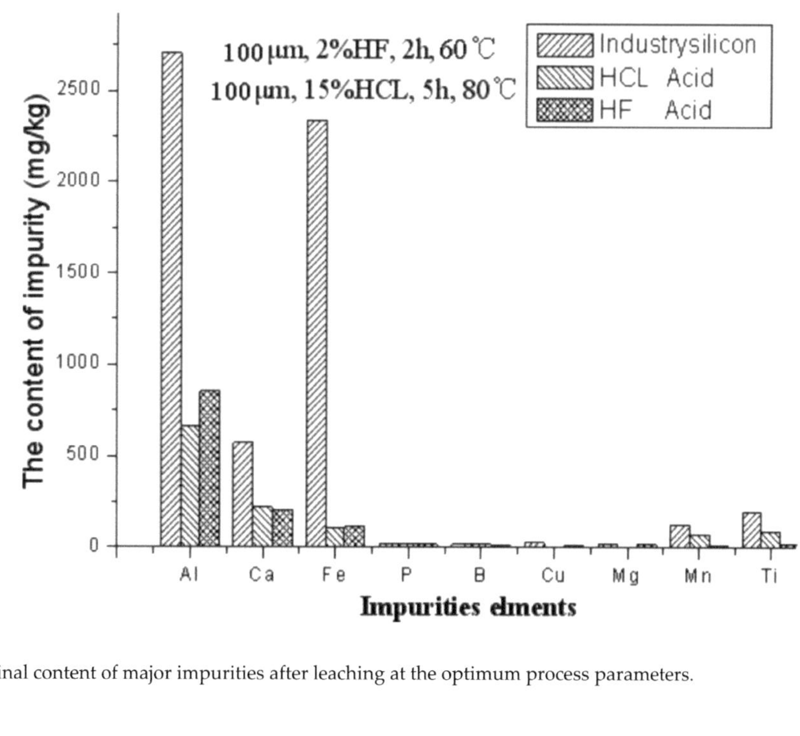

Figure 7. Final content of major impurities after leaching at the optimum process parameters.

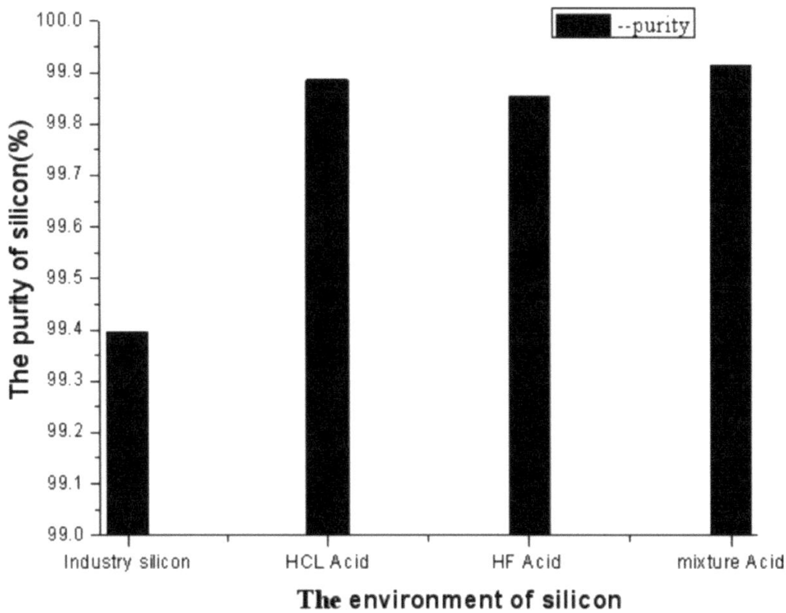

Figure 8. Final purity of silicon powders after leaching at the optimum process parameters.

The removal efficiencies of aluminium, iron and calcium impurities are up to 70.9, 94.82 and 82.69%, respectively. (2) The formation of insoluble fluorides is a rejection for the impurity removal under hydrofluoric acids, and it is dangerous under the mixture of hydrofluoric and hydrochloric acids. In order to avoid the above situation, it is better to select hydrochloric acid as the acid leaching solution. (3) The content of major impurities increases as the particle sizes of MG-Si powders decrease, because the more friable phases contained impurities concentrate in the finer fraction during sieving.

Silicon of 99.9% purity was obtained by leaching with hydrochloric acid alone from MG-Si with iron (0.234%), aluminium (0.271%) and calcium (0.058%) under the optimum process parameters. On the other hand, silicon of above 99.9% purity could be also obtained after leaching with hydrochloric acid and hydrofluoric acid. The significant improvement achieved by mixture solutions was due to the Fe-Si and Fe-Si-Ti phases, which were not leached by hydrochloric acid. Although the silicon powders of 99.9 wt% could be obtained after the acid leaching, the major metallic impurities in MG-Si could not be got rid of fully only with the acid leaching because there was a critical factor in the acid leaching process.

There are two main reasons as follows: the distribution of impurities in MG-Si is non-uniform and their distribution can be divided into two situations: one is light-coloured inclusions, which are confined to the grain boundaries and can be clearly defined at grain boundaries; the other one is dark-colour inclusions, which are distributed clutter-like particles in MG-Si. In order to separate these impurities from MG-Si as much as possible in the wet pre-treatment process, MG-Si needs to be crushed. During the course of this processing, the impurities existed in MG-Si powders by three forms. The first form is a single particle alone in the

MG-Si powders, the second form is partly enchased in the silicon and partly naked to the air, the third form of impurities is completely coated by silicon with no contact with the outside world. When the MG-Si powders are treated with the leaching agent, the first forms of impurity can quickly react with the leaching agent and go into the solution. Leaching rate can be improved through strengthening stir, increasing the leaching agent concentration and reaction temperature and so on. The second form of impurities contacted with the outside world only partially, the reaction between the agent and the impurities needs to keep from outside to inwards, and this reaction is controlled by a chemical reaction process, and leaching rate can be improved by increasing the reaction temperature, the leaching agent concentration and reducing the original particle radius. The third form of impurities in silicon is coated completely without contact with the outside world. Because silicon does not react with the leaching agent and the reaction between the leaching agent and the impurities is difficult, the reaction process is controlled by internal diffusion. Increasing the reaction temperature, reducing the original particle radius and the thickness of solid membrane can improve their leaching rate. Under the wet pre-treatment process, the first forms of impurity can quickly react with the leaching agent and thus they can be removed commendably. The second form of impurities can also be removed after a period of reaction time. The third form of impurities coated in the silicon is difficult to contact and react with the leaching agent as silicon acts as a good barrier coating for impurities [4].

3. Refining of metallurgical grade silicon by electron beam melting

3.1. Research status of silicon purification by electron beam melting

The global PV industry has been rapidly developed over the past 10 years, which led to a large demand for silicon materials. Silicon was divided into four categories according to the purity of itself that is metallurgical grade silicon (MG-Si), chemical grade silicon (CG-Si), solar grade silicon (SOG-Si) and electronic grade silicon (EG-Si). The solar cell is the key technology for the solution of the energy problem in our planet. In order to search a steady material supply for solar cell, a new process to produce silicon with low cost is necessary. A lot of studies had been explored for the production of high purity silicon by EBM processes [5]. The silicon sample with purity of 99.999% was obtained from an initial purity of 99.91% by EBM [6]. Takashi et al. found that 75% aluminium, 89% calcium, 90% carbon and 93% phosphorus were removed by EBM under 10^{-2} Pa and 30 min; in the meantime, first-order rate equation was used for carbon, calcium and aluminium, and second-order equation was fit for the dephosphorization [7]. Kazuhiro et al. [8, 9] carried out industrial scale experiment for EBM purification process. MG-Si with 25 ppmw of phosphorus was successfully purified to below 0.1 ppmw. Osokin V.A. et al. [10] indicated that the phosphorus content decreased from 0.001 to 0.0003% after EBM under a scanning electron beam in an 8.9×10^{-3} Pa vacuum. Silicon ingot with the conversion efficiency of 14.1% was successfully produced by a new technology. The technology includes electron beam, plasma beam and directional solidification prepared SOG-Si, which was invented by Japan's JET Company [11]. Above studies indicated that electron beam melting has been studied in-depth in foreign countries, but it is still at its initial stages in China due to some reasons.

3.2. Impurity removal process and effect analysis of electron beam melting

In order to study preparation of SOG-Si by EBM, in this section, a new technology of purification process that combined EBM and directional solidification would be researched. The content and distribution of impurity at different locations of silicon disk were also discussed by electron probe microanalysis (EPMA). Impurity contents and surface analysis were confirmed by inductively coupled plasma mass spectrometry (ICP-MS) and scanning electronic microscope (SEM). The schematic diagram of structure and principle for EBM is shown in **Figure 9**. Silicon particles were washed with acetone in an ultrasound cleaner with the objective of removing possible solid residues from the surface before EBM and silicon with 0.4 kg was placed in the copper crucible for every time experiment. Silicon material was melted with electron beam until the whole silicon was melted. After refining process, the powder was reduced tardily until the beam disappeared at the centre of the copper crucible, which provides a temperature gradient from outside to inside in the copper crucible. **Table 3** gives the process parameters.

3.2.1. The macro-morphology of silicon disk

Figure 10 shows the top and bottom views of silicon disk obtained after EBM with a diameter of 130 and 25 mm thickness, respectively. The bottom of silicon disk was very coarse because

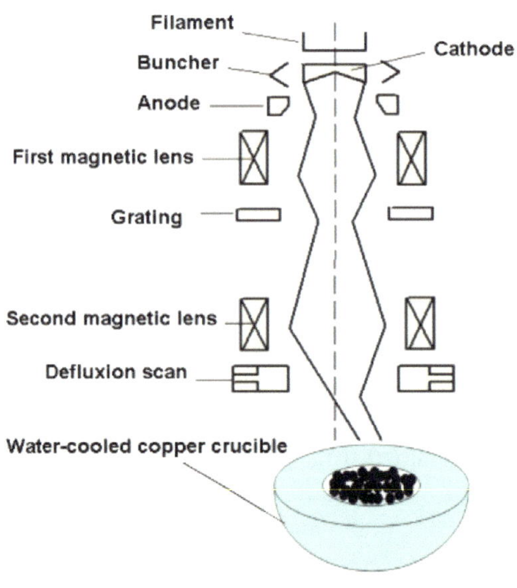

Figure 9. Schematic diagram of structure and principle for electron beam furnace.

Melting time (min)	**20**
Beam power (kW)	10–15
Internal pressure of the furnace (Pa)	2.5×10^{-3} to 5.0×10^{-3}

Table 3. Experimental parameters used in the electron beam melting.

the bottom of the disk contacted with the refrigerated copper crucible, which led to it not fully melting under EBM.

Figure 11 shows the schematic representation of the regions from silicon disk for impurity profile analysis.

3.2.2. Distribution of impurity elements along axial

Distributions of metal impurity and non-metal elements along axial are shown in **Figures 12** and **13**, respectively. It could be seen that metal impurity was dragged from bottom to top along the direction of solidification. The distribution of non-metal impurity was not affected. The location with the lowest impurity content was at about 10 mm above silicon sample bottom. The metal impurities gathered at the top could be explained by segregation coefficients. Metal impurities were pushed aside the liquid region during solidification process because the effective segregation coefficient of metal impurities in silicon was less than 1. Thus, the content of metal impurities increased gradually along the ingot height direction.

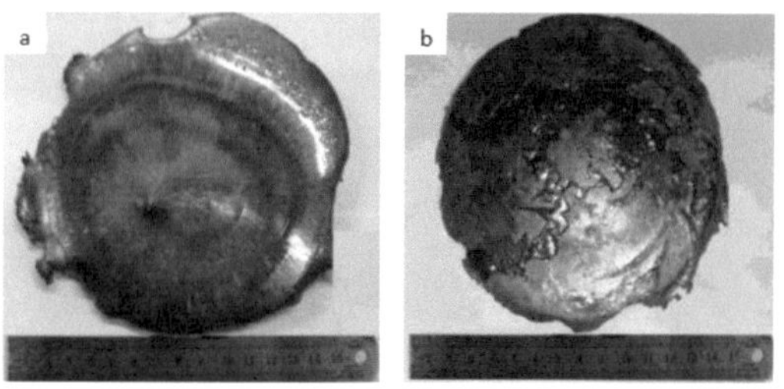

Figure 10. Top view (a) and bottom view (b) of silicon disk after electron beam melting.

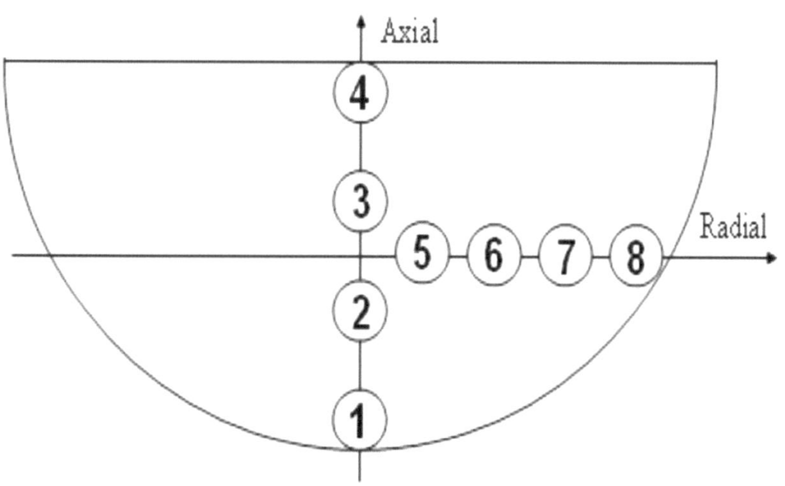

Figure 11. Schematic representation of the regions of silicon disk for impurity profile analysis.

Figure 14 shows the diagram of silicon cake solidification process. There were two temperature gradients during silicon solidification by the special structure of water-cooled copper crucible; they were the axial and the radial temperature gradient. Metal elements were dragged to the remaining liquid region during solidification. Thereby, the solidified part was purified. Elements like Cu, Al, Ti and Fe were segregated effectively to the liquid In the process of solidification because they have low segregation coefficients ($k \ll 1$). However, the non-metal elements like C, O, B and P were little affected because the segregation coefficient of them is close to unit [12]. One noteworthy fact in **Figure 4** is that metal impurities concentrations at the bottom are greater than the area just over it. Non-metallic impurity elements also have the same distribution at the bottom besides B. This could be explained by the fact that the bottom of the silicon disk contacts with the water-cooled copper crucible, so the melting was not sufficient. The bottom of silicon disk has already solidified before metallic impurities had enough time to diffuse fully.

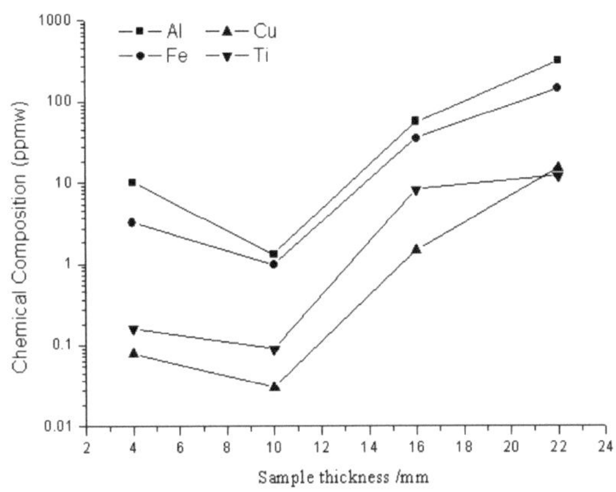

Figure 12. Distribution of metal impurity elements along axial.

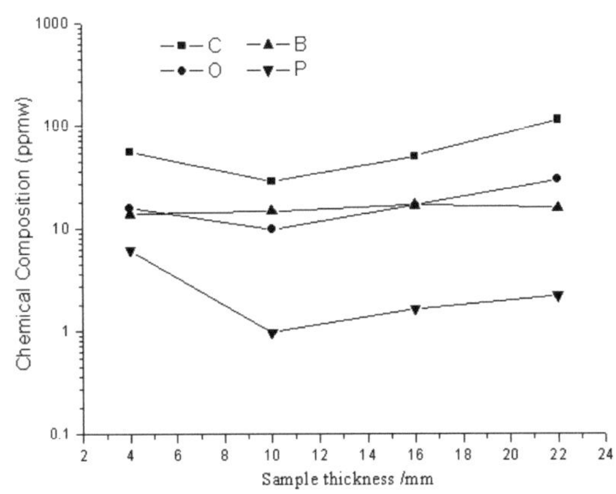

Figure 13. Distribution of non-metal impurity elements along axial.

3.2.3. Distribution of impurity elements along sample radial

Figures 15 and **16** show the distribution of impurity along radial. Metal impurities were also dragged from the edge to the centre in the water-cooled copper crucible. There was almost no change of the non-metal impurities contents from the edge to the centre of the silicon disk. We also found that C element presented the same distribution trends in **Figures 15** and **16**. Carbon could reverse diffusion and could be dragged in the opposite direction of solidification because the segregation coefficient of it was nearer to the unit and the atomic radius of it was also smaller relative to silicon [13]. Silicon carbides precipitation also formed when the carbon concentration was over saturated.

Figure 14. Schematic diagram of the solidification process of silicon cake: (a) the initial stage, (b) the intermediate stage and (c) the end stage.

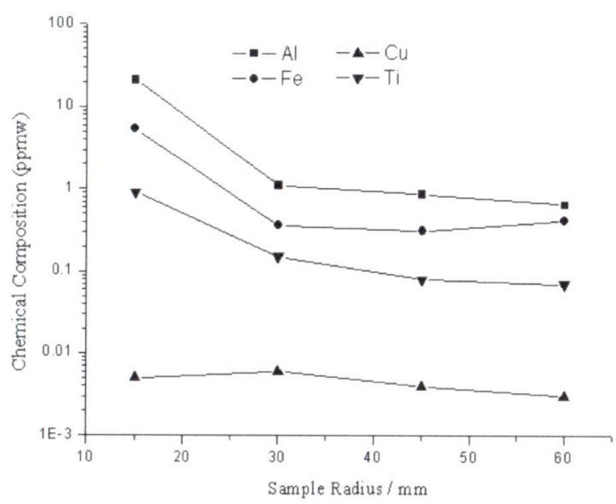

Figure 15. Distribution of metal impurity elements along sample radial.

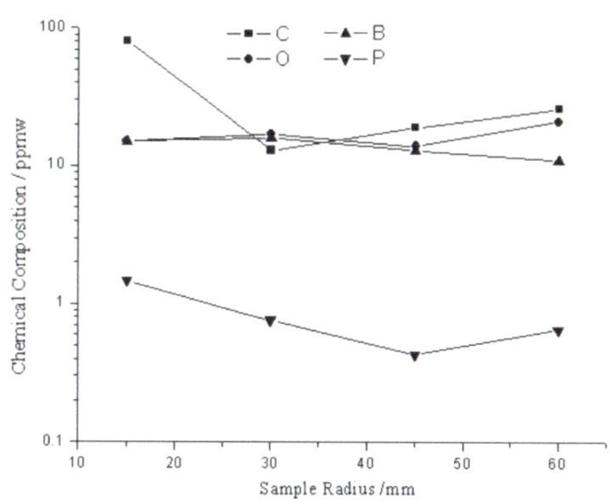

Figure 16. Distribution of non-metal impurity elements along sample radial.

3.2.4. The distribution of impurity elements along axial by electron microprobe analysis

Figure 17(a)–(c) shows the distribution of impurity elements by EPMA corresponding to the metal elements of 4, 2 and 1 sample, respectively.

The distribution of metal elements such as Al, Fe and Ca at the bottom of silicon was very evenly. In the meantime, there was no segregation at the grain boundaries because the bottom of disk contacted the copper crucible, which caused silicon at this position to not fully melt from **Figure 17(c)**. Thus there was no time to spread. From **Figure 17(b)**, it could be seen that metal impurities generated obvious segregation at grain boundaries with the solidification process. At the same time, the number of grain boundaries was very few because the grain size was comparatively large. Most of the inclusions aggregated at the final stage of solidification (the top of solidification portion), due to segregation effects in **Figure 17(a)**. Meanwhile, under the effect of the vertical and horizontal temperature gradient, the grain size decreased and the number of grain boundaries increased. Most of the metal impurities distributed at the grain boundary of silicon crystals owing to the segregation effects and a small number of them was uniformly distributed in grains from the EPMA of the samples. The regularities of distribution for non-metallic elements (such as B and P) were basically the same at the different locations of silicon disk; they scattered equally between grain boundary and grain. The above analysis results were alike along the axial and radial of silicon disk.

3.2.5. Effect of vapour pressure on impurities removal efficiency

Figures 18 and **19** show the changes of metallic and non-metallic impurities in the silicon dealing with different process conditions, respectively. We used sample no. 2 in **Figure 3** to represent the impurity content of silicon cake after electron beam treatment, because the impurities contents in silicon were lowest at this position. From **Figures 18** and **19**, it could be

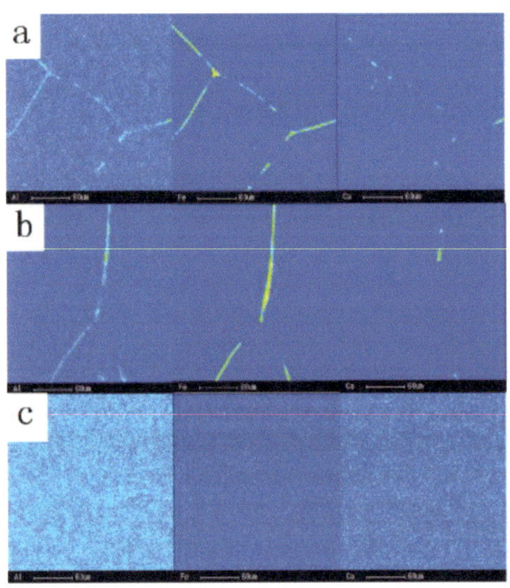

Figure 17. Electron microprobe analysis of metal impurity elements.

seen that metal impurities in silicon had been greatly reduced; however, non-metallic impurities had no change after acid treatment. The contents of metal and non-metallic impurities had been reduced again adequately after EBM. After leaching process, many kinds of impurities elements also remained in MG-Si, but only eight types of impurities still remained in silicon sample after EM because the vapour pressure of Mn, Mg and Ca was higher than that of silicon, therefore, they could be easily removed from silicon by evaporation. The content of Ti, Fe, Al and Cu could also be reduced significantly, although the vapour pressures of them were closer than that of silicon [14]. But the contents of boron did not change along the thickness and radius of silicon after EBM. Because the vapour pressure of this element was very low in relation to that of silicon (10^{-4} Pa of boron and 10^{-1} Pa of silicon) so it was very difficult to remove boron. Therefore, the vacuum process had no effect on its removal. Another reason was the segregation coefficient of boron in silicon was near unit; therefore, unidirectional solidification process for removing boron was also invalid. Plasma melting in an oxidizing atmosphere (O_2, CO_2 or H_2O) for removing boron was effective, boron was transformed into the oxide form (vapour pressure is changed) in this process [15].

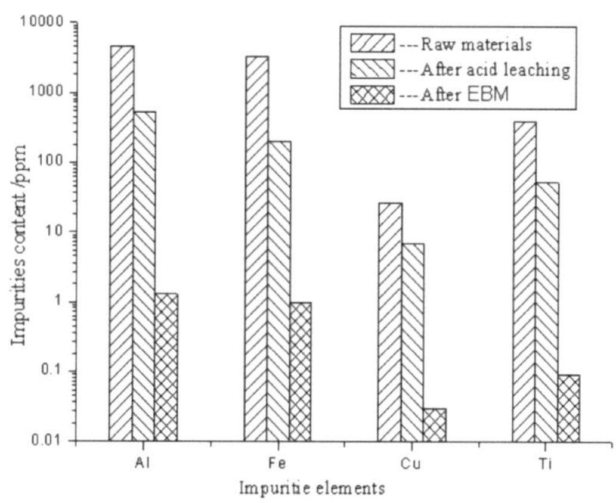

Figure 18. The changes of metal impurities after different treatment processes.

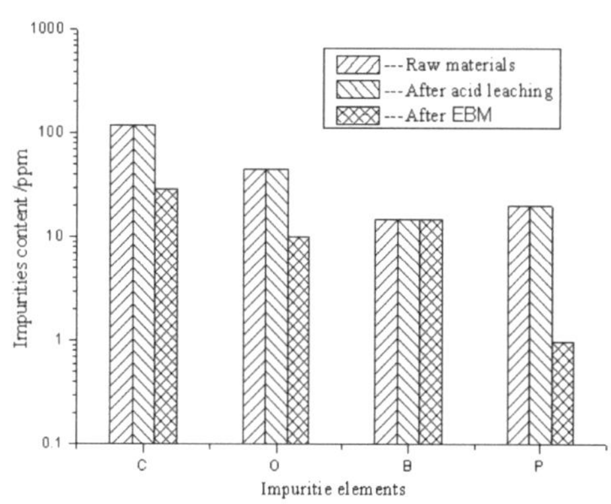

Figure 19. The changes of non-metallic impurities after different treatment processes.

3.3. Limit of impurity removal by electron beam refining

Based on the analysis of the electron beam melting process and the distribution of impurity, we can conclude the following conclusions: (1) Silicon disk (99.995% in mass) was obtained from MG-Si (99.88% in mass) on the centre of the disk after electron beam melting, which had been considered to be a process technically capable of purifying polycrystalline silicon. (2) The distribution characteristics of impurity along sample thickness and sample radius in silicon disk were different. Because metal elements like Cu, Al, Ti, Fe and so on had very low segregation coefficients so they were segregated to the liquid part effectively during solidification. But the non-metal elements like C, O, B, P and so on were a little affected by directional solidification because the segregation coefficient of them was close to unit. (3) The impurities with higher vapour pressure could be removed from silicon easily by evaporation. The impurities like Ti, Fe, Al and Cu with a closer vapour pressure than that of silicon could also be reduced significantly. But it was very difficult to remove the boron because its vapour pressure was very low in relation to the vapour pressure of silicon (10^{-4} Pa for boron and 10^{-1} Pa for silicon) and its segregation coefficient in silicon was near unit.

4. Removal of boron from metallurgical grade silicon (MG-Si) by electromagnetic induction slag melting

4.1. Characteristics and their removal methods of boron impurities in silicon

Scrap of SEG-Si is the main supply for solar cells, but it is difficult to afford a steady supply. Therefore, alternative production processes are needed, refining of metallurgical grade silicon is a very feasible process [16]; however, removing boron by vacuum processes is very difficult because the vapour pressure of boron is very low in relation to that of silicon (10^{-4} Pa for boron and 10^{-1} Pa for silicon). Another difficulty regarding boron extraction is that its segregation coefficient in silicon is near unit which results that there is no removal effect by unidirectional solidification process difficult [17]. A process usually used for the boron removal is plasma melting in an oxidizing atmosphere (O_2, CO_2 or H_2O). In this process, boron is transformed into the oxide form, increasing its vapour pressure [18, 19]. However, it needs a very large initial investment using plasma equipment. Slag refining is very effective method for the removal of boron from molten silicon. The technical viability of the slag refining process has been demonstrated by many researches, but all of them were based on laboratory scale, static theoretical research. Meanwhile, refining time was very long in their study. The values of L_B obtained from their experiments were mostly between 1.5 and 2.5 [20–24]. Meantime, the partition ratio of boron between SiO_2-CaO-Na_2O and SiO_2-CaO-Al_2O_3 slag systems and liquid Si melt at 1823 K were also simulated using the new assessed thermochemical databank together with the FACT oxide thermodynamic database by Bale et al. [25]. The calculated values were approximately two times higher than the experimental values, which indicated that the reaction kinetic barriers might play important roles in the refining processes. In the present study, pilot scale and dynamic slag experiments were carried out by using electromagnetic induction slag melting. The partition ratio L_B, which is the ratio of the content of impurities in the slag to the content of silicon, has been measured, and the removal effects of some metallic impurities are also analysed. The mechanism of boron by EISM was discussed. Additionally, the thermodynamics and kinetics of boron removal were also carried out. Microanalysis and impurity contents were analysed by inductively coupled plasma atomic emission spectrometer (ICP-AES) and electron probe microanalysis (EPMA).

4.2. Experiment results and analysis of electromagnetic induction slag

4.2.1. The testing process and parameters of electromagnetic induction slagging

The schematic diagram of experimental devices is shown in **Figure 20**. An inductive furnace with 200 kW and 3000 Hz was used for melting. The silicon was loaded in a quartz ceramic crucible with an inner diameter of 120 mm and height of 220 mm. The crucible was surrounded by a graphite heater since because at room temperature it was not a conductor. The induction melting of silicon was carried out through water-cooled copper coil under protective gas. Three kinds of slag systems were chosen according to previous studies and the CaO/SiO_2 ratios in all slag systems was 1.21. Silicon material has already been treated by acid leaching, metal impurities have been reduced significantly but non-metallic impurities such as B and P did not change. In each experiment, 3 kg silicon and 0.3 kg slag (10 wt% of silicon) were used. The boron and metal impurities contents in the silicon phase and the slag phase were analysed by ICP-AES. The partition ratio of boron was defined as follows:

$$L_B = \frac{(\text{ppm\% B})}{[\text{ppm\% B}]} \quad (1)$$

where (ppm% B) and [ppm% B] are the concentrations of boron in slag and silicon, respectively. Experimental conditions and parameters were shown in **Table 4**.

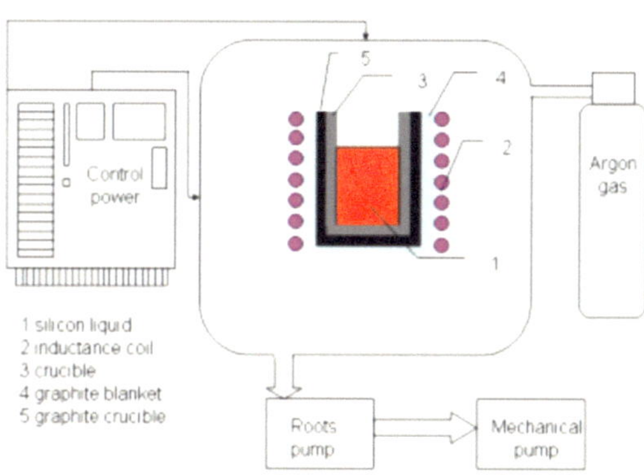

Figure 20. The schematic diagram of experimental devices.

4.2.2. The test parameters of electromagnetic induction on slagging boron removal effect

4.2.2.1. The effect of the third oxide and refining time on boron removal

Four types of typical slag systems were chosen for comparative study according to the previous research [20–23]. The photograph of silicon ingots after test is shown in **Figure 21**. From **Figure 21(a)**, it could be found that the most slag froze at the bottom of silicon ingots after test and a small amount of them covered at the side of silicon ingot. Slag deposited to the bottom of the crucible and solidified here because the density of slag and silicon melt is different, the former is larger. So, they can be well separated. The internal silicon ingot was very bright, and no slag agent in the inner was found from **Figure 21(b)**. This proved again that the slag and silicon could separate very well.

The values of L_B after melting test under different types of slag at 1773 K for 1 h are shown in **Figure 22**. It could be seen that boron removal was significantly improved by using EISM compared to previous static test results [19–21]. Although CaO/SiO_2 ratios were same in different slag systems, the values of L_B were very different by adding different third oxide to the CaO/SiO_2 slag system. The values of L_B significantly reduced by adding CaF_2 but significantly enhanced by adding Na_2O and Al_2O_3. Therefore, this chapter focused on the removal of boron at different refining parameters under SiO_2-CaO-Na_2O and SiO_2-CaO-Al_2O_3 slag systems. The reasons for above phenomenon would be presented in the thermodynamic analysis section of this article.

Parameters	**Conditions**
Initial boron content in silicon	15 ppm
Sorts of used slags	SiO_2-CaO; SiO_2-CaO-(10 wt%)CaF_2
SiO_2-CaO-(10 wt%)Na_2O; SiO_2-CaO-(10 wt%)Al_2O_3	Experimental times
0.5–2 h	Experimental temperatures
1723–1873 K	Atmosphere
High purity Ar(99.999%)	

Table 4. Experimental conditions in the slag treatment.

Figure 21. Photos of silicon ingot after electromagnetic induction slag melting.

Figure 23 shows the change of boron removal with refining time at 1823 K under SiO_2-CaO-Na_2O and SiO_2-CaO-Al_2O_3 slag systems. It could be found that L_B was significantly improved compared to previous studies (L_B value was in the range1.5–2.5) by EISM though change situation of L_B in the two kinds of slag systems was different. Diffusion resistance of impurity elements at the boundary layer could be reduced and the diffusion of them also enhanced by increasing the melt flow rate under EISM. So the removal rate of boron could be greatly sped up and the required refining time was also reduced substantially. Impurity removal effect was obviously better under SiO_2-CaO-Al_2O_3 slag systems than that under SiO_2-CaO-Na_2O systems. And L_B value gradually increased with the increasing times under the SiO_2-CaO-Al_2O_3 systems. This showed that the increase in refining time was conducive to the removal of boron. But L_B value increased and then reduced with the increasing times because the use of Na_2O might increase the basicity and also decrease the melting point of the slag. However, sodium oxide could be easily reduced to the silicon phase and became volatile at high temperatures, which made it difficult to work for a prolonged period [22].

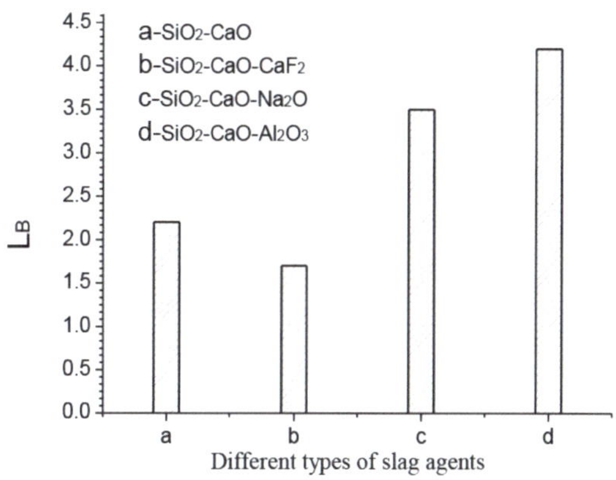

Figure 22. Partition ratio L_B under different types of slag systems.

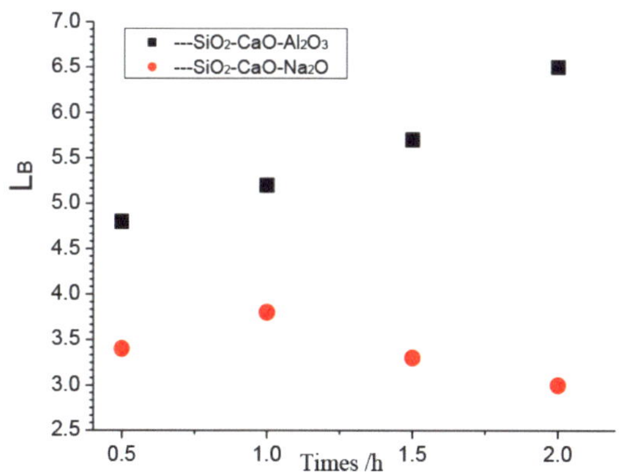

Figure 23. Partition ratio L_B at different times at 1823 K.

4.2.2.2. The effect of refining temperature and slag composition on boron removal

Figure 24 shows the change of boron removal with refining temperatures for 1 h under SiO_2-CaO-Na_2O and SiO_2-CaO-Al_2O_3 slag systems. The value of L_B gradually increased with the increase of refining temperatures under SiO_2-CaO-Al_2O_3 slag systems. The removal effect of boron was raised along the refining temperatures and is increased owing to the viscosity of slag agent debased along with the temperature increased. at the same time, the silica activity in slag and the mobility of slag both enhanced. The above conditions were both conducive to the removal of boron. However, under the SiO_2-CaO-Na_2O systems the removal effect of boron was better at low temperatures than that at high temperatures. The reason for it was also because sodium oxide could be easily reduced to the silicon phase and became volatile at high temperatures [22].

Figure 25 shows the changes of boron removal with the content of Na_2O and Al_2O_3 in slag. L_B had the same change trend with the content of Na_2O and Al_2O_3 in SiO_2-CaO-Na_2O and SiO_2-CaO-Al_2O_3 slag systems. It increased as the proportion of Na_2O and Al_2O_3 in slag increased. The better partition ratio and a higher value for capacity could be obtained if compounds with higher basicity than that of CaO or CaF_2 were selected because both the basicity of slags and the activity of silica in slag could be greatly improved by adding alkaline oxide into slag [22].

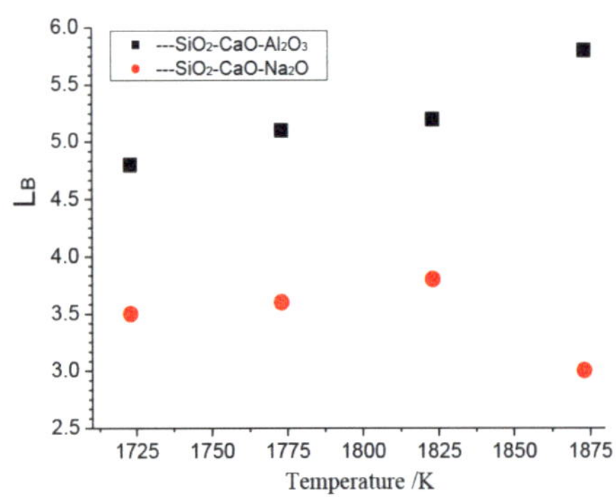

Figure 24. Partition ratio L_B at different temperatures for 1 h.

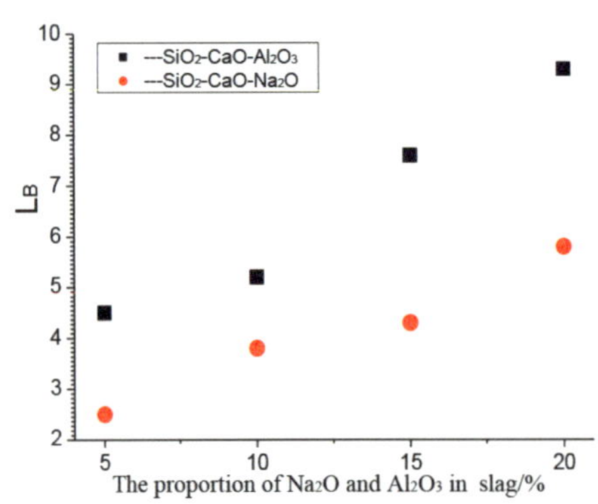

Figure 25. Partition ratio L_B with different contents of Na_2O and Al_2O_3 in slag at 1823 K for 1 h.

4.2.2.3. Removal of other impurity elements

Not only the boron elements could be easily removed by EISM but also other elements could be removed such as Ca, Al and Mg. **Figure 26** shows that Al, Ca and Mg elements could be removed by slag refining; meanwhile their removal rates were 85.0, 50.2 and 66.7%, respectively. **Figure 27** shows the analysis of metal impurity in silicon and slag by EPMA after refining for 2 h at 1823 K under SiO_2-CaO-Al_2O_3 systems. It could be clearly seen that the distribution of Ca, Al and Mg elements was same on both sides of the interface between silicon and slag. They all concentrated in slag phase. The Ca and Al in the slag did not contaminate silicon melt because the state of them existed in slag was relatively stable and the form of them contacted with silicon liquid was molten oxide. Therefore silicon liquid could not be contaminated. Al and Mg elements were more inclined to gather at slag phase because the oxygen content in slag phase was relatively high and the affinity of these two elements with slag was greater than that of silicon.

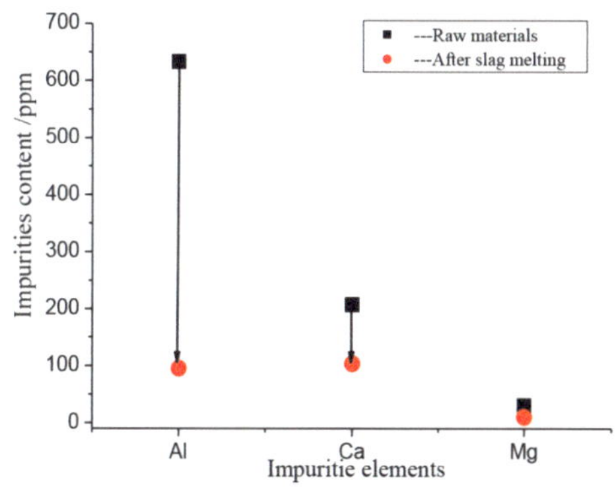

Figure 26. Impurity elements content of silicon before and after slag refining at 1823 K for 2 h.

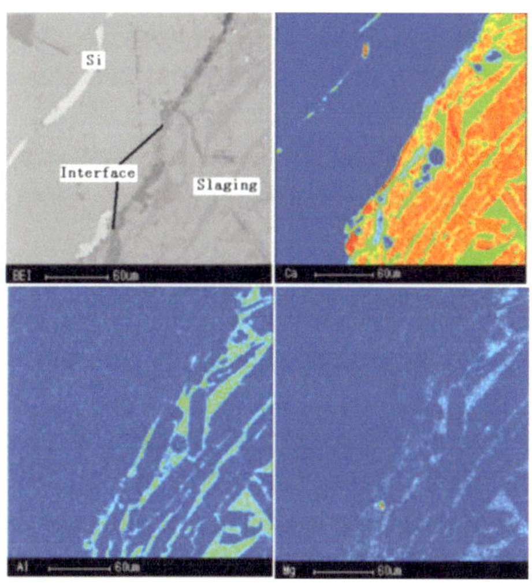

Figure 27. EPMA of metal impurity elements in silicon and slag after refining at 1823 K for 2 h.

4.3. Thermodynamic and Kinetic analysis of boron removal

4.3.1. Thermodynamic analysis of boron removal

The reaction of removal of boron from Si by using basic fluxes could be expressed by Eq. (2):

$$\underline{B}+\frac{3}{4}(SiO_2)=\frac{3}{4}Si+(BO_{1.5}) \quad (2)$$

The boron dissolved from silicon liquid to the slag at the interface and silica got reduced and entered the liquid melt, boron element got oxidized and entered the slag due to chemical reaction. The distribution coefficient of boron was given as the ratio by Eq. (3):

$$L_B = \frac{(\%B)}{[\%B]} = \frac{Kf_B}{\gamma_{BO_{1.5}} k_{x\to\%}}\left(\frac{a_{SiO_2}}{a_{Si}}\right)^{3/4} \quad (3)$$

where the square brackets denote boron dissolved in the silicon and parentheses denote boron in the slag. From **Figure 3**, it could be known that removal effect of boron under SiO_2-CaO-Na_2O and SiO_2-CaO-Al_2O_3 slag systems was better than that under SiO_2-CaO-CaF_2. The reason was as follows: the activity coefficient of boron oxide was calculated for CaO/SiO_2-based slag at 1723–1873 K and could be approximately expressed as Eq. (4):

$$\ln\gamma_{BO_{1.5}} = -4.00(CaO/SiO_2)+3.67 \quad (4)$$

When CaF_2 was added in the slag, CaO could be generated after CaF_2 reacted with silica, thereby increasing the basicity of slags. Although the addition of CaF_2 provided a broader basicity range compared to the binary system, the content of SiO_2 in slag reduced and the oxygen partial pressure also significantly reduced. Meanwhile, the ratio of CaO/SiO_2 in Eq. (4) increased and resulted $\gamma_{BO_{1.5}}$ also reduced. But the change degree $\gamma_{BO_{1.5}}$ was small relative to $(a_{SiO_2}/a_{Si})^{3/4}$. Therefore, L_B would reduce according to Eq. (3) [23]. It was apparent that when Na_2O and Al_2O_3 were added in slag, a better partition ratio and a higher value for capacity might be obtained. This was because not only the basicity of slag could be greatly improved but also the oxygen partial pressure of slag could also be greatly increased by adding alkaline oxide into slag. In this case, the oxygen partial pressure of $(a_{SiO_2}/a_{Si})^{3/4}$ also significantly increased. Meanwhile, the ratio of CaO/SiO_2 in Eq. (4) increased and resulted $\gamma_{BO_{1.5}}$ also reduced. So L_B would increase according to Eq. (3). In addition, the use of Al_2O_3 might decrease the melting point of the slag and the mobility and affinity of slag were also improved markedly [24]. L_B value gradually increased with temperatures under EISM. This was because the activity coefficient of boron (f_B^0) in dilute solution silicon was expressed by Eq. (5) between 1723 and 1923 K [22]:

$$\log f_B^0 = -\frac{1.11\times10^4}{T}+5.82 \quad (5)$$

increased with temperature. Meanwhile, the equilibrium constant K also increased. The effect of temperature on $\gamma_{BO_{1.5}}$ was very small, so it could be ignored. The activity coefficient of boron oxide was affected by the incorporation between boron and silicate network. The further decreased on $\gamma_{BO_{1.5}}$ as the silica content increased was not necessarily a function of the amount of boron incorporated in the network but might be related to the degree of polymerization of the slag structure. So, L_B would increase according to Eq. (3) under SiO_2-CaO-Al_2O_3 slag system. But L_B value would reduce with the increasing temperatures under the SiO_2-CaO-Na_2O slag system because Na_2O could easily become volatile at high temperatures [22]. Because the boron was micro in the silicon, boron diffuson from silicon into slag phase needed certain time. This will be discussed in detail in the following section.

4.3.2. Kinetic analysis of boron removal

Following five steps were needed in refining impurity element by slag refining: (1) impurity transfers from the bulk metallic phase to the metal boundary layer $[\%X]_b \rightarrow [\%X]_\delta$; (2) impurity diffuses through the metal boundary layer $[\%X]_\delta \rightarrow [\%X]_i$; (3) metal impurity was oxidized at the interface between metal melt and slag melt $[\%X]_i \rightarrow (\%X)_i$; (4) impurity spread through the slag boundary layer $(\%X)_i \rightarrow (\%X)_\delta$ and (5) impurity transferred from the slag boundary layer to slag bulk phase $(\%X)_\delta \rightarrow (\%X)_b$. Above process depended on stirring and mixing between silicon fluid and slag. We often use gas bubbling or mechanical devices to increase the mass transfer in the bulk phases. Therefore, the viscosity of slag properties was important. The low velocities in the slag and the low mass transfer of the impurity element could be obtained if the slag had a high viscosity; meanwhile, the diffusivity of the impurity element was also lowered. The mass transfer coefficients β (in the metal) and β_S (in the slag) determined processes for Steps 2 and 4, respectively. A serious difficulty with refining silicon by extraction to a second (slag) phase was the problem of mixing in the slag phase. Often, the slag phase was relatively viscous, so that it was difficult to mix the impurity element throughout the slag.

In this section, we used the electromagnetic induction slag refining method. The melt movement which was produced by electromagnetic force in the crucible is shown in **Figure 28**. Electromagnetic force was formed due to the interaction function by the induction eddy current in melt and the magnetic field in the medium frequency induction coil. Slag agent followed the silicon fluid movement and they could contact fully in crucible because the electrical conductivity of them was different. Electromagnetic induction melting was better relative to gas mixing and mechanical agitation because Electromagnetic induction melting process did not contact with silicone melt and did not lead into other impurities. The effective boundary layer thickness (units) is as follows according to boundary layer theory:

$$\delta = \frac{c - c^*}{\left(\frac{\partial c}{\partial x}\right)_{x=0}} \tag{6}$$

where c^* is the concentration for interface and c is the concentration for liquid bulk concentration outside the boundary layer ($c^* > c$). In addition, the mass transfer coefficient of impurity could be expressed as follows by the surface renewal theory:

$$\beta = \sqrt{DS} \tag{7}$$

where D is the diffusion coefficient for impurity element and S is the surface renewal rate.

The boundary layer thickness was thinner as the concentration gradient near interface was larger by Eq. (6). Thickness of boundary layer could be reduced if the concentration gradient became larger by increasing fluid flow. Diffusion resistance did not exist finally when the flow rate increased (run up to critical velocity) until the thickness of boundary layer achieved zero. Diffusion resistance of impurity elements at the boundary layer could be reduced and the diffusion of them could be enhanced with the increase in the melt flow rate by electromag-

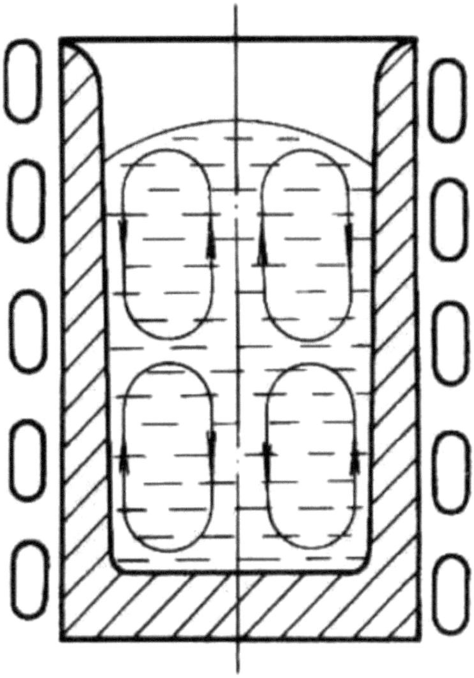

Figure 28. The flow lines of silicon melt under electromagnetic induction melting.

netic stirring. At the same time, the diffusion coefficient of impurity elements also could be enhanced. The mass transfer coefficient β and β_S both had been increased according to Eq. (7). Therefore, the removal of boron could be improved and sped up.

The above is why the boron impurities in silicon could be significantly reduced by the EISM.

4.3.3. Conclusion

The following conclusions can be drawn from the above experimental process and analysis process: (1) MG-Si with 15 ppm of boron was successfully purified to 2 ppm after EISM at 1823 K for 2 h under SiO_2-CaO-Al_2O_3 slag systems. Meanwhile, Al, Ca and Mg elements in MG-Si were also well removed and their removal efficiency reached 96.4, 91.8 and 76.2%, respectively. (2) The value of L_B increased with temperature and refining time owing to the mobility of slag is enhanced and the viscosity of slag agent is reduced; meanwhile, the silica activity in slag increases along with increase in temperature. (3) The boron content of silicon could be significantly improved and the level of it achieved the SOG-Si requirement after EISM because diffusion resistance of impurity at the boundary layer could be reduced and the diffusion of impurity could be enhanced by increasing the melt flow rate. At the same time, the mass transfer coefficient β and β_S both had been enhanced; therefore, the removal of boron could be improved and sped up. Above discussion showed that EISM was a very effective method for removal of boron from silicon.

Acknowledgements

This research was supported by the National Natural Science Foundation of China (No. 50674018); young- and middle-aged backbone teachers training program of Chengdu University of Technology (No. 10912-JXGG201516) and Key Natural Science Foundation of Sichuan Provincial Department of Education (No. 15ZA0079).

Author details

Dawei Luo

Address all correspondence to: luodawei2013@cdut.cn

College of Materials, Chemistry & Chemical Engineering, Chengdu University of Technology, Sichuan, China

Collaborative Innovation Centre of Panxi Strategic Mineral Resources Multi-Purpose Utilization, Chengdu University of Technology, Chengdu, Sichuan, China

References

[1] Dietl J. Hydrometallurgical purification of metallurgical grade silicon. Solar Cells, 1983, 10:145–154. DOI:10.1007/s12598-009-0043-1.

[2] Kantos I.C., Gonalves A.P., Santos C.S., et al. Purification of metallurgical grade silicon by acid leaching. Hydrometallurgy, 1990, 23(2–3): 237–246. DOI:10.1016/0304-386X (90)90007-O.

[3] Kotval P.S. Process for the production of refined metallurgical silicon, U.S. Patent 4195067, 1997. ID: lens.org/125-322-381-086-850.

[4] Zhan L.Y. Pre-treatment experimental study of direct preparation solar grade silicon from metallurgical grade silicon [thesis]. Kunming University of Science and Technology, Kunming, 2007.

[5] Maijer D.M., Ikeda T., Cockcroft S.L., Maeda M., Rogge R.B. Mathematical modelling of residual stress formation in electron beam remitting and refining of scrap silicon for the production of solar-grade silicon. Materials Science and Engineering, 2005, 390:188–201. DOI:10.1016/j.msea.2004.08.036.

[6] Pires J.C.S., Braga A.F.B., Mei P.R. Profile of impurities in polycrystalline silicon samples purified in an electron beam melting furnace. Solar Energy Materials and Solar Cells, 2003, 79:347–355. DOI:10.1016/S0927-0248(02)00471-3.

[7] Takashi I., Masafumi M. Purification of metallurgical silicon for solar grade silicon by electron beam button melting. Iron and Steel Institute of Japan International, 1992, 32:635–642. DOI:10.2355/isijinternational.32.635.

[8] Kazuhiro H., Noriyoshi Y., Yoshiei K. Evaporation of phosphorus in molten silicon by an electron beam irradiation method. Materials Transactions, 2004, 45:844–849. DOI:10.2320/matertrans.45.844.

[9] Noriyoshi Y., Kazuhiro H., Yoshiei K. Removal of metal impurities in molten silicon by directional solidification with electron beam heating, materials transactions, 2004, 45:850–857. DOI:10.2320/matertrans.45.850.

[10] Osokin V.A., Shpak P.A., Ishchenko V.V., Panibratskii V.A., Piyuk E.L. Electron-beam technology for refining polycrystalline silicon to be used in solar power applications. Metallurgist, 2008, 52:121–127. DOI:10.1007/s11015-008-9018-y.

[11] Yuge N., Abe M., Hanazawa K., Baba H., Nakamura N., Kato Y., Sakaguchi Y., Hiwasa S., Aratani F. Purification of metallurgical grade silicon up to solar grade. Progress in Photovoltaics: Research and Applications, 2001, 9:203–209. DOI:10.1002/pip.372.

[12] Möller H.J. Semiconductors for solar cell applications. Progress in Materials Science, 1991, 35:205–206. DOI:10.1016/0079-6425(91)90001-A.

[13] Barbosa L.C., Zone refining of silicon. [thesis], Faculdade de Engenharia de Campinas, Universidad Estadual de Campinas, Campinas, Sǎo Paulo, Brazil, 1981.

[14] Pires J.C.S., Otubo J., Braga A.F.B., Mei P.R. The purification of metallurgical grade silicon by electron beam melting. Journal of Materials Processing Technology, 2005, 169:16–20. DOI:10.1016/j.jmatprotec.2004.03.035.

[15] Ikeda T., Maeda M. Elimination of boron in molten silicon by reactive rotating plasma arc melting. Materials Transactions Japan Institute of Metals, 1996, 37:983–987. DOI:10.2320/matertrans1989.37.983.

[16] Braga A.F.B., Moreira S.P., Zampieri P.R., Bacchin J.M.G. Mei P.R. New processes for the production of solar-grade polycrystalline silicon: A review. Solar Energy Materials & Solar Cells, 2008, 92:418–424. DOI:10.1016/j.solmat.2007.10.003.

[17] Ciszek T.F., Schwuttke G.H., Yang K.H. Solar-grade silicon by directional solidification in carbon crucibles. IBM journal Of research Develop, 1979, 23:270–277. DOI:10.1147/rd.233.0270.

[18] Takashi I., Masafumi M. Elimination of boron in molten silicon by reactive rotating plasma arc melting. Materials Transactions Japan Institute of Metals, 1996, 37:983–987. DOI:10.2320/matertrans1989.37.983.

[19] Wu J.-J., Ma W.-H., Yang B., Dai Y.-N., Morita K. Boron removal from metallurgical grade silicon by oxidizing refining. Transactions of Nonferrous Metals Society of China, 2009, 19: 463–467. DOI:10.1016/S1003-6326(08)60296-4.

[20] Liaw H.M., Secco F. Purification of metallurgical grade silicon by slagging and impurity redistribution. Solar Cells, 1983, 10:109–118. DOI:10.1016/0379-6787(83)90012-1.

[21] Ryouji N., Kichiya S., Fumitaka T., Nobuo S. Thermodynamics of boron in silicon melt. Metallurgical and materials transactions B, 1994, 25:903–907. DOI:10.1007/BF02662772.

[22] Torsten W., Klaus S. Chemical equilibria between silicon and slag melt. Metallurgical and Materials Transactions B, 1994, 25:497–504. DOI:10.1007/BF02650071.

[23] Leandro A., Viana T., Kazuki M. Removal of boron from molten silicon using $CaO-SiO_2$ based slags. Iron and Steel Institute of Japan International, 2009, 49:783–787. DOI:10.2355/isijinternational.49.783.

[24] Leandro A., Viana T., Yomei T., Toshinobu Y., Kazuki M. Behavior and state of boron in $CaO-SiO_2$ slags during refining of solar grade silicon. Iron and Steel Institute of Japan International, 2009, 49:777–782. DOI:10.2355/isijinternational.49.777.

[25] Bale C.W., Chartrand P., Degterov S.A., Eriksson G., Hack K., Mahfoud R.B., Melançon J., Pelton A.D., Petersen S. FactSage thermochemical software and databases, Calphad, 2002, 26:189–228. DOI:10.1016/j.calphad.2008.09.009.

Chapter 5

Non-Vacuum Process for Production of Crystalline Silicon Solar Cells

Abdullah Uzum, Seigo Ito, Marwan Dhamrin and Koichi Kamisako

Additional information is available at the end of the chapter

http://dx.doi.org/10.5772/intechopen.68170

Abstract

Existing technologies for conventional high-efficient solar cells consist of vacuum-processed, high cost, sophisticated, and potentially hazardous techniques ($POCl_3$ diffusion, SiNx deposition, etc.) during crystalline silicon solar cell manufacturing. Alternative research studies of non-vacuum and cost-efficient processes for crystalline silicon solar cells are in continuous demand. However, there is not a well understanding of utilizing schemes and the achievable performances for such applications and techniques in solar cell fabrication. This chapter addresses the non-vacuum processes and applications for crystalline silicon solar cells. Such processes including spin coating and screen-printing phosphorus and boron diffusions for the formation of n+ and p+ emitter or back surface fields, spin coating and spray-deposited antireflection coatings for silicon solar cells. Application techniques were explained by combining and comparing experimental results with the calculation and simulations. Consequently, the aim of this chapter is to provide a good understanding of the non-vacuum processes for crystalline solar cells both with simulation and with experimental proves.

Keywords: low-cost solar cell, non-vacuum process, single diffusion selective, screen-printing, crystalline silicon

1. Introduction

Solar energy is one of the most interesting and practical alternative source of energy against the conventional fossil fuels. Crystalline silicon-based solar cells are dominant by far in photovoltaic industry and shares about 90% solar cell production worldwide. Low cost is still one of the main interests for the photovoltaic industry while stabilizing high conversion efficiency. Utilized materials and processing techniques are the major components that need to be considered for cost reduction. A conventional p-type solar cell fabrication process basically comprises of texturing, n-type emitter formation (phosphorus diffusion), bulk passivation, antireflection coating, back surface field, and back/front metallization steps. Although, high-efficiency solar cells greater than 20% including record efficiencies [1] provide a good performance, they cannot demonstrate the feasibility for low-cost scale due to the complex structures and long and costly fabrication processes. Therefore, the efficiency of standard industrial cells still remains in the range of 15–20%. In order to meet industrial requirements with simple low-cost technologies with high throughput, cost-effective methods need to be investigated and adapted to the solar cell manufacturing process. By adaptation of such simple processing techniques including spin coating, spray deposition, and screen-printing methods with proper materials, high-efficiency solar cells that can be compatible with the commercial counter parts should be achievable. This study determines the ground for totally vacuum-free, low-cost crystalline silicon solar cell manufacturing process and applications.

2. Basic low-cost methods for film deposition

2.1. Spin-on deposition technique

The spin-on deposition (SOD) process generally applied by placing the substrate on the spinner and then the liquid source is dropped on the center of the wafer as shown in **Figure 1 (a)**. An optimized rotation speed is needed which depends on the size of the wafer and viscosity of solution, to avoid contamination on rear side of wafer. The coated wafer is usually baked on a hot plate in order to evaporate solvents (**Figure 1 (b)**) and to achieve a concrete layer. Then, the subsequent procedures such as thermal annealing at high temperatures, etc. might follow if necessary.

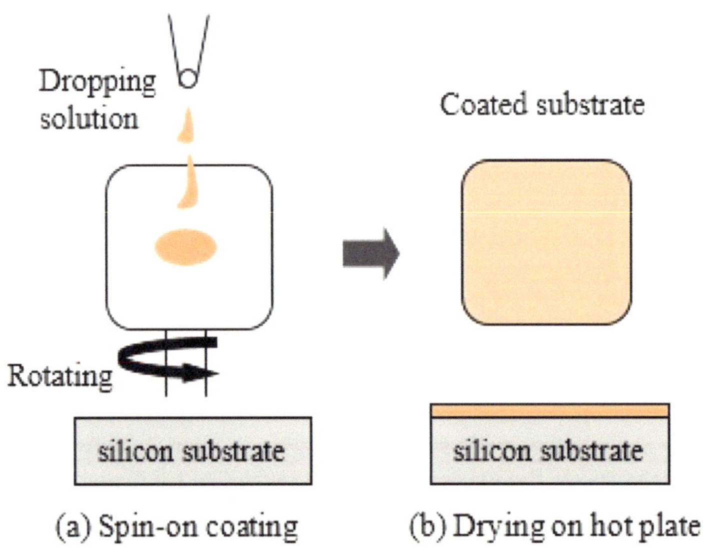

Figure 1. Spin coating technique. (a) Spin-on coating and (b) drying on hot plate.

Possibility of in-line processing for dopant sources to form semiconductor junctions, low energy consumption, simplicity, and low cost are the main advantages of SOD. For most of the SOD sources, low viscosity leads to homogeneous coatings even on textured samples but optimizations are usually necessary for a specific task. Depending on the SOD sources even storing in a refrigerator might be required, warming up to room temperature is usually necessary before the deposition process.

2.2. Screen-printing technique

Screen-printing is a fast, reliable, and cost effective technology which is used mainly for metallization purposes for silicon solar cell industry. The quality of the printing depends mainly on the parameters including the quality of printing mask, viscosity of the paste, snap-off distance, squeegee speed, scrapper pressure, squeegee pressure, and squeegee angle to the mask. For printing process, paste is forced through the openings of the emulsion layer onto the surface of the wafer. The printing mask consists of a frame that holds a stretched fabric with a photo stencil attached to the mesh with the required design of the grid pattern. That allows the transference of a diffusion ink or paste through a screen in a designed pattern on a silicon substrate located below and aligned with the pattern to be transferred as described in **Figure 2**.

The distance between the designed mask and the solar cell surface is called "snap-off" distance. It concerns with the vertical printing properties and by changing the snap-off distance, the pressure on the mask can be changed. The printing speed is also an important parameter and should be adjusted for each printing task because it affects the printed thickness/amount of the paste and the quality of the printing. Due to the increased applied force with increased printing speed, the print through process becomes easier. In order to achieve high throughput rates, a fast printing speed is required. Because the squeegee is the tool which contacts

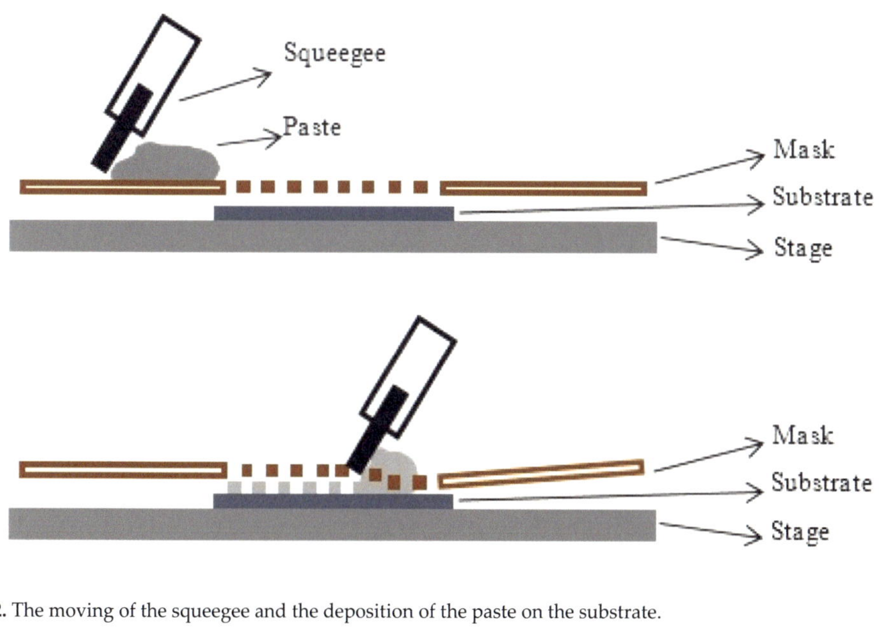

Figure 2. The moving of the squeegee and the deposition of the paste on the substrate.

directly to the mask and plays a key role on paste transfers through the mask. Hardness and the shape of the squeegee are needed to be considered. An excessive pressure on the squeegee results in a limited printing quality of fine-line printed contacts. It leads to an increased abrasion, and thus the lifetime of the screen-printing form and the squeegee is shortened. After the deposition of the paste on the surface, wafers are usually dried before the thermal annealing in case of the requirement of high temperature process. Besides the availability to form homogenous structures, the possibility of forming any designed structures in a simple way maybe the most attractive advantage for using screen-printing technique.

2.3. Spray deposition technique

Spray deposition technique is one of the low cost thin film-coating processes for solar cells. This technique is based on spraying particles of a precursor solution utilizing a spraying nozzle which can be controlled manually by a hand-pump or by automated pumping system. Fast processing, simplicity and low cost are the main advantages of spray deposition technique. In **Figure 3**, given an example setting schematic of spray deposition system, including spray nozzle, beaker for precursor solution, pipe which connects the pump to the nozzle and a manual or automated pump. The thickness of the coated films is usually controlled by the amount of precursor. And the uniform coating can be achieved by optimizing the height and facing angle of the nozzle to the substrate and the pressure of pumping. Generally, the homogeneity of the film can be altered by rotating the substrate in between each spraying pace. Considering the different materials and precursors, the substrate temperature is crucial. Therefore, pace of spraying in mL s^{-1} also needs to be taken into account.

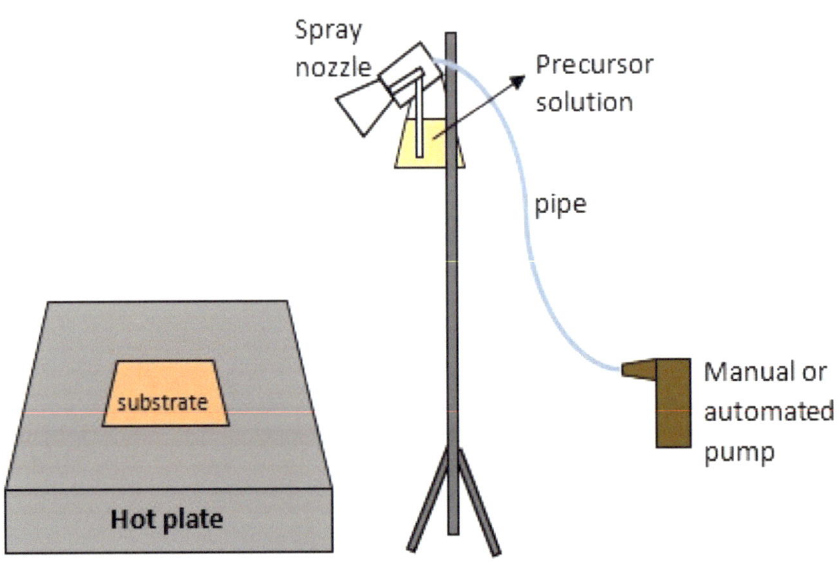

Figure 3. A schematic of a simple spray coating system.

3. Antireflection coating films for low-cost solar cells

In conventional crystalline silicon solar cells micrometer scale alkaline or acidic surface texturing is the primary method for light trapping into the solar cell. Additionally, antireflection coatings (ARC) are used to reduce the reflectivity. Various material and thin films have been used for ARC purposes including SiO_2, SiN_x, TiO_2, Al_2O_3 thin films, etc. SiN_x is the chief ARC film in conventional silicon solar cells industry with very good antireflective properties [2, 3]. SiN_x and its stack layers (e.g., SiN_x/SiO_x) are also used for passivation effect which leads higher carrier lifetimes [3]. These films are deposited by well-known plasma-enhanced chemical vapor deposition (PECVD) technique [4, 5] which has drawbacks including the need of toxic and hazardous gases such as SiH_4 and NH_3 with vacuum processing for CVD operation and considerably high costs. In order to meet industrial requirements with simple low-cost technologies with high throughput, cost-effective methods such as spin coating and spray deposition techniques are required to be investigated for crystalline silicon solar cell fabrication. Primarily, TiO_2 with good optical characteristics, found to be an attractive antireflection coating for solar cells [6, 7]. A considerable amount of literature has been published on sol-gel processes for silicon solar cells including TiO_2, TiO_2-SiO_2 [6, 7] and some Al_2O_3-based Ti doped mixed sol-gel sources as well have been introduced [8, 9]. On the other hand, ZrO_2 can be considered as a promising material for an antireflection coating of solar cells with high refractive index and thermal strength [10]. However, more research and analysis are needed to carry out for better understanding of this kind of material for adapting to the solar cell manufacturing.

3.1. Spin coating ZrO_2 film and TiO_2 films for ARC

3.1.1. Material optimization

In case of spin coating applications, development of spin coated material for a quality application while keeping desired electrical properties in a high-quality level is crucial for applications. Generally, nanoparticle processing makes it possible to utilize the solution processes for solar cells. Proper control of the material compositions and the applicability of low temperature processes are the main advantages. Investigating such kind of broad range of possibility of material and processes is important for solar cell manufacturing in terms of new material development, enhancement of the properties of a designed solar cell, especially the absorption of photons, and for cost reduction as well. This study consists investigation of colloidal suspension of ZrO_2 nanoparticles in water (ZR-30AH, provided by Nissan Chemical Industry Co. Ltd.), surface-modified colloidal suspension of TiO_2 nanoparticles in water (nanoUSE-Ti, Nissan Chemical Industry Co. Ltd.), and diluted TiO_2 nanoparticles (PST-18NR, provided by JGC Catalysts and Chemicals Ltd.). The average diameter of particles and the rate of solid components of ZrO_2 nanoparticles, surface-deactivated TiO_2-nanoparticles and TiO_2-nanoparticles are 50 nm (30 wt%), 13 nm (30 wt%), and 18 nm (17 wt%), respectively. Transmission electron microscopy (TEM) images of each nanoparticle were given in **Figure 4**.

Solutions for composite films were prepared by mixing one of the nanoparticles with ethanol and organic polymer of ethyl cellulose. For the preparation of ethyl cellulose, mixture of

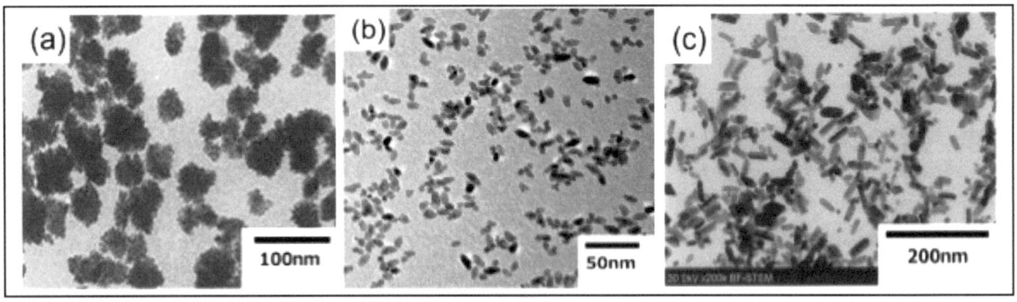

Figure 4. TEM images of nanoparticles of; (a) ZrO_2 (ZR-30AH, Nissan Chemical Industrial Co. Ltd.), (b) TiO_2 with surface deactivation (nanoUSE-TI, Nissan Chemical Industrial Co. Ltd.), and (c) TiO_2 (PST-18NR, JGC-CCIC Co. Ltd.) [11].

E0265 and E0266 (provided by Tokyo Chemical Industry Co. Ltd.) [11] was used. 50 g of E0265 and 50 g of E0266 were dissolved in 1 L of ethanol prior to mixing.

When solution process materials are aimed to be used for a specific purpose, the adaptation of the material for the process while achieving a quality physical and electrical properties are crucial. Considering the solar cell manufacturing process, textured surfaces are common for a solar cell to improve light trapping. Therefore, an ARC film needs to be applied smoothly to such surfaces for better spectral response. The nanoparticle-polymer composite layers formed on flat or alkaline textured crystalline silicon (c-Si) substrates and evaluation of the coating films were carried out as an antireflection coating (ARC) films. Based on this consideration, zirconium bare solution was formed by spinning on the surface of the textured wafer without any additives with spin speed of 4000 rpm for 20 s (acceleration time is 4 s). Between the textured pyramids detrimental cracks were observed after drying, as shown in **Figure 5 (a)**. These cracks need to be avoided to form a quality film. A mixed solution with ethanol was used [11] in a volume of 1:1; however, cracks could not be avoided totally. A total crack-free forming of the films was achieved by using of ethyl cellulose [11].

The ratio of the mixture was adjusted to 1:32:2 (in volume) of ethyl cellulose, ethanol, and zirconium sol, respectively. Ethyl cellulose is a well-known and widely used polymer with good material properties which is suitable to utilize for quality films [12]. It dissolves well in ethanol that is crucial in this study [12]. Spin coating conditions was similar with that of previous experiments which were without ethyl cellulose. As shown in **Figure 5 (d)**, quality film with smooth surface could be achieved with ethyl cellulose included solution. (This solution is used for the rest of the study of ZrO_2-P composite films). The crack-free surface may be due to the contribution of ethyl cellulose in terms of elasticity of the film. As well as the good light permeability of the ethyl cellulose is a great advantage when using for ARC materials where the transparency is so important.

As another alternative solution-processed ARC material, TiO_2 was also investigated and adapted to solar cell processing. Similar optimizations were carried out for TiO_2 solution as that of previously explained for ZrO_2 solution. TiO_2 solution was mixed with various amount of ethanol and surface condition of textured silicon was examined. 2, 4, 6, or 8 mL of ethanol mixed into 1 mL TiO_2 solution (spin speed of 5000 rpm for 25 s). It was confirmed that TiO_2

Figure 5. SEM surface images of antireflection coating on textured Si surface formed by ZrO_2 sol (with various amount of ethanol or ethylcellulose in 1 mL ZrO_2 sol); (a) pure ZrO_2 sol without ethanol or ethyl cellulose, (b) with 1 mL ethanol, (c) with 9 mL ethanol, and (d) with 8 mL ethanol + 0.25 mL, 10% ethyl cellulose solution in ethanol [11].

solution mixed with 8 mL of ethanol combination provided crack-free surface as shown in **Figure 6 (d)**.

3.1.2. Sample preparation and characterization of thin films

Square-shaped p-type Si wafers (25 × 25 mm^2) were cut out from 6-inch CZ-Si p-type wafers for optical-based experiments. First, wafers were dipped into 20% HF for 1 min and soaked in ionized water. Then, UV/O_3 cleaning was carried out to achieve a completely cleaned surface. Optimizations of spin coating process were carried out with various spin speed-time profiles. After optimizations, 5000 rpm for 25 s with acceleration of 5 s found to be optimum in most cases for this work. After the deposition of the films by spin coating, annealing was carried out 125°C for 5 min. Scanning electron microscope (SEM) measurements (JSM-6510, JEOL),

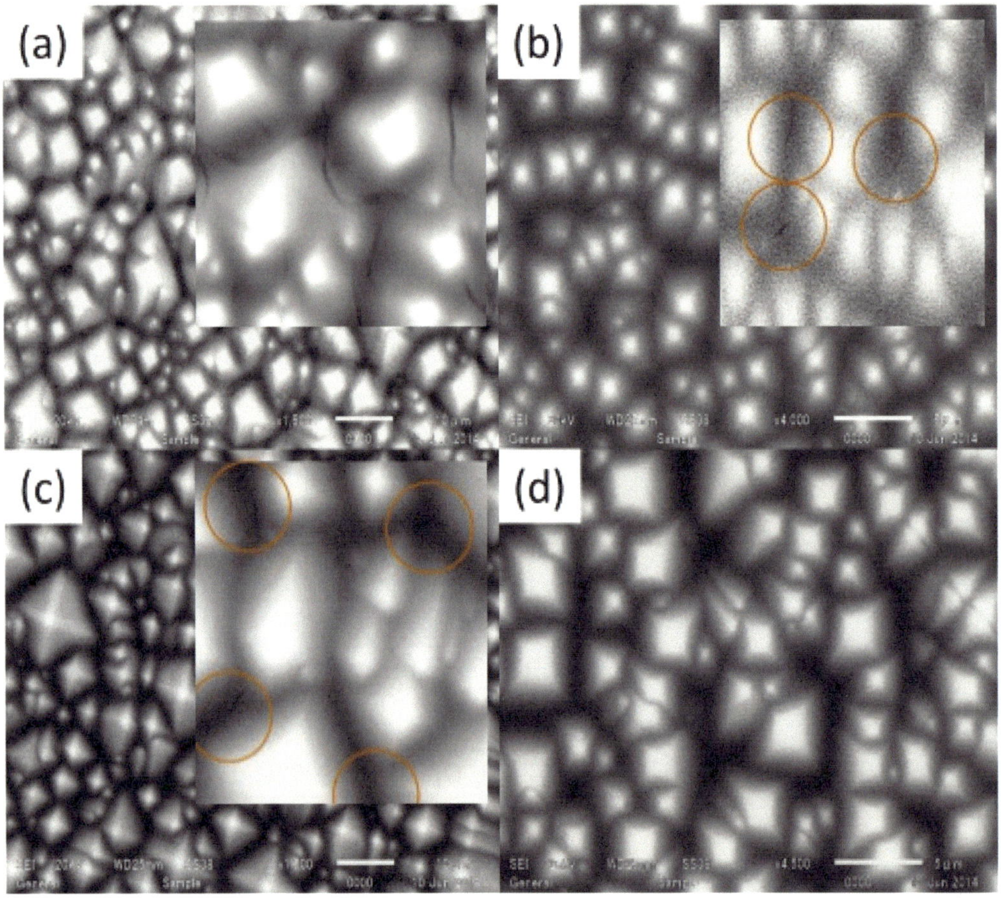

Figure 6. SEM surface images of antireflection coating on Si surface formed by surface-eliminated TiO_2 sol (various amount of ethanol in 1 mL TiO_2 sol); (a) with 2 mL ethanol, (b) with 4 mL ethanol, (c) with 6 mL ethanol, and (d) with 8 mL ethanol. The circles indicate the position of cracks [11].

reflection analysis (Lambda 750 UV/VIS Spectrometer, Perkin Elmer) and ellipsometer measurements (UviselErAgms-nds, Horiba Jobin Yvon) were mainly used for the analysis. C-Si solar cells were fabricated and analyzed as well.

Applied spin speed for thin film formation is important for a homogeneous formation of the thin film. The control of the film thickness is mainly established by the spin speed for spin coating applications. Reflection spectrometry is used for absorption, reflectivity, and transmission analysis which measures the absorbed portion of the photons as a function of wavelength. In case of an ARC material for silicon solar cells, it is important to analyze the characteristics of the ARC film by comparing the reflection on bare flat silicon surface, on bare textured silicon surface and such structured surfaces with applied ARC film. **Figure 7 (a)** compares the reflectivity of flat silicon substrates with and without ZrO_2-P layer coated on the surface with spin speed of 1000–4000 rpm for 20 s (acceleration time is 4 s). The lowest average reflectance (the reflectance average between the wavelengths of 300–1100 nm) of 28% was achieved by ZrO_2 coating with a spin speed of 1000 rpm, a steady reflectance tendency from wavelength of around 700–1000 nm can also be confirmed. Considering the reflection of bare silicon surface as 40%, decrease of surface reflection down to 28% owing to the formed ZrO_2 film on the flat silicon surface, can be confirmed.

The reflectivity of textured silicon wafers covered by ZrO_2-P composite film including different amount of ethanol from 6 to 10 mL in the solution, was measured. Average reflectance of 6.5% was observed as the lowest value, when film formed with 0.5 mL zirconium solution (5.71%vl) + 0.25 mL ethyl cellulose (2.85%vl) + 8 mL (91.4%vl) combined solution. **Figure 7 (b)** shows the reflectance dependence of alkaline textured silicon substrates coated with ZrO_2-P composite film for spin speed in a range of 500–3000 rpm. Decrease of reflectance can be confirmed when increasing spin speed. Average reflectance reaches below 7% at spin speeds of 1500 rpm. Similar optimizations were performed for the films coated by SD-TiO_2 solution, both on flat or textured silicon substrates. Amount of ethanol was changed from 4 to 10 mL, in 1 mL SD-TiO_2 solution + 0.1 mL ethyl cellulose mixture. First, the mixed solution was formed on flat silicon surface by spin coating and annealing was carried out at 125°C for 5 min on a hot plate. In **Table 1**, the average reflectance of flat silicon surface coated with surface deactivated TiO_2-ethylcellulose film is summarized for each amount of ethanol in the solution.

Films formed with precursor solution contains 8 mL of ethanol provides the lowest average reflectances. Average reflectances of 19.92 and 20.04% were observed when the films were formed by for spin speed of 5000 and 8000 rpm, respectively. The reflectivity of textured silicon surface coated with SD-TiO_2-P film by varying the amount of ethanol from 4 to 10 mL, is

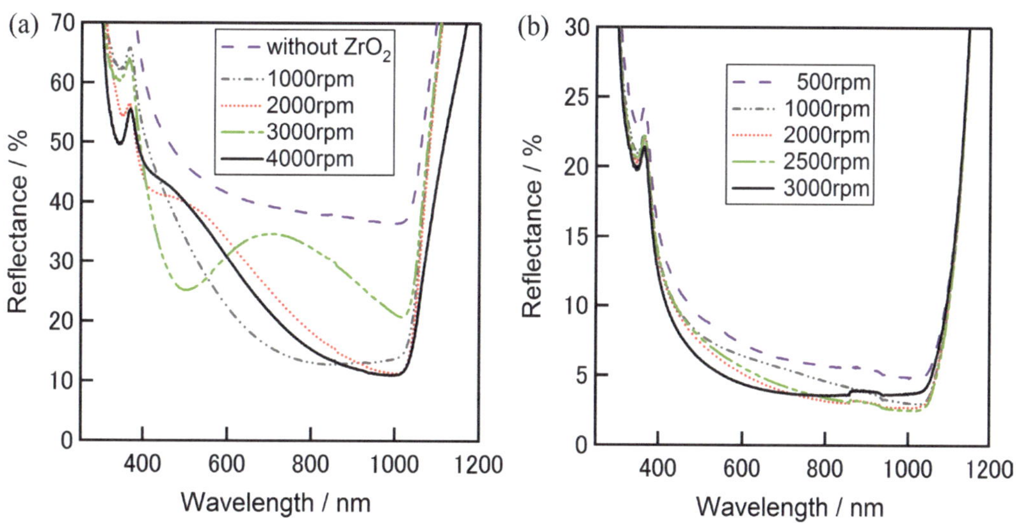

Figure 7. Reflectance spectra of flat surface silicon substrate with ZrO_2-polymer composite ARC; (a) by changing the spin-coat rotating speed and (b) reflectance of texture surface silicon substrate with ZrO_2-polymer composite ARC by changing the spin-coat rotating speed [11].

Speeds	EtOH (mL)			
	4	6	8	10
5000 rpm	30.00	24.00	19.92	22.66
8000 rpm	25.78	20.83	20.04	24.15

The volume of surface-deactivated TiO_2–sol was 1 mL [11].

Table 1. Average reflectance of flat silicon surface with surface-deactivated TiO_2-polymer ARC coating by different spin coating speeds.

given in **Figure 8**. Similarly as on flat silicon surface, the lowest average reflectance of 5.6% was achieved with the precursor contains 8 mL of ethanol.

3.1.3. Simulation and formation of ZrO_2-polymer/surface-passivated TiO_2-polymer composite multilayer film

Due to the difference on reflection of ambients and materials, photons come toward to surface of a solar cell experience three basic phenomenon: reflection, absorption, and transmission. In order to improve the efficiency of solar cells reflection must be minimized while absorption is maximized. It is known that multilayer structure of ARCs based on the gradual increase of each layer in the structure reflectance can be reduced significantly [13]. According to this fact, ⟨air/ZrO_2-polymer/surface-passivated TiO_2-polymer/Si⟩ multilayer structure was built with gradual of refractive indexes sequential order, as in $n_{air} < n_{ZrO2\text{-}polymer} < n_{TiO2\text{-}polymer} < n_{silicon}$.

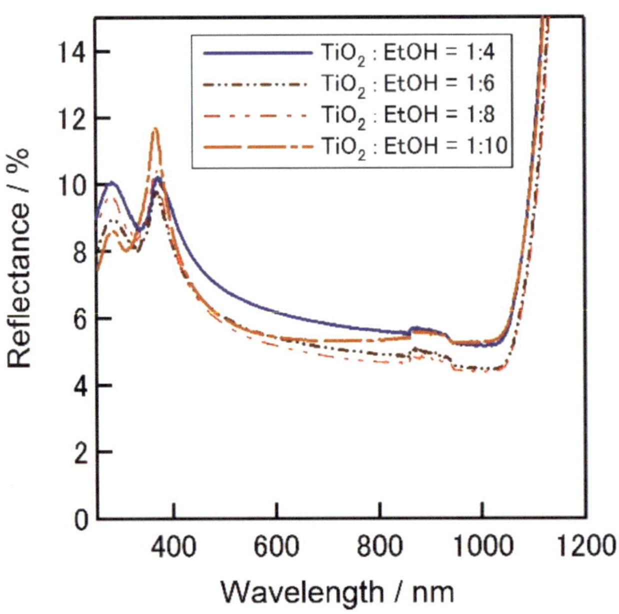

Figure 8. Reflectance spectra of textured silicon surface with surface-deactivated TiO_2-polymer antireflection coating [11].

http://dx.doi.org/10.5772/intechopen.68170

Refractive indexes and extinction coefficients of ZrO_2-P and SD-TiO_2-P composite layers were achieved via fitting of ellipsometry measurements [11] as can be seen in **Figure 9 (a)** and **(b)**, where the refractive indexes of air and silicon are 1 and 4.3, respectively.

Only the absorbed photons can contribute on generation of electron-hole pairs where the reflected or transmitted photons are considered as energy loss. As the absorbed portion of total photons increases, the electrons moving to the conduction band increase thus contributes to the short-circuit current density (J_{sc}). Based on this basic fact, measured absorption-reflection spectra can be used to estimate J_{sc}. In order to determine this estimation Fresnel approach can be used. Fresnel equations describe the reflection and transmission behavior of photons regarding to a multilayer medium with different refractive indices [14]. Considering the refractive index, thickness of each layer and the transmission characteristics of a multilayer stack, it can be reduced down to a single imaginary layer by using Fresnel approach.

Follows a simulation study for the ⟨air/ZrO_2-polymer (ZrO_2-P)/surface-deactivated TiO_2-polymer (SD-TiO_2-P)/Si⟩ structure in order to estimate the short-circuit current density (J_{sc}). Thickness of the ZrO_2-P and SD-TiO_2-P composite layers were changed from 10 to 100 nm, and "incident photon to current conversion efficiency (IPCE)" was set to 100% for the calculation for the sake of achieving a comparable data. Eq. (1) was used to calculate J_{sc} to estimate optimum thicknesses of ZrO_2-P and SD-TiO_2-P composite films for a good multilayer ARC film.

$$J_{sc,max} \leq \frac{eS}{hc}\int_{300}^{1200} P_{in} \times Abs(\lambda)\ d\lambda \qquad (1)$$

where $J_{sc,max}$ is maximum short-circuit current density (mA cm^{-2}), e is charge of an electron (1.602×10^{-19} A s), h is Plank's constant (6.626×10^{-34} eV s), c is speed of light (2.998×10^{8} m s^{-1}), P_{in} is the incident power in W m^{-2}, and λ is to denote wavelength (nm). Calculated J_{sc} values for ZrO_2-P/SD-TiO_2-P composite multilayer ARC films are given in **Figure 10**, and written down in detail in **Table 2**.

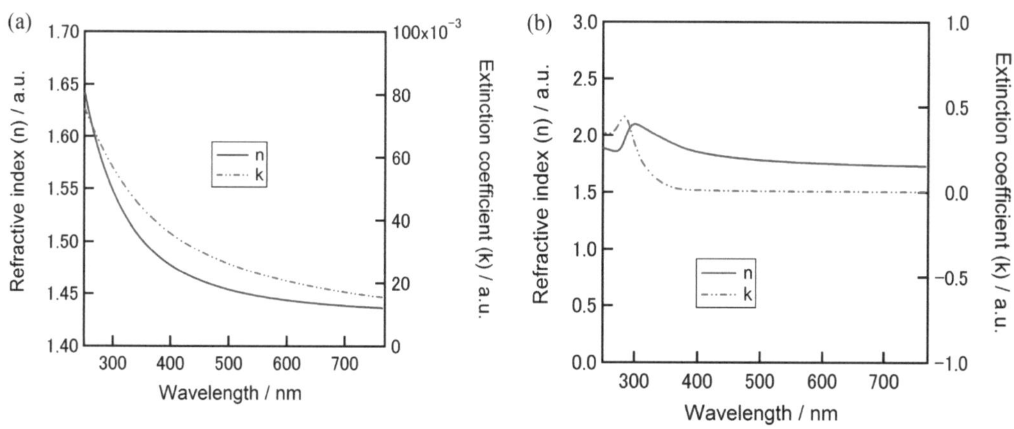

Figure 9. Fitting of ellipsometry measurements of (a) nano-colloid ZrO_2 (ZR30-AH)-polymer composite and (b) surface-deactivated TiO_2-polymer composite [11].

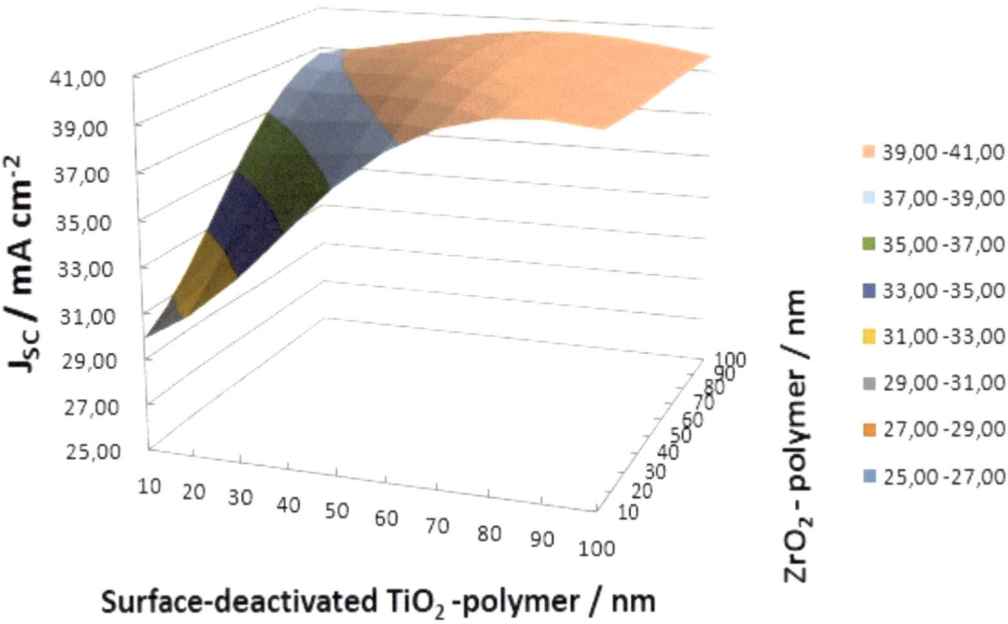

Figure 10. Estimated short-circuit current density with thickness variations of ZrO_2-polymer and surface-deactivated TiO_2-polymer films on flat-surface silicon solar cells [11].

ZrO_2/nm	TiO_2/nm									
	10	**20**	**30**	**40**	**50**	**60**	**70**	**80**	**90**	**100**
10	29.90	31.07	32.82	34.95	37.02	38.65	39.72	40.27	40.37	40.15
20	30.66	32.12	34.07	36.18	38.03	39.36	40.15	40.46	40.39	40.06
30	31.74	33.42	35.40	37.33	38.88	39.92	40.46	40.58	40.38	39.98
40	33.06	34.81	36.66	38.31	39.55	40.33	40.66	40.65	40.37	39.94
50	34.48	36.13	37.73	39.07	40.03	40.59	40.78	40.68	40.37	39.93
60	35.84	37.27	38.56	39.60	40.33	40.73	***40.83***	40.69	40.36	39.93
70	37.00	38.13	39.13	39.92	40.47	40.76	40.82	40.66	40.35	39.92
80	37.87	38.72	39.45	40.04	40.47	40.71	40.76	40.60	40.30	39.89
90	38.45	39.04	39.56	40.01	40.37	40.60	40.64	40.50	40.21	39.82
100	38.76	39.13	39.50	39.87	40.20	40.43	40.47	40.35	40.08	39.72

Table 2. Estimated short-circuit current density with thickness variations of ZrO_2-polymer and surface-deactivated TiO_2-polymer films on flat-surface silicon solar cells [11].

Highest J_{SC} of 40.83 mA cm^{-2} was confirmed with ⟨60 nm ZrO_2-P/70 nm SD-TiO_2-P⟩ composite multilayer ARC film. In experimental side of this investigation, such a multilayer film experimentally, first SD-TiO_2-P film was spin coated on the textured silicon surface and annealed at 125°C for 5 min, then the reflectance of the surface was measured. To form and characterize such a multilayer film experimentally, first SD-TiO_2-P film was spin coated on the textured silicon surface and then annealed at 125°C for 5 min, then the reflectance measurement was carried out. After the reflectance measurement of SD-TiO_2-P coated surface, ZrO_2-P film was spin coated on it and annealing was done similarly. **Figure 11** presents the comparison of reflectance spectra of ARC films on textured silicon surface of SD-TiO_2-P or ZrO_2-P/SD-TiO_2-P composite. Average reflectance of around 5.5% is achieved owing to the multilayer-structured ARC where the average reflectance of the surface without ARC was 16.3%.

3.1.4. Fabrication of p-type crystalline silicon solar cells

p-type CZ-Si solar cells with surface area of 25 × 25 mm^2, with/without ZrO_2-P or SD-TiO_2-P composite ARC layers and with/without ZrO_2-P/SD-TiO_2-P composite multilayer ARC, were fabricated. For textured solar cells, alkaline texturing was carried out [11] based on KOH solution in addition to Alka-Tex (provided by GP Solar) supporting agent at 80°C for 30 min. After the etching, all wafers were rinsed into 20% HF solution and soaked with ionized water.

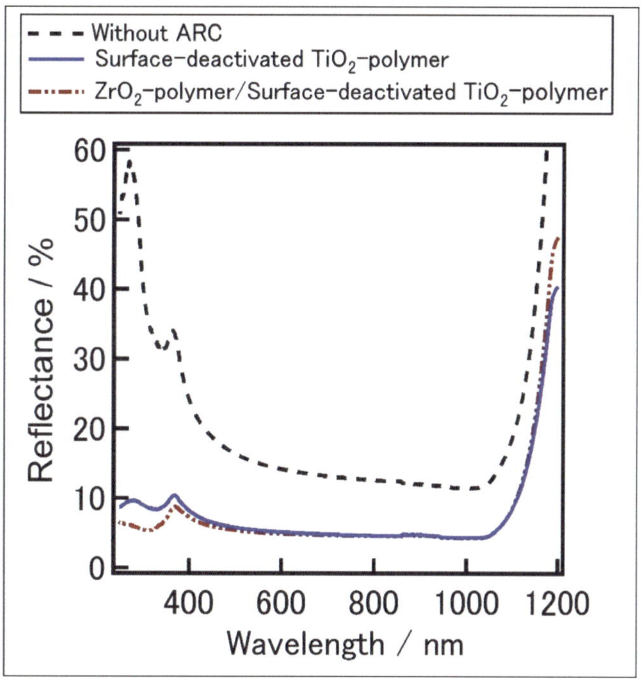

Figure 11. Reflectance spectra of textured silicon surface without ARC, with surface-deactivated TiO_2-polymer, ZrO_2-polymer/surface-deactivated TiO_2-polymer multilayer ARC [11].

For further cleaning, RCA cleaning process [15, 16] was carried out using a $NH_4OH/H_2O_2/H_2O$ (1:1:5 in volume) solution, at 80°C for 10 min. Follows a dip into 20% HF solution, and into $HCl/H_2O_2/H_2O$ (1:1:5 in volume) for 10 min at 80°C for a complete cleaning process. Prior to phosphorus diffusion, back side of the wafers were coated by polysilazane to avoid diffusion of phosphorus on back side. Coating of polysilazane was performed at 1500 rpm for 20 s (two times subsequent), and annealed at 600°C for 60 min in ambient O_2. $POCl_3$ diffusion was carried out at drive in temperature of 900°C for 30 min in ambient N_2. Post diffusion phosphorus silica glass was removed by 10% HF solution and wafers were cleaned in ionized water. Screen-printing Ag and Al was used for front and back contacts, respectively and cofired at 780°C for 1 min. ARC films were formed after the firing step by spin coating at 5000 rpm for 25 s) and annealing was carried out at 125°C for 5 min. For the cells with ZrO_2-P/SD-TiO_2-P composite multilayer structure ARC, first SD-TiO_2-P composite film was deposited and annealed and then ZrO_2-P composite film was deposited and annealed. The electrical characteristics of the fabricated solar cells with or without ZrO_2-P composite ARC are summarized in **Table 3**.

Best efficiency on flat surface silicon solar cell without ARC was 7.25% with J_{sc} of 24.03 mA cm^{-2}. After applying ZrO_2-P composite ARC, conversion efficiency increases up to 9.78% with J_{sc} of 30.22 mA cm^{-2} due to the formed ARC layer. In case of textured silicon solar cells, the conversion efficiency of solar cells without ZrO_2-P ARC increases from 10.2 up to 10.9%, where the J_{sc} were 28.74 and 30.62 mA cm^{-2}, respectively. This change is mainly due to the combined antireflection effect of textured surface and ZrO_2-P composite ARC layer.

On the other hand, considering the fabricated solar cells with SD-TiO_2-P ARC and ZrO_2-P/SD-TiO_2-P composite multilayer ARC, improvement of J_{sc} up to 32.44 and 33.00 mA cm^{-2} can be confirmed comparing to the J_{sc} of those of solar cells without ARC which was 30.93 mA cm^{-2}, respectively (values are average of three cells). Comparing the average efficiencies of the cells without ARC (11.99%) with those cells with ZrO_2-P/SD-TiO_2-P composite multilayer

ARC	Surface structure	*Jsc* (mA cm^{-2})	V_{oc} (V)	*FF* (%)	η (%)
Without ARC	Flat	24.03	0.527	57.1	7.25
With ZrO_2-polymer composite	Flat	30.22	0.537	60.2	9.78
Without ARC (textured ref 1)	Textured	28.74	0.511	69.4	10.20
With ZrO_2-polymer composite	Textured	30.62	0.510	69.7	10.89
Without ARC (textured ref 2)	**Best**	**29.33**	**0.570**	**71.7**	**11.99**
	Average	30.93	0.569	63.1	11.06
With surface-deactivated TiO_2 polymer	**Best**	**30.90**	**0.574**	**72.3**	**12.84**
	Average	32.44	0.575	63.7	11.84
With ZrO_2-polymer/surface-deactivated TiO_2 polymer	**Best**	**31.42**	**0.575**	**71.5**	**12.91**
	Average	33.00	0.574	62.7	11.85

Table 3. Photovoltaic characteristic of fabricated silicon solar cells on flat or textured silicon surface, with or without ZrO_2-polymer ARC.

ARC (12.85%), the improvement with an increase of J_{sc} (2.07 mA cm^{-2}) is worthfull. The best cell with ZrO_2-P/SD-TiO_2-P composite multilayer ARC is stand alone with the electrical characteristics of; J_{sc} of 31.42 mA cm^{-2}, V_{oc} of 575 mV, fill factor (*FF*) of 71.5%, and efficiency of 12.91%. **Figure 12** presents the *J-V* curve comparisons for best cells from each group of **Table 3**. From these results, the potential of spin coating ZrO_2-P, SD-TiO_2-P ARC and ZrO_2-P/SD-TiO_2-P based ARC on enhancing the optical performance of solar cells could be confirmed which can be a promising ARC candidate for the aim of low-cost silicon solar cells. For further analysis Ref. [11] can be referred.

It is worthy to note that better performances can be achieved especially on V_{oc} and on *FF*, with using high lifetime high quality wafers, a good bulk passivation and with a proper edge isolation.

3.2. Spray deposited TiO_2 and spin coating ZrO_2 films for ARC

In this part, spray deposited compact TiO_2 layer as a single layer ARC and spin-coated ZrO_2/spray deposited TiO_2 films as a double-layer ARC will be analyzed based on our previous work [17]. A compact TiO_2 film is commonly used for dye synthesized solar cells and perovskite solar cells between the front contact and porous TiO_2 layer to block the electron recombination which boost the current [18, 19]. This phenomena and adaptation of compact TiO_2 film to the silicon solar cells as an ARC will be discussed.

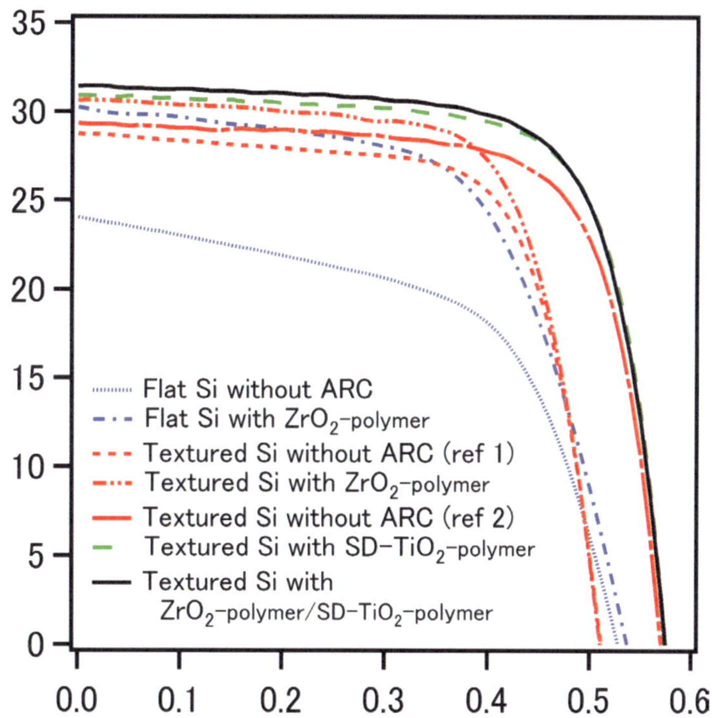

Figure 12. *J-V* curve of the best fabricated solar cells without any ARC film and with ZrO_2-polymer composite, with surface-deactivated TiO_2-polymer composite or ZrO_2-polymer composite/Surface-deactivated TiO_2 polymer-composite multilayer structure ARC films.

3.2.1. Sample preparation

p-type silicon wafers (25 × 25 mm^2) were cut out from 6-inch CZ-Si p-type wafers for the optical measurements. First, all wafers were dipped into the 20% diluted HF for 1 min and rinsed in distilled water. Afterwards, UV/O_3 surface treatment was carried out for a complete cleaning process. ZrO_2-polymer composite film was deposited on silicon substrate by spin coating using the zirconium sol (ZR-30AH, provided by Nissan Chemical Industry Co. Ltd., Chiba, Japan). Optimized ZrO_2 solution as explained in title of 3.1.1 composed by ethyl cellulose, ethanol, and ZR-30AH in ratio of 2:16:1 (in volume) was used in this work as well. Deposition of the ZrO_2 film was carried out by spin coating with a spin speed of 1500 rpm for 25 s (acceleration time is 5 s). In case of TiO_2 compact film deposition, spray pyrolysis was utilized and homogenous films were established by spraying precursor solution using a glass atomizer. Prior to spraying process crystalline silicon wafers were set on a conventional hot plate and heated up until the surface temperature of the substrate reaches 450°C. The TiO_2 precursor solution was composed of titanium bis-isopropoxide bis-acetylacetonate (TAA) and ethanol (1:10%vl). Annealing temperature was optimized for TiO_2 as 450 and as 125°C for ZrO_2, respectively. Analyses were mainly carried out by reflection analysis (by ultraviolet-visible spectroscopy, Lambda 750 UV/VIS Spectrometer, Perkin Elmer, Waltham, MA, USA) and ellipsometer (UviselErAgms-nds, Horiba Jobin Yvon, Kyoto, Japan) and finally by the performance of the fabricated silicon solar cells with or without ARC.

3.2.2. Characterization of the films

In order to evaluate the reflectivity performance of single layer TiO_2-compact film, at first flat surface silicon wafers were used. The thickness of the TiO_2 film was mainly controlled by the spraying amount of TiO_2 precursor solution in a range of 0–12 mL as previously mentioned. **Figure 13 (a)** shows the comparison of the reflectivity of flat samples with or without TiO_2-compact film on silicon surface. Depending the film thickness, minimum of the reflectance (<2%) shifts toward to the longer wavelengths. A steep decrease before the minimum can be confirmed for all TiO_2 film coated surfaces. The minimum of the reflectance shifts to the wavelengths as high as around 1000 nm for the films coated with 10 and 12 mL TiO_2 precursors due to the thickening of the film. The average reflectivity (the reflectance average between the wavelength of 300–1100 nm) was plotted by changing the spraying amount of the TiO_2 precursor as 0, 2, 4, 6, 8, 10, and 12 mL, respectively (**Figure 13 (b)**). A deposition rate of TiO_2 is defined as 10.1 nm mL^{-1} according to ellipsometry measurements. The lowest average reflectance of 21.41% was achieved by spray-deposited TiO_2 coating with a thickness of 60.6 nm. These values are in a relative agreement of calculated value of 56.8 at the wavelength of 550 nm for the refractive index of TiO_2 as 2.2 [20]. The bottom of reflectance valley shifted to higher wavelength with increasing the solution volume from 2 to 12 mL, gradually which shows the thickness dependence of reflection as well. Using the 8 mL precursor solution, the second valley arouse at the lower wavelength which was close to the valley bottom of 2 mL precursor solution. The second valley at the lower wavelength was also shifted to the larger wavelength with increasing the amount of precursor solution.

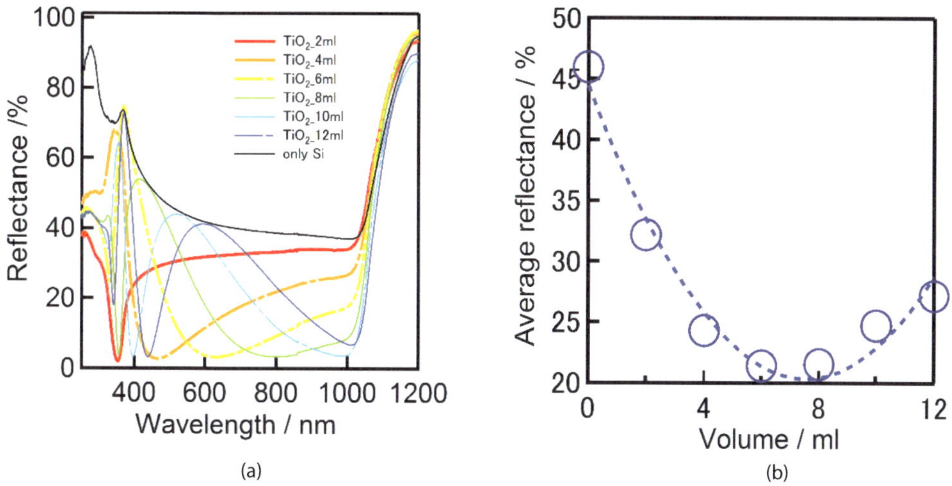

Figure 13. Comparison of the reflectivity of the flat samples with TiO_2 antireflection coatings by various amount TiO_2 precursor solution; (a) reflectance spectra depending to spraying amount and (b) average reflectance (300–1100 nm wavelength) depending to spraying amount [17].

In case of reflectivity measurement for ZrO_2-polymer composite/TiO_2-compact multilayer, first, TiO_2-compact single layer was formed and measured, then ZrO_2-polymer composite was formed on TiO_2 covered surface and remeasured. The comparison of reflectivity of TiO_2-compact single layer with the ZrO_2-polymer composite/TiO_2-compact multilayer formed on textured silicon wafers is given in **Figure 14**. The average reflectance was improved further with applying ZrO_2-polymer composite on the TiO_2-compact film, offering between 5 and 12% reflectance gain from 300 to 450 nm and a steady lower reflectance tendency from wavelength of around 700–1000 nm can also be confirmed.

3.2.3. Evaluation of the fabricated p-type crystalline silicon solar cells

Silicon solar cells were fabricated with a variety of final surface condition; without ARC on textured surface, with TiO_2-compact film on textured surface and with ZrO_2-polymer composite/TiO_2-compact multilayer ARC film on textured surface. For the fabrication of silicon solar cells, 25 mm × 25 mm p-type CZ-Si wafers were used as well. In this work, alkaline texturing process, RCA cleaning, UV/O_3 cleaning process, diffusion barrier of polysilazane forming were carried out in a similar way as explained in Section 3.1.4. $POCl_3$ diffusion was carried out at 930°C for 35 min in ambient N_2. Post diffusion PSG glass was removed by diluted HF and wafers were cleaned by ionized water. SiO_2 film was formed by thermal oxidation process at temperature of 800°C for 10 min under O_2 gas. Before the metallization process, wafers were set on a hot plate heated at deposition temperature (450°C) and deposition of TiO_2 films were carried out by spray pyrolysis. Then, front and back contacts were formed by screen-printing Ag and Al, respectively. Cofiring was carried out at 780°C for 1 min in an oven. After evaluating the solar

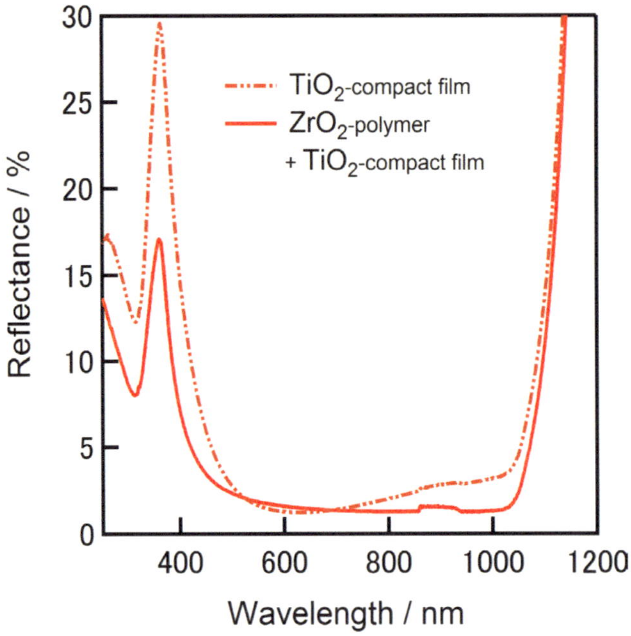

Figure 14. Comparison of the reflectivity of the textured silicon wafers with TiO_2 single layer and spin-coated ZrO_2/spray-deposited TiO_2 double-layer [17].

cells on with a single layer TiO_2-compact films on the wafer surfaces, ZrO_2-polymer composite films were deposited on the finished cells and compared to those of the cells without ARC and with single layer TiO_2-compact films. Deposition of ZrO_2-polymer composite ARC films were carried out by the annealing step after spin coating process (1000 rpm on flat surface, 1500 rpm on textured surface, 5 s acceleration+25 s, annealing at 125°C, 5 min).

Conversion efficiency of the cells improved from 15.2 to 15.9%, where the J_{sc} also increased further from 35.3 to 37.2 mA cm^{-2} owing to the ZrO_2/TiO_2 multilayer ARC film (the J_{sc} of the cell without ARC was 28.7 mA cm^{-2}). The electrical characteristics of the cells are summarized in **Table 4**.

A total of 70 mV increase of V_{oc} of the cells with ARC can be confirmed. Though, SiO_2 film was formed by thermal oxidation process at temperature of 800°C for 10 min under O_2 gas, the photovoltaic results with/without SiO_2 layer have not been compared in this study. Hence, the increase of V_{oc} is due to the compact TiO_2 layer, which can prevent excessive fire though of Ag into p-n junction. The greater *FF* of the cells with ARC layer can be attributed to the lower contact resistances in which the effect of additional SiO_2 layer can be considerable in order to avoid internal shunts. *J-V* characteristics of the silicon solar cells fabricated with TiO_2-compact single layer and ZrO_2-polymer composite/TiO_2-compact multilayer ARC on surface of the cells is given in **Figure 15**.

One can conclude that a significant improvement on J_{SC} and η could be confirmed owing to the ZrO_2-polymer composite/TiO_2-compact multilayer ARC when compared to the textured

Surface structure	ARC	J_{sc} (mA cm^{-2})	V_{oc} (V)	FF (%)	R_{series} (Ω cm^2)	R_{shunt} (Ω cm^2)	η (%)
Textured	–	28.7	0.511	69.4	0.71	250	10.2
Textured	TiO_2-compact	35.3	0.581	74.1	0.40	655	15.2
Textured	ZrO_2-polymer/TiO_2-compact	37.2	0.583	73.4	0.45	535	15.9

Table 4. Photovoltaic characteristics of fabricated crystalline silicon solar cells with various surface structures with/without antireflection coating on textured surface [17].

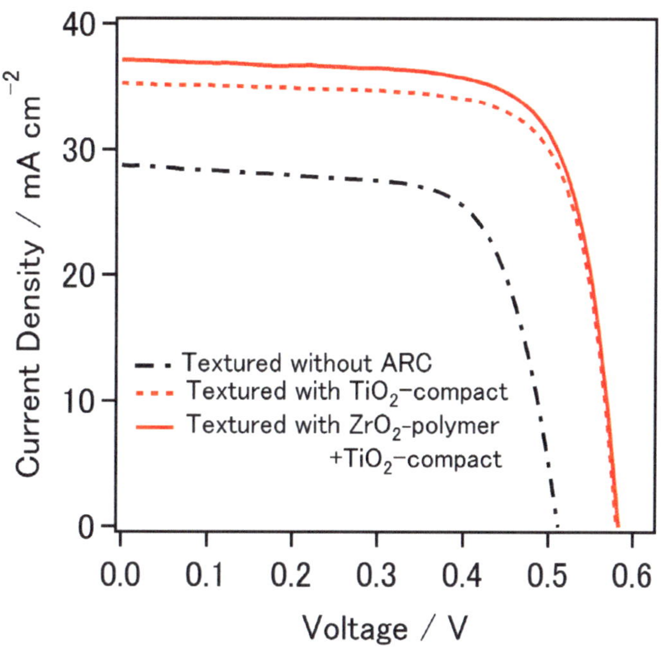

Figure 15. *J-V* characteristics of the silicon solar cells fabricated with TiO_2-compact single layer and ZrO_2-polymer composite/TiO_2 -compact multilayer ARC [17].

cells without ARC. The increase of the J_{SC} was related to minimizing the reflectance loses which can boost the performance of the cell. These results suggest that ZrO_2/TiO_2-based multilayer ARC films formed by spray pyrolysis deposition technique and spin coating technique could be an attractive alternative as a low-cost, simple, and vacuum-less process for ARC coating for silicon solar cells.

3.3. Spray deposited TiO_2 and Al_2O_3 films

3.3.1. Sample preparation

Compact TiO_2 and Al_2O_3 films formed by spray deposition technique are another attractive low cost ARC alternative for crystalline silicon solar cells. In this part, TiO_2 and Al_2O_3 film formations on flat crystalline silicon substrates, evaluation of the coating films was carried out. And textured silicon solar cell fabrication with Al_2O_3/TiO_2 double-layer antireflection coating films was explained. The detailed analysis of the modeling to show the effects of organics can be found in Ref. [21]. For the experimental of optical studies, 25 × 25 mm^2 Cz-Si p-type wafers were used. All wafers were cleaned in 20% HF solution and soaked in ionized water. Then, UV/O_3 cleaning was applied in order to achieve a complete clean surface without mobile ions. Deposition of TiO_2 and Al_2O_3 films were established by spray pyrolysis. Prior to spraying process crystalline silicon wafers were set on a conventional hot plate and heated up until the surface temperature of the substrate reaches 450°C. The TiO_2 precursor solution was composed of titanium bis-isopropoxide bis-acetylacetonate (TAA) and ethanol (1:10%vl). The Al_2O_3 precursor solution was 0.03 M of aluminum (III) acetylacetonate (Al(acac)$_3$) in ethanol solution. Deposited film thicknesses were mainly controlled by the amount of sprayed precursor solutions. Each layer was analyzed by ellipsometer (Uvisel ErAgms-nds, Horiba Jobin Yvon).

3.3.2. Characterization of the films

At first, in order to analyze and optimize TiO_2 and Al_2O_3 spray on deposited layers for the processing of Al_2O_3/TiO_2 double-layer film, single-layer TiO_2 and Al_2O_3 films were analyzed. In depth analysis was carried out with fitting models of "flat layer" as the main body of the coated layer and "rough layer" as the surface residue when depositing a spray-deposited TiO_2 and Al_2O_3 films [21] and fitted according to the real SEM images. As an optimum value of 60 mL TiO_2 and 100 mL Al_2O_3 precursor solutions were considered. The analysis was carried out improved by considering the "void" and "organic" parts in the models. The existence of the organics and their contributions were confirmed by the FTIR measurements and by modeling [21]. For the optimization process of $\langle Al_2O_3/TiO_2 \rangle$ double-layer, thickness of TiO_2 film was varied between 20 and 120 nm (precursor solution 2–12 mL) while the thickness of Al_2O_3 film was changed from 68 to 135 nm (precursor solution 75–150 mL). After the reflectance measurements of the cells with ARC, the maximal short-circuit current densities from the reflectance results were calculated using Eq. (1) in order to observe the optimum Al_2O_3 and TiO_2 film thicknesses to obtain efficient Al_2O_3/TiO_2 double-layer ARC film.

Figure 16 presents the short-circuit current density ($J_{sc,max}$) calculated by (Eq. (1)) using experimental reflectance spectra of silicon substrate with Al_2O_3/TiO_2 ARC film for various thicknesses of Al_2O_3 and TiO_2 films. According to the calculations using experimental absorption spectra, the maximum value of short-circuit current density ($J_{sc,max}$) of 38.9 mA cm^{-2} was achieved with 90 nm Al_2O_3/40 nm TiO_2 double-layer ARC film (the data was not shown in the manuscript). Based on the experimental founding, 90 nm Al_2O_3/40 nm TiO_2 double-layer ARC film was applied when fabricating silicon solar cells which will be explained in following section. Comparison of reflectance of silicon surface with/without ARC and/or with/without texturing is given in **Figure 17**. As known texturing of silicon surface have a significant effect on reducing the reflectivity, as also proved in this study by the decrease of reflectance from 40 to around 20% after texturing. Considering the surface condition of a silicon substrate, reflectivity of 20.1 and 9.71% were observed on flat and textured surfaces, respectively. The effect of texturing on J_{SC} can be expected [22]. Owing to the double-layer ARC film the reflectivity was decreased down to the minimum of 0.4% at around 600 nm.

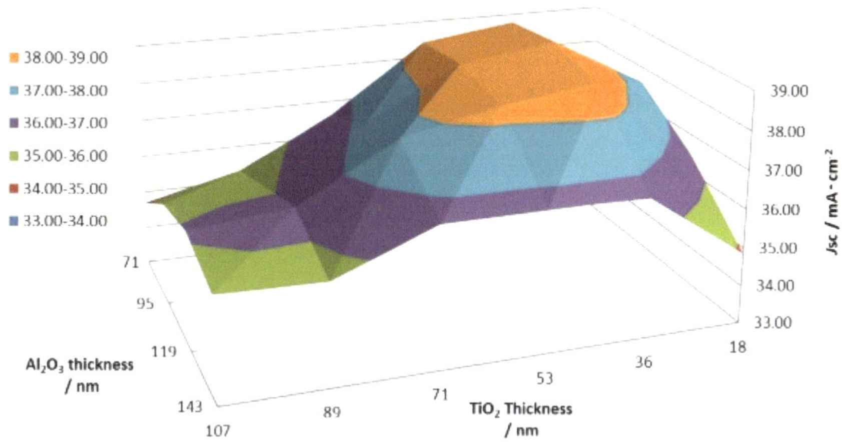

Figure 16. Estimated J_{sc} calculated by the experimental reflectivity data using ⟨$Al_2O_3/TiO_2/Si$⟩ layers with various thicknesses of Al_2O_3 and TiO_2 films [21].

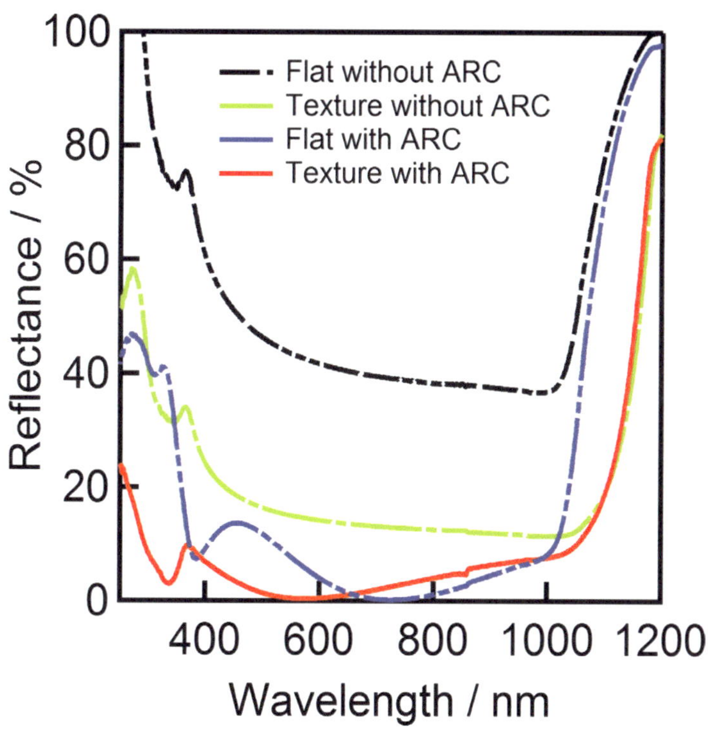

Figure 17. Reflectance spectra of silicon wafers with various surface structures [21].

3.3.3. Solar cell fabrication

Electrical characteristics of fabricated solar cells with various surface structure and ARC film were given in **Figure 18**, and summarized in **Table 5**. Comparing the flat or textured surface solar cells without ARC, J_{SC} of 27.6 mA cm^{-2} improved up to 33.5 mA cm^{-2} owing to the textured surface. In case of solar cells with Al_2O_3/TiO_2 double-layer ARC J_{SC} of solar cells were improved from 34.3 (flat surface) up to 37.0 mA cm^{-2} (textured surface). Consequently, the CZ-Si p-type solar cells with Al_2O_3/TiO_2 double-layer ARC on textured surface reached a η of 15.5% with a V_{oc} of 590 mV and *FF* of 71.2. A significant improvement on J_{SC} and η could be confirmed owing to the Al_2O_3/TiO_2 ARC when compared to the textured cells without ARC.

The ARC effect on the cells can be seen by the increase of J_{SC} (**Figure 18 (a)**) and can be confirmed by the EQE coverage over the spectrum (**Figure 18 (b)**). The suppression of EQE below 400 nm may be attributed to the absorption of TiO_2 layer. The ripples on EQE in the range of 500–1000 nm may take attention which is similar for all measurements at the same measurement points, can be considered as non-significant measurement errors. The decrease of open circuit voltage was attributed to the degradation of carrier lifetime of the bulk due to the over-annealing (>350°C) [23, 24]. In order to avoid this deterioration, an additional ultra-thin passivation layer-like silicon dioxide may be formed between silicon substrate and TiO_2 layer to boost open circuit voltage. A J_{SC} calculation was performed for single layer of TiO_2 and SiN_x with theoretical reflectance and (Eq. (1)) which was resulted as 39.50 and 40.67 mA cm^{-2}, respectively. The J_{SC} of cells with TiO_2/Al_2O_3 double-layer ARC were estimated as 43.46 mA cm^{-2} which is above these values. In some references, also, the advantage of double-layer ARC was shown with low reflectance less than single layer ARC [25–28]. Moreover, another advantage of TiO_2/Al_2O_3 is the non-vacuum spray pyrolysis deposition, which can realize the

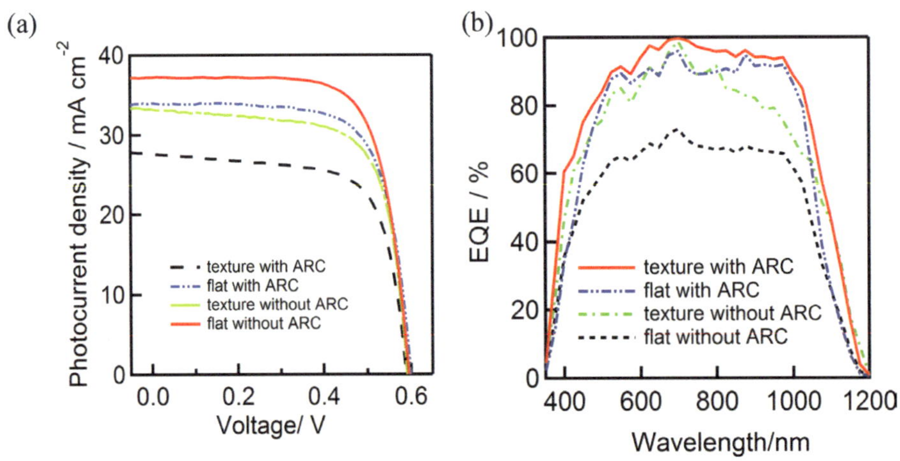

Figure 18. Current-voltage curves (a) external quantum efficiency (EQE) spectra and (b) of four fabricated sets of solar cells with/without texture and antireflection coatings (ARC) [21].

Surface structure	ARC	Jsc/mA cm^{-2}	V_{oc}/V	FF	η/%
Flat	Without	27.6 ± 0.1	0.588 ± 0.02	0.708 ± 0.012	11.5 ± 0.2
Flat	With	34.3 ± 0.8	0.596 ± 0.03	0.706 ± 0.014	14.4 ± 0.5
Textured	Without	33.5 ± 0.6	0.588 ± 0.05	0.702 ± 0.004	13.8 ± 0.2
Textured	With	37.0 ± 0.2	0.590 ± 0.09	0.712 ± 0.026	15.5 ± 0.8

The average data and the errors have been calculated using three different samples [21].

Table 5. Average data of photovoltaic characteristics of fabricated Si solar cells with various surface structures with/without surface texturing and antireflection coating (ARC).

lower cost production system than batch vacuum process like a SiN_x in terms of installation and running costs [29].

4. Screen-printing phosphorus diffusion for low-cost solar cells

A range of methods are in use to form n+ emitters or back surface fields for p-type silicon solar cells. Phosphorus oxychloride ($POCl_3$) is the main technique in the industry, while there are other alternatives as orthophosphoric acid (H_3PO_4) by spray or liquid sources by spin-on deposition techniques. Some have industrial application difficulties, high process cost, difficulty in wafers handling, or feasibility for batch or inline diffusion processes. Considering these techniques, screen-printing is a well-known and appealing technique with the advantages of low-cost, simplicity, both inline or batch mode diffusion possibility and industrial availability. In this study, screen-printing phosphorus diffusion is presented as an alternative to conventional dopant deposition techniques. Special attention was given to the impact of diffusion time, temperature variation, or the ambient gas effect during diffusion on the quality of screen-printed phosphorus diffusions on mc- and CZ-Si wafers. Diffusion tube with a gas input and output gates at the head and the end of the tube were used by changing the ambient during the diffusion process. By applying such a diffusion process we aimed to use the ambient effectively on diffusion process and adding to the diffusion time, diffusion temperature, the effect of ambient gas was also investigated.

4.1. Phosphorus oxychloride diffusion

Phosphorus diffusion using $POCl_3$ (phosphorus oxychloride) liquid source bubbled by N_2 is the dominant diffusion n+ emitter forming in silicon solar cell manufacturing process. Overall diffusion process occurs in two main steps: predeposition and drive-in. During predeposition, liquid $POCl_3$ source bubbled by N_2 gas flow and, evaporated source go through the chamber to react externally introduced O_2. This reaction takes place according to Eq. (2) in order to form P_2O_5 on the surface of wafers while Cl_2 is aired out during the reaction. The schematics of the $POCl_3$ diffusion process can be seen in **Figure 19**.

$$4\,POCl_3 + 3\,O_2 \rightarrow 2\,P_2O_5 + 6\,Cl_2 \tag{2}$$

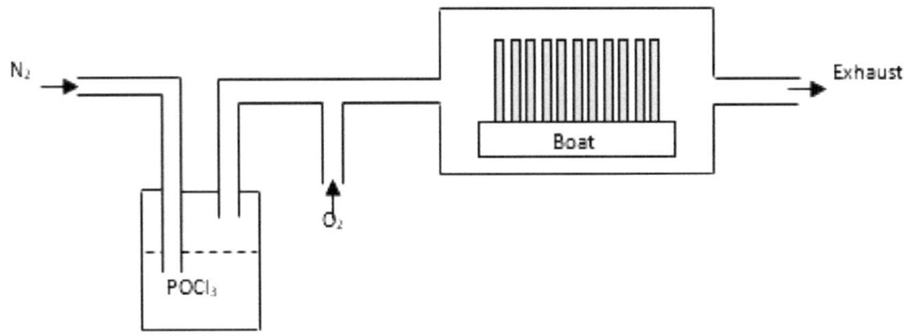

Figure 19. The schematic of $POCl_3$ diffusion system.

After predeposition step, $POCl_3$ source is halted and drive-in process starts to take place via the deposited phosphorus rich layer on the surface. During the drive-in step, diffusion of phosphorus atoms into silicon occurs through the reduction of P_2O_5 by silicon according to Eq. (3).

$$2\,P_2O_5 + 5Si \rightarrow 5\,SiO_2 + 4P \tag{3}$$

Generally, predeposition step is applied at lower temperatures than the drive-in step where the formation of diffused phosphorus layer establishes when the drive-in comes up. The surface concentration of diffused layer is controlled by increasing the O_2 gas flow during the drive-in process.

In silicon solar cell manufacturing process, wafers are stacked vertically in a boat in a furnace as in **Figure 19** prior to $POCl_3$ diffusion. If the wafers are set in order one by one, both side of wafers achieved similar diffusion of phosphorus. By that, gettering of impurities in silicon can also accomplish owing to the advantage of two side diffusion. However, back-to-back setting order also widely used for manufacturing in order to diffuse as much samples in a single process. One of the disadvantages of the $POCl_3$ diffusion system is that the same amount of source has to be used even for small size wafers because of the batch nature of the system. Additionally, long ramp-up and ramp-down cycles may be required which leads to time consumption even though the independent control of the predeposition and drive-in steps is an advantage.

4.2. Screen-printing phosphorus diffusion

As explained under Section 2.2, screen-printing is a fast, reliable and cost effective technology which is mainly used for metallization purposes for silicon solar cell industry. Screen-printable dopant sources are also introduced to the photovoltaic industry and are under research. As an attractive technology to deposit dopant sources on the surface of the substrates makes inline diffusion process possible to form dopant-diffused layers. After the deposition, diffusion process is usually performed by using either belt furnaces or quartz tube at high temperatures. Applied dopant paste usually is needed to be dried before the thermal annealing. Similarly, to the other diffusion techniques, after the deposition of the dopant paste by screen-printing, phosphorus atoms are released from the dried paste during the diffusion according to reduction of P_2O_5, as given in Eq. (3). Among other advantages, the availability to form homogenously diffused layers, the possibility of forming selective structures in any desired pattern are some of the most attractive points of using screen-printing technique dopant diffusion.

4.3. Diffusion process equipment

Both for solar cell manufacturing in industry or research and development of dopant diffusion process are mainly performed by using quartz tube or belt furnaces. A brief introduction of each equipment and technique will be given.

4.3.1. Diffusion in quartz tube

For diffusion process in quartz tube, wafers are set to a loading boat and are placed in the quartz tube. Diffusion process takes place according to the settled thermal cycle profile. The scheme of the quartz tube and heating system is given in **Figure 20**. The ambient inside the tube can be varied by changing the gas introduced into the tube from the end-side of the tube. Diffusion by solid sources of for $POCl_3$ and BBr_3 bubbled by N_2 needs to be carried out by such a tube. For other deposition techniques including screen-printing, spin-on, or spray deposition, it is possible to diffuse wafers from a single side or both sides. Depending on the type of quartz tube, gas ambient inside the tube can be controlled by introducing more than one gas separately.

4.3.2. Diffusion in belt furnace

Diffusion by a belt furnace offers a fast ramp-up and cool-down processes usually by infrared heating systems which can shorten the total process time significantly. These kinds of systems usually have a couple of separated heating zones. The control of the diffusion profile is usually carried out by setting different temperatures for each zone and by the speed of the belt. Also, gases can be introduced through the openings to change the ambient. Single or double side diffusion is both possible for belt furnace diffusion. For single-side diffusion, SOD method, spraying, or screen-printing technique are usually used. The most important drawback of belt furnace diffusion is the contamination by metallic impurities due to the conveyer belt.

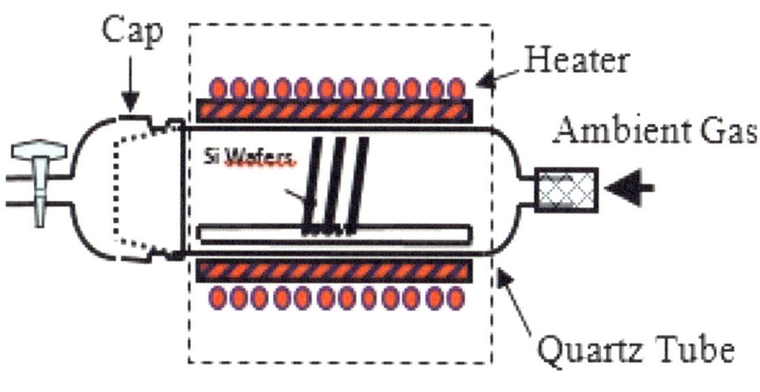

Figure 20. Quartz tube diffusion system.

5. Screen-printing phosphorus diffusion on P-type c-Si

5.1. Sample preparation and experimental

A total of 1.2 Ωcm mc-Si and 2.5–3 Ωcm CZ-Si p-type wafers with a surface area of 50 × 50 mm^2 were used for the quality evaluation experiments of the screen-printing phosphorus diffusion using Aquanyl phosphorus source provided by Nippon Gohsei Co. For the texturing of mc-Si p-type wafers, HF: HNO_3 (1:10 vl%) was used. CZ-Si p-type wafers were alkaline textured in KOH bath at 80°C. After printing the phosphorus paste on both side of the wafers by a screen-printing machine, wafers were dried at 150°C for 3 min. Diffusion process was carried out in a conventional quartz tube furnace in various ambients (Ar, N_2, O_2, and N_2+5% O_2) at peak temperatures of 875 and 900°C. For the sheet resistance evaluation, phosphorus silica glass was removed first by 5% HF and sheet resistances were measured by four-point-probe method. In case of lifetime studies, $n^+/p/n^+$ symmetric structures were formed on mc- and CZ p-type wafers and the effect of impurity gettering was evaluated after phosphorus diffusion. In order to estimate the carrier lifetime of the bulk, n+ layers were etched away using HF:HNO_3 (1:10 vl%) and minority carrier lifetimes were measured using the QSSPC [31, 32] technique after 3% iodine-ethanol chemical passivation.

5.2. Impact on sheet resistance

Diffused layers are basically characterized by sheet resistance, R_{sheet}, which is the ratio of resistivity, ρ, to thickness of the sample, d, $R_{sheet} = \rho / d$. In this study, ambient gas, diffusion temperature and the diffusion time are important parameters that affect the value of sheet resistance after diffusion. In order to minimize front recombination losses high sheet resistances are advantageous for n^+ emitter crystalline silicon solar cells. However, high sheet resistances limit to obtain a good Ω contact for metallization. On the other hand, these recombination losses can be reduced by lowering the phosphorus concentration which can lead an enhancement of blue response and thus the efficiency of solar cells.

Figure 21 shows the effect of the sheet resistance of emitter on internal quantum efficiency (IQE) in short wavelength range simulated by PC-1D. As can be seen, the higher the sheet resistance emitter, the better the blue response. High-efficiency devices using lithography or any advanced metallization contact technology can use high sheet resistance emitter potential. In case of $POCl_3$ diffusion, the control of sheet resistance is managed by temperature profile and diffusion ambient [33]. In this work, screen-printing phosphorus diffusion was carried out at peak temperature of 875 and 900°C in different ambients. **Figure 22** shows the average sheet resistances of phosphorus-diffused mc- and CZ-Si p-type wafers for each diffusion condition. For the mc- and CZ-Si p-type wafers sheet resistances of phosphorus diffused at 875°C in Ar, N_2, and N_2 + 5% O_2 ambient resulted in a range of from 80 to 90 and from 60 to 70 Ω sq^{-1}, respectively. At the peak diffusion temperature of 900°C, the range of sheet resistances were from 40 to 50 and from 30 to 40 Ω sq^{-1}, for mc and CZ-Si wafers, respectively.

Higher sheet resistances were observed at both 875 and 900°C in ambient O_2 for both types of wafers. The phenomenon of diffusion of screen-printed phosphorus into the silicon wafers can be explained by the reduction reaction of P_2O_5 with silicon as in Eq. (3). Surface of the

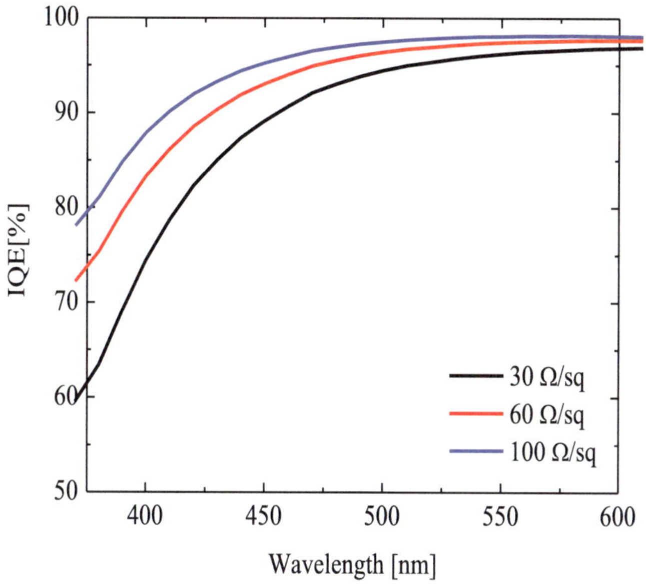

Figure 21. The simulated internal quantum efficiency (IQE) for different sheet resistances by PC-1D.

silicon oxidies are faster with the existence of O_2 ambient. Due to the heavy O_2 flow during the diffusion may cause reoxidation of phosphorus atoms. Thus, limits the amount of phosphorus atoms to be diffused which results high sheet resistance [30]. As a conclusion of that, 100% O_2 ambient may not be suitable for quality emitter formation. These results shows that proper emitter sheet resistances can be controlled by adjusting the diffusion temperature, diffusion time, and ambient gas in a similar manner as in the $POCl_3$ diffusion process when adapting a proper screen-printing dopant paste.

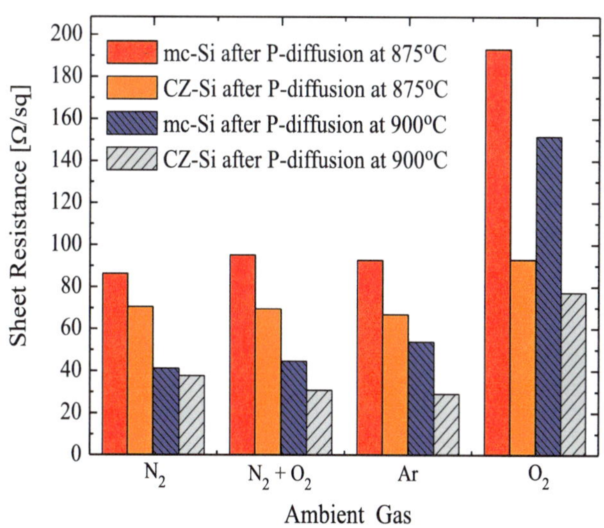

Figure 22. Sheet resistances measured on mc- and CZ-Si wafers after phosphorus diffusion at 875 and 900°C for 30 min [30].

5.3. Impact on carrier lifetime improvement

Metallic impurities have a lifetime killing effect for crystalline silicon solar cells, which becomes more crucial especially when using low solar grade silicon to grow mc-Si ingots. Especially top and bottom of the ingots contains high concentrations of metal impurities [34], which requires additional gettering before or during solar cell processing [35]. It is well known that the detrimental effect of these impurities can be reduced by can be reduced by phosphorus gettering [36, 37]. Owing to the phosphorus gettering, bulk lifetime of the silicon wafers can be improved by removing the impurities. In this work, $n^+/p/n^+$ symmetric structure of mc- and CZ-Si p-type wafers were formed by phosphorus diffusion in order to examine the effect of phosphorus gettering for our screen-printing diffusion process via estimation of bulk lifetimes. As a reference, initial lifetimes of the wafers were measured prior to phosphorus diffusion after removing the saw damages by acidic etching and then by using chemical passivation with a 3% iodine–ethanol solution for the measurement. These initial lifetimes were compared by post diffusion lifetimes after removing n^+ layers by acidic etching and passivating the bulk by the 3% iodine–ethanol chemical passivation as that used for the initial lifetime measurements. **Figure 23 (a)** and **(b)** shows the average carrier lifetime dependence of CZ and mc-Si p-type wafers to the ambient gas and diffusion temperature. After diffusion in ambient N_2, N_2+5%O_2, and Ar, improvement of carrier lifetimes were observed for both type of wafers. After diffusion at 900°C in ambient N_2, carrier lifetimes of CZ-Si p-type wafers reaches up to 350 μs from initial value of 200 μs [30].

The carrier lifetimes of mc-Si p-type wafers increased to up to 70 μs, about a tenfold increase from the initial carrier lifetime, as given in **Figure 23 (b)**. Considering that all wafers went through the same procedures before and after the diffusion processes, carrier lifetime improvements can be attributed to the gettering effect. The effect of oxygen on carrier lifetime was clearer for mc-Si wafers that show a good gettering effect even for shallow emitters.

The excess carrier dependent lifetimes for the CZ and mc-Si wafers improved and higher carrier lifetimes were realized at low-injection levels after phosphorus diffusion as shown in **Figure 24 (a)** and **(b)**.

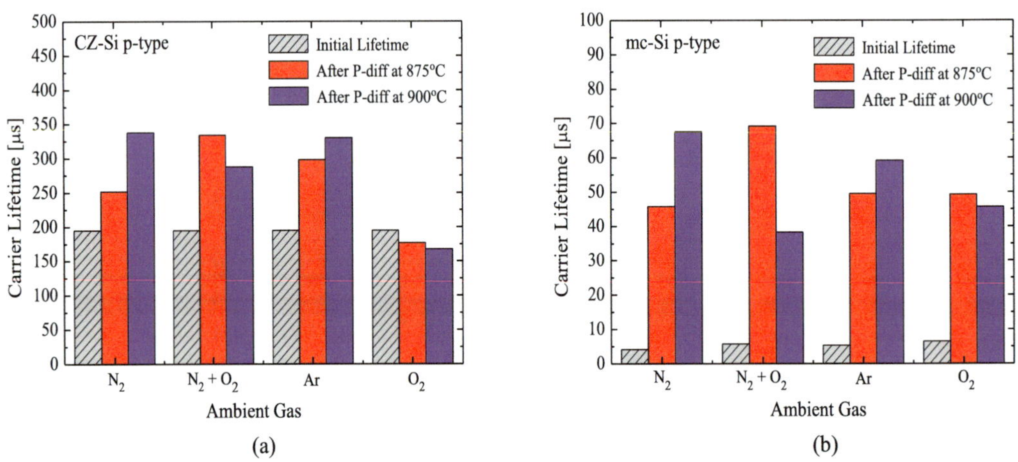

Figure 23. Increase in carrier lifetime after phosphorus diffusion on (a) CZ-Si wafers and (b) mc-Si wafers at different temperatures and in various ambient gases [30].

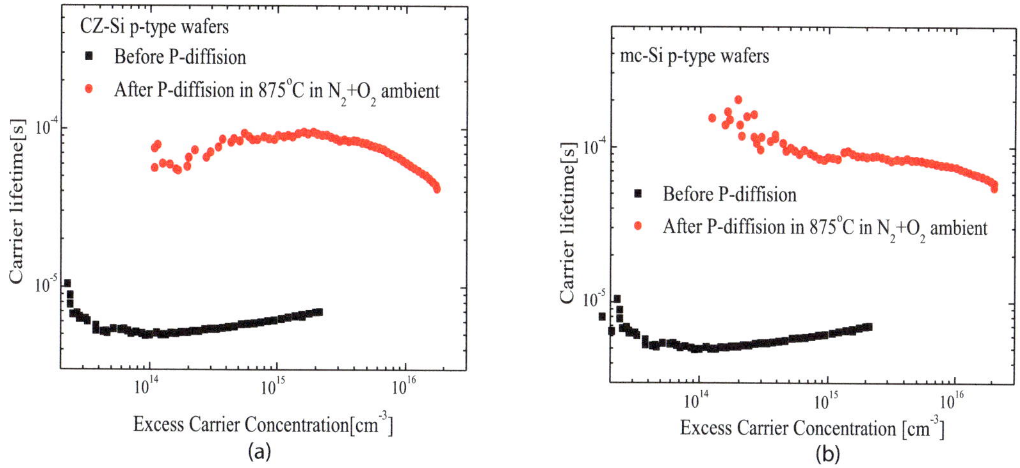

Figure 24. The carrier lifetime as a function of excess carrier concentration in (a) CZ-Si wafers and (b) mc-Si wafer before and after phosphorus diffusion [38].

6. Selectively screen-printing and single diffusion process

A lowly doped shallow emitter is required for a good blue response and a high quantum yield [39, 40]. However, a highly doped deep emitter is required for a good Ω contact for metallization. A selective emitter structure is an optimum tradeoff to combine these both emitter conditions with highly doped low sheet resistance regions under the printed contact fingers and lowly doped high sheet resistance regions between the fingers [41]. By the advantage of that, selective structures lower the surface recombination velocity of minority carriers [42], and leads to reduced contact series resistances as well [43, 44]. Additionally, blue response of solar cells can be improved [45, 46], high open circuit voltage and fill factor [47] can be achieved owing to the selective emitter structures. There are various techniques and applications to form selective emitter structures [48]. Double-diffusion with masking [49, 50], selective diffusion barrier process [51], oxide masking process and etch back [45, 52], implantation process [53] are among these techniques. Depending on the technique, need of laser ablation, additional steps like masking, repeated diffusions, etch back processes are among the drawbacks of such techniques. A screen-printable ink based on Si-nanoparticles was also provided which are selectively deposited prior to phosphorus diffusion to form heavily doped areas of the selective emitter structure [54]. This technology is already commercially available and includes an extra step of diffusion process for mass production. For instance, in laser doping through PSG [55], a light $POCl_3$ diffusion is performed first and resulted PSG acts as a doping source and highly doped regions are formed by laser ablation. Another laser-based process [47] offers a formation of the passivation layer simultaneously with structuring a highly doped region by a laser beam processing. There can be find some other processes for selective emitter solar cells mostly at research levels [56–59]. On the other hand, as a simple process for the formation of selective structures with single diffusion [60], based on the auto-diffusion mechanism, highly and lowly doped regions are formed at a single diffusion step. In this work, a single diffusion screen-printable diffusion technique is discussed as a simple and cost-effective method for industrial selective emitter cell processing. Applicability of forming selectively diffused structures using screen-printing phosphorus paste on p-type mc-Si and CZ-Si wafers will be explained.

6.1. Sample preparation and experimental

A total of 50 × 50 mm^2 alkaline-textured CZ- and acid-textured mc-Si p-type wafers were used for the characterization. Phosphorus diffusion paste was selectively screen-printed on the surface of the wafers using a 200 μm finger pattern and dried for 3 min at 150°C. Phosphorus diffusion was performed in a conventional furnace for 30 min at a peak temperature of 875°C in N_2+5%O_2 ambient. After phosphorus diffusion, PSG glass and oxide layers were removed by 5% HF and the selective emitter sheet resistance was determined using the specially optimized four-point-probe machine to estimate the sheet resistance of the selective emitter structures. Since phosphorus dopant paste was screen-printed selectively on CZ- and mc-Si p-type wafers with the same pattern as metal contacts, beneath the screen-printed phosphorus lines, the silicon wafer was doped heavily. The regions between the lines were doped lightly by doping atoms diffusion from the printed source via the gas atmosphere. **Figure 25** shows the principle of single screen-printing phosphorus diffusion to form the selective emitter structure.

6.2. Sheet resistances of selective emitter structure

In order to confirm that this simple process ability to form selective emitter structures, sheet resistances of CZ, and mc-Si p-type wafers were measured after diffusion of selectively printed phosphorus source. To examine the pattern of the formed selective emitter, a line scan was performed by four-point-probe measurements. The distribution of measured sheet resistance after the phosphorus diffusion for 30 min at 875°C in N_2+5%O_2 ambient on selectively screen-printed CZ-Si p-type substrates is shown in **Figure 26**.

Lower sheet resistance under the printed fingers and higher sheet resistance between the fingers can be confirmed. It should be noted that the accurate values of sheet resistances of the structure may differ from these measured values due to tight geometry of the selective emitter structure which can affect the measurement. However, highly diffused regions beneath the printed area and shallow regions between the printed areas can still be confirmed after the single screen-printed phosphorus diffusion process. One can conclude selective emitter structure can

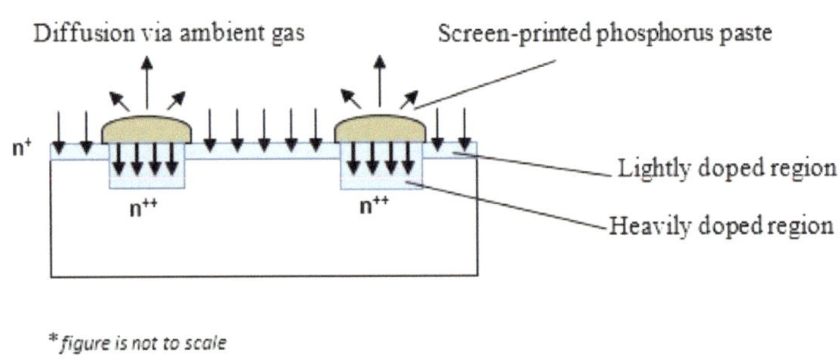

Figure 25. Principle of selective emitter formation process by single screen-printed phosphorus diffusion [30].

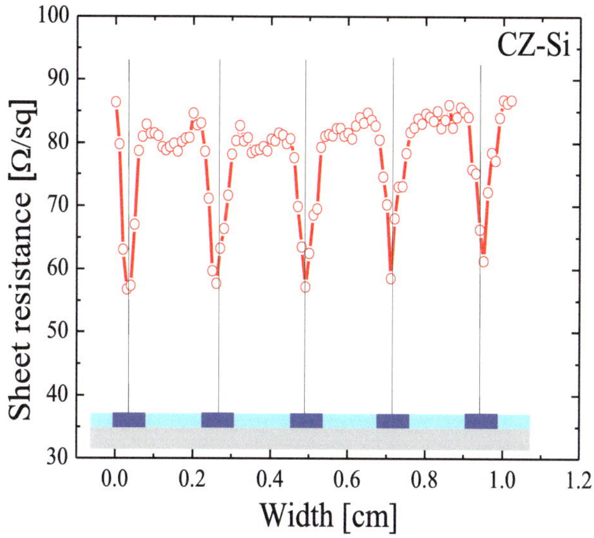

Figure 26. The distribution of measured sheet resistance after the phosphorus diffusion on selectively screen-printed CZ-Si p-type wafer [30].

be achieved with a single diffusion of a screen-printing phosphorus diffusion paste taking the advantage of auto-diffusion. Shallow regions under the illuminated area make it possible to have a better current collection properties and a low dark saturation current can be achieved by proper passivation applications [61, 62]. In order to confirm the diffusion profile of selectively screen-printed phosphorus, secondary ion mass spectrometry (SIMS) analysis was performed [30]. **Figure 27** shows SIMS profiles measured at two different region; under the printed finger area (highly doped region) and between the printed fingers (lowly doped region) of a CZ-Si p-type wafer went under single phosphorus diffusion process (875°C for 30 min in N_2+5%O_2 ambient).

Surface dopant concentration of the finger area was realized as a value of 1.4 × 10^{21} cm^{-3} and the concentration of the area between the fingers was realized as a lower value of 4.7 × 10^{20} cm^{-3} which shows the effect of auto-diffusion to achieve selective emitter structure. The actual sheet resistances were calculated using the SIMS data and estimated as 15 and 45 Ω sq^{-1}, under and between the fingers, respectively.

The sheet resistances base on the SIMS profile were calculated as follows; in order to calculate the carrier mobility first, μ_{max} and μ_{min} were set to 1414 and 68.5 cm^2 Vs^{-1}, respectively. N_0 was 9.2 × 10^{16} cm^{-3} and absorption coefficient α is set as 0.711 and N is the actual concentration taken from the SIMS measurement. Mobility of carriers was calculated according to Eq. (4).

$$\mu_n = \mu_{min} + \frac{(\mu_{max} + \mu_{min})}{1 + \left(\frac{N}{N_0}\right)^{\alpha}} \quad (4)$$

After determination of the carrier mobility, resistivity was calculated using Eq. (5).

$$\rho = \frac{1}{qN\mu_n} \quad (5)$$

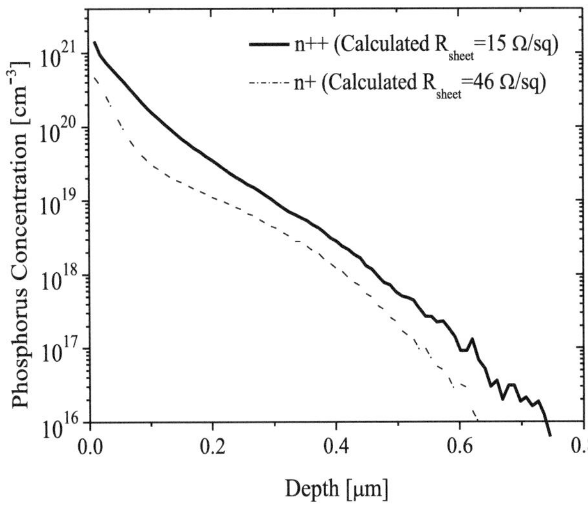

Figure 27. SIMS profile of highly doped regions under the fingers and lowly auto-doped regions between the fingers diffused at 875°C for 30 min [30].

Sheet resistance was calculated per each layer according to the following Eq. (6).

$$R_{shpl} = \frac{\rho_1 + \rho_2}{2(x_2 - x_1)} \tag{6}$$

Finally, total sheet resistance was calculated as the sum of inverse values of the sheet resistances per layer, as defined in Eq. (7).

$$R_{sh\cdot total} = \sum (R_{shpl})^{-1} \tag{7}$$

7. Single diffusion selective emitter solar cell fabrication

7.1. Sample preparations and experimental

Industrial-type selective emitter monocrystalline silicon solar cells with area of 156 × 156 mm^2 were fabricated by the single screen-printed diffusion process and compared to the homogenous emitter solar cells fabricated with standard $POCl_3$ diffusion. The process sequence used in this experiment is shown in **Figure 28 (a)**.

The screen-printing patterns of the printed dopant paste and the metal fingers as in **Figure 28 (b)** were similar in shape. However, the width of the fingers of printed paste was wider than the width of the metal finger to ensure the alignment. In case of the fabricated solar cells, the width of the phosphorus printed fingers were 400 μm where the finger width of the metal contacts was 70 μm.

Figure 29 shows the SIMS profiles of the selective structure formed on a CZ-Si p-type by single phosphorus diffusion at optimized conditions for the cell fabrication process at 870°C for 10 min in ambient N_2+5%O_2. However, note that the SIMS analyses of carbon, oxygen, and phosphorus concentrations were measured in cycles 10 nm, where the first 20 nm were used

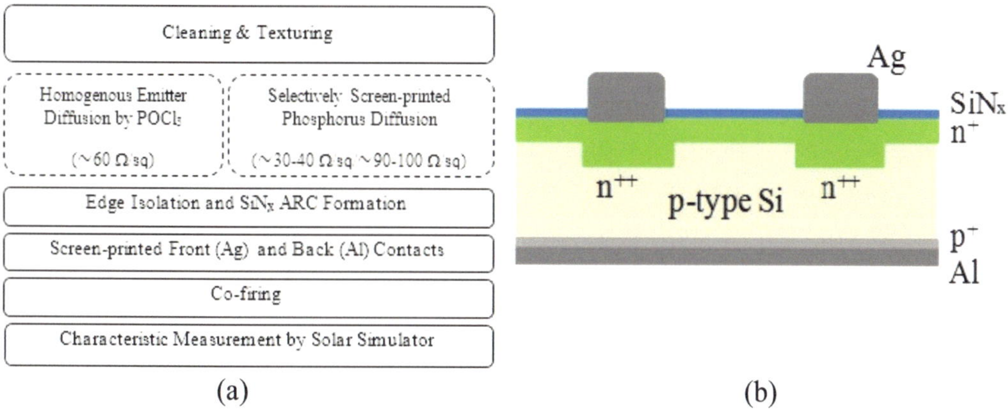

Figure 28. (a) Process sequence of cell fabrication process and (b) schematic of selective emitter solar cell with highly doped regions under the contacts and lowly doped regions between the contacts.

for the carbon and oxygen analyses and the phosphorus was measured starting from the 30 nm cycle. Therefore, the data of the first 20 nm depth was not obtained, and the actual surface concentrations are higher than the maximum values shown in **Figure 29**.

7.2. Fabricated homogenous and selective emitter solar cells

Table 6 summarizes the I–V measurement results of the homogenous emitter and selective emitter solar cells as measured by a calibrated solar simulator: the short-circuit current densities (J_{sc}), open-circuit voltages (V_{oc}), fill factor (*FF*), and conversion efficiency (*Eff*).

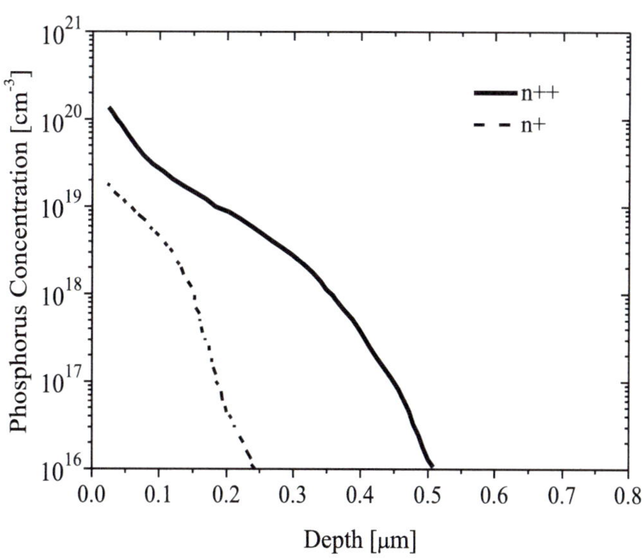

Figure 29. SIMS profiles of highly doped regions under the contacts and out-diffused lowly doped regions between the contacts diffused with the similar profile used for solar cell fabrication at 870°C for 10 min.

		V_{oc} (mV)	J_{sc} ($mAcm^{-2}$)	FF (%)	Eff (%)	I_{sc} (A)	V_{pm} (mV)	I_{pm} (A)	P_{max} (W)	R_s (Ωcm^2)	R_{sh} (Ωcm^2)
Homog. emit.	Avg.	618	36.4	77.9	17.5	8.7	518	8.1	4.2	1.1	3250
	Best	**617**	**36.6**	**78.3**	**17.7**	**8.7**	**518**	**8.2**	**4.2**	**1.1**	**2440**
Select. emit.	Avg.	625	36.3	79.2	17.9	8.7	528	8.1	4.3	1.1	5020
	Best	**625**	**36.5**	**79.2**	**18.1**	**8.7**	**529**	**8.2**	**4.3**	**1.0**	**21,700**

Table 6. Electrical characteristics of homogenous emitter and selective emitter solar cells [30].

The improvements in V_{oc} and *FF* by the selective emitter structure were confirmed. As shown in **Table 6**, the V_{oc} of the selective emitter cells was clearly improved by about 7 mV. The *FF* of selective emitter solar cells all exceeded 79%, which is an average increase of 1–2% compared to the homogenous emitter solar cells. However, no clear improvement in the short-circuit current was observed. The short-circuit currents seem to have been limited by the higher carrier recombination velocity due to the larger finger widths of the phosphorus-dopant paste and/or lower surface passivation effects. If surface passivation can be further improved, higher *FF* and V_{oc} can be also realized owing to the reduced defects and carrier recombination in the emitter region [63]. Thinner finger widths of phosphorus dopant paste can be considered to enhance current collection on the surface.

Figure 30 shows a comparison of the I–V curves of the best homogenous emitter solar cell and selective emitter solar cell. These results demonstrate the first trial of selective emitter formation using our screen-printed phosphorus source to prove the validity of the concept of single diffusion process. Further optimizations are in progress to realize higher efficiencies.

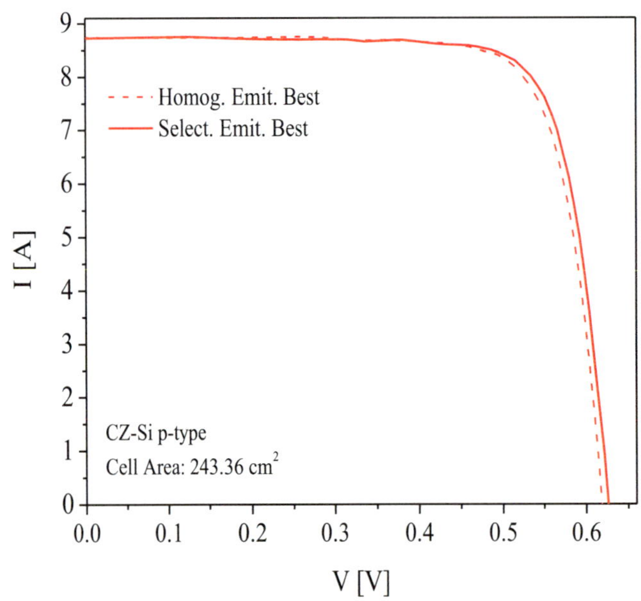

Figure 30. *I-V* curve comparison of the best homogenous emitter and selective emitter solar cells [30].

8. Conclusion

Non-vacuum processing techniques for cost-effective and simple crystalline silicon solar cell manufacturing were introduced. A special attention was given to the low-cost spin coating and spray deposited ARC alternatives instead of the main stream SiN_x ARC and to the low-cost screen-printing phosphorus doping process with the application of selective emitter solar cells accomplished by a single deposition and single diffusion step. About the introduced materials and films, first evaluation of the material and processes were carried out and then silicon solar cells were fabricated. ZrO_2-polymer composite/surface-deactivated TiO_2-polymer composite films were introduced by spin coating deposition method and best efficiency of 12.91% was achieved with *J*sc of 31.42 mA cm^{-2}, V_{oc} of 575 mV, and fill factor *(FF)* of 71.5%. An improvement of 0.86% with an increase of *J*sc of 2.07 mA cm^{-2} was confirmed when compared to those of fabricated cells without the ARC. TiO_2-compact and ZrO_2-polymer composite antireflection coating layers were also introduced by simple and cost-effective spray pyrolysis deposition and spin coating process. Decrease in surface reflectance below 2% was confirmed owing to the ZrO_2-polymer composite/TiO_2-compact multilayers from around 550 –1050 nm of the spectrum. Efficiency of the fabricated cells increased up by a factor of 0.8% and reaches 15.9% with further increase of *J*sc 2 mA cm^{-2} owing to the applied ZrO_2-polymer composite/TiO_2-compact multilayer ARC. In case of spray pyrolysis deposition-based Al_2O_3/TiO_2 double-layer ARC, a reflectivity lower than 0.4% could be confirmed at 600 nm. Average conversion efficiency of fabricated cells (average of four samples) were 15.5% with a J_{SC} of 37.0 mA cm^{-2}, where the improvement was about 3.5 mA cm^{-2} and 1.7% on J_{SC} and on efficiency, respectively. Considering the future lowcost solar cells with non-vacuum process, introduced ARC films with good optical properties can be a promising alternative for ARC of solar cells. On the other hand, a cost effective screen-printing phosphorus diffusion process was introduced for an easy forming of selective emitter solar cells. Owing to the phosphorus diffusion carrier lifetimes of CZ-Si p-type wafers could be improved from 195 up to 350 μs through phosphorus gettering. By a single print and single diffusion process lowly and highly doped regions were achieved for selective emitter solar cell applications. Large-area monocrystalline solar cells were fabricated successfully by this method and more than 18% conversion efficiency was achieved. As an attractive and cost effective method, adapting such simple methods including spin coating, spray deposition, or screen-printing methods to all solar cell manufacturing sequences are believed to be play a key role for future low-cost solar cells.

Author details

Abdullah Uzum[1,2], Seigo Ito[1]*, Marwan Dhamrin[3] and Koichi Kamisako[3]

Address all correspondence to: itou@eng.u-hyogo.ac.jp

1 University of Hyogo, Himeji, Japan

2 Karadeniz Technical University, Trabzon, Turkey

3 Tokyo University of Agriculture and Technology, Tokyo, Japan

References

[1] Deng W, Chen D, Xiong Z, Verlinden P J, Dong J, Ye F, Li H, Zhu H, Zhong M, Yang Y, Chen Y, Feng Z, Altermatt P. 20.8% PERC solar cell on 156 mm × 156 mm P-Type multicrystalline silicon substrate. IEEE Journal of Photovoltaics. 2016;**6**:3–9. DOI: 10.1109/JPHOTOV.2015.2489881

[2] Richards B S. Comparison of TiO_2 and other dielectric coatings for buried-contact solar cells: a review. Progress in Photovoltaics: Research and Applications. 2004;**12:**253–281. DOI: 10.1002/pip.529

[3] Dekkers HF, Beaucarne G, Hiller M, Charifi H, Slaoui A. Molecular hydrogen formation in hydrogenated silicon nitride. Applied Physics Letters AIP. 2006;**89**:211914. DOI: 10.1063/1.2396900

[4] Aberle A, Hezel R. Progress in low-temperature surface passivation of silicon solar cells using remote-plasma silicon nitride. Progress in Photovoltaics. 1997;**5**:29–50. DOI: 10.1002/(SICI)1099-159X199701/02)(5:1<29::AID-PIP149>3.0.CO;2-M

[5] Wan Y, McIntosh KR, Thomson AF. Characterisation and optimisation of PECVD SiNx as an antireflection coating and passivation layer for silicon solar cells. AIP Advances. 2013;**3**:032113. DOI: 10.1063/1.4795108

[6] Brinker CJ, Harrington MS. Sol-gel derived antireflective coatings for silicon. Solar Energy Materials. 1981;**5**:159–172. DOI: 10.1016/0165-1633(81)90027-7

[7] Richards BS. Single-material TiO_2 double-layer antireflection coatings. Solar Energy Materials and Solar Cells. 2003;**79**:369–390. DOI: 10.1016/S0927-0248(02)00473-7

[8] Vitanov P, Loozen X, Harizanova A, Ivanova T, Beaucarne G. A study of sol-gel deposited Al_2O_3 films as passivating coatings for solar cells application. In: Proceedings of the 23rd European Photovoltaic Solar Energy Conference and Exhibition; 1–5 September 2008; Valencia, Spain. pp. 1596–1599. DOI: 10.4229/23rdEUPVSEC2008-2CV.5.5

[9] Liang Z, Chen D, Feng C, Cai J, Shen H. Crystalline silicon surface passivation by the negative charge dielectric film. Physics Procedia. 2011;**18**:51–55. DOI: 10.1016/j.phpro.2011.06.056

[10] Wang MT, Wang TH, Lee JY. Electrical conduction mechanism in high-dielectric-constant ZrO_2 thin films. Microelectronics Reliability. 2005;**45**(5–6):969–972. DOI: 10.1016/j.microrel.2004.11.014

[11] Uzum A, Kuriyama M, Kanda H, Kimura Y, Tanimoto K, Ito S. Non-vacuum processed polymer composite antireflection coating films for silicon solar cells. Energies. 2016;**9**:633–649. DOI: 10.3390/en9080633

[12] Koch W. Properties and uses of ethylcellulose. Industrial and Engineering Chemistry. 1937;**29**(6):687–690. DOI: 10.1021/ie50330a020

[13] Born M, Wolf E. Principle of Optics. 7th ed. Cambridge University Press; Cambridge, UK, 1999. p. 67. ISBN: 0521642221

[14] Kobiyama M. Basic Theory of Thin Film Optics (in Japanese). Optronics Co. Ltd.; 2002. ISBN-10: 4-900-47496-7, ISBN-13: 978-4900474963

[15] Kern W, Puotinen DA. Cleaning solutions based on hydrogen peroxide for use in silicon semiconductor technology. RCA Review. 1970;**31**:187–206.

[16] Reinhardt K, Kern W, editors. Handbook of Silicon Wafer Cleaning Technology. 2nd ed. William Andrew; 2008, Norwich, USA. ISBN: 978-0-8155-1554-8

[17] Uzum A, Kuriyama M, Kanda H, Kimura Y, Tanimoto K, Fukui H, Izumi T, Harada T, Ito S. Sprayed and spin-coated multilayer antireflection coating films for non-vacuum processed crystalline silicon solar cells. International Journal of Photoenergy. 2017;3436271, 5 pages. DOI: 10.1155/2017/3436271

[18] Haque SA, Palomares E, Cho BM, Cho ANM, Hirata N, Klug DR, Durrant JR. Charge separation versus recombination in dye-sensitized nanocrystalline solar cells: The minimization of kinetic redundancy. Journal of the American Chemical Society. 2005;**127**: 3456–3462. DOI: 10.1021/ja0460357

[19] Yu H, Zhang S, Zhao H, Will G, Liu P. An efficient and low-cost TiO_2 compact layer for performance improvement of dye-sensitized solar cells. Electrochimica Acta. 2009;**54**:1319–1324. DOI: 10.1016/j.electacta.2008.09.025

[20] Hocine D, Belkaid MS, Pasquinelli M, Escoubas L, Torchio P, Moreau A. Characterization of TiO_2 antireflection coatings elaborated by APCVD for monocrystalline silicon solar cells. Physics Status Solidi C. 2015;**12**:323–326. DOI: 10.1002/pssc.201400085

[21] Kanda H, Uzum A, Harano N, Yoshinaga S, Ishikawa Y, Uraoka Y, Fukui H, Harada T, Ito S. Al_2O_3/TiO_2 double layer anti-reflection coating film for crystalline silicon solar cells formed by spray pyrolysis. Energy Science and Engineering. 2016;**4**:269–276. DOI: 10.1002/ese3.123

[22] Meiners BM, Holinski S, Schafer P, Hohage S, Borchert D. Investigation of anti-reflection-coating stacks for silicon heterojunction solar cells. In: Proceedings of 29th European PV Solar Energy Conference and Exhibition; 2014. Amsterdam, Netherlands. DOI: 10.4229/EUPVSEC20142014-2AV.3.22

[23] Yu IS, Wang YW, Cheng HE, Yang ZP, Lin CT. Surface passivation and antireflection behavior of ALD on n-type silicon for solar cells. International Journal of Photoenergy. 2013;**431614**:7. DOI: 10.1155/2013/431614

[24] Jhaveri J, Avasthi S, Man G, McClain WE, Nagamatsu K, Kahn A, Schwartz J, Sturm JC. Hole-blocking crystalline-silicon/titanium-oxide heterojunction with very low interface recombination velocity. In: Proceedings of IEEE 39th Photovoltaic Specialists Conference (PVSC); 2013 June; pp. 3292–3296. DOI: 10.1109/PVSC.2013.6745154

[25] Ali K, Khan SA, Jafri MZM. Effect of double layer (SiO_2/TiO_2) anti-reflective coating on silicon solar cells. International Journal of Electrochemical Science. 2014;**9**:7865–7874.

[26] Moradi M, Rajabi Z. Efficiency enhancement of Si solar cells by using nanostructured single and double layer anti-reflective coatings. Journal of Nanostructures. 2013;**3**:365–369. DOI: 10.7508/jns.2013.03.013

[27] Wright DN, Marstein ES, Holt A. Double layer anti-reflective coatings for silicon solar cells. In: Proceedings of the 31st IEEE Photovoltaic Specialists Conference; January 2005; pp. 1237–1240. DOI: 10.1109/PVSC.2005.1488363

[28] Wang WC, Tsai MC, Yang J, Hsu C, Chen J. Efficiency enhancement of nanotextured black silicon solar cells using Al_2O_3/TiO_2 dual-layer passivation stack prepared by atomic layer deposition. ACS Applied Materials & Interfaces. 2015;**7**:10228–10237. DOI: 10.1021/acsami.5b00677

[29] Yang CH, Lien SY, Chu CH, Kung CY, Cheng TF, Chen PT. Effectively improved SiO_2-TiO_2 composite films applied in commercial multicrystalline silicon solar cells. International Journal of Photoenergy. 2013;**823254**:8. DOI: 10.1155/2013/823254

[30] Uzum A, Hamdi A, Nagashima S, Suzuki S, Suzuki H, Yoshiba S, Dhamrin M, Kamisako K, Sato H, Katsuma K, Kato K. Selective emitter formation using single screen-printed phosphorus diffusion process. Solar Energy Materials and Solar Cells. 2013;**109**:288–293. DOI: 10.1016/j.solmat.2012.11.013

[31] Sinton RA, Cuevas A, Stuckings M. Quasi-steady-state photoconductance, a new method for solar cell material and device characterization. In: Proceedings of 25th IEEE Photovoltaic Specialist Conference; Washington DC, USA. 1996. pp. 457–460. DOI: 10.1109/PVSC.1996.564042

[32] Sinton RA, Cuevas A. Contactless determination of current–voltage characteristics and minority-carrier lifetimes in semiconductors from quasi-steady-state photoconductance data. Applied Physics Letters. 1996;**69**:2510–2512. DOI: 10.1063/1.117723

[33] Lossen J, Mittelstadt L, Dauwe S, Lauer K, Beneking C. Making use of silicon wafers with low lifetimes by adequate $POCl_3$ diffusion. In: Proceedings of 20th European Photovoltaic Solar Energy Conference; Barcelona, Spain. 2005. pp. 1411–1414.

[34] Davis Jr, Rohatgi A, Hopkins RH, Blais PD, Rai-Choudhury P, McCormick JR, Mollenkopf HC. Impurities in silicon solar cells. IEEE Transection of Electron Devices. 1980;**27**(4):677–687. DOI: 10.1109/T-ED.1980.19922

[35] Cuevas A, Stocks M, Armand S, Stuckings M, Blakers A, Ferrazza F. High minority carrier lifetimes in phosphorus-gettered multicrystalline silicon. Applied Physics Letters. 1997;**70**(8): 1017–1019. DOI: 10.1063/1.118469

[36] Macdonald D, Cuevas A, Kinomura A, Nakano Y, Geerligs LJ. Transition-metal profiles in a multicrystalline silicon ingot. Journal of Applied Physics. 2005;**97**:033523. DOI: 10.1063/1.1845584

[37] Macdonald D, Cuevas A, Samundsett C, Ferrazza F. Response to phosphorus gettering of different regions of cast multicrystalline silicon ingots. Solid-State Electronics. 1999; **43**(3): 575–581.

[38] Hamdi A, Uzum A, Nagashima S, Suzuki S, Suzuki H, Yoshiba S, Dhamrin M, Kamisako K, Sato H, Katsuma K, Kato K. Phosphorus diffusion impact on carries lifetime of P-type multi- and mono-crystalline silicon wafers using novel screen printed diffusion source. In: Proceedings of 21th Photovoltaic Science and Engineering Conference; December 2011; Fukuoka, Japan. DOI: 10.1016/S0038-1101(98)00283-4

[39] King RR, Sinton RA, Swanson RM. Studies of diffused phosphorus emitters: Saturation current, surface recombination velocity, and quantum efficiency. IEEE Transection of Electron Devices. 1990;**37**(2):365–371. DOI: 10.1109/16.46368

[40] Zhao J, Wang A, Green MA. Emitter design for high-efficiency silicon solar cells. Progress in Photovoltaics: Research and Applications. 1993;**1**(3):193–202. DOI: 10.1002/pip.4670010303

[41] Misiakos K, Lindholm FA. Toward a systematic design theory for silicon solar cells using optimization techniques. Solar Cells. 1986;**17**(1):29–52. DOI: 10.1016/0379-6787(86)90057-8

[42] Haverkamp H, Dastgheib-Shirazi A, Raabe B, Book F, Hahn G. Minimizing the electrical losses on the front side: Development of a selective emitter process from a single diffusion. In: Proceedings of 33th IEEE Photovoltaic Specialists Conference; San Diego, CA, USA; 2008. pp. 430–433. DOI: 10.1109/PVSC.2008.4922443

[43] Meier DL, Schroder DK. Contact resistance: Its measurement and relative importance to power loss in a solar cell. IEEE Transections on Electron Devices. 1984;**31**:647–653. DOI: 10.1109/T-ED.1984.21584

[44] Bähr M, Dauwe S, Mittelstadt L, Schmidt J, Gobsch G. Surface passivation and contact resistance on various emitters of screen-printed crystalline silicon solar cell. In: Proceedings of 19th European Photovoltaic Solar Energy Conference; Paris, France; 2004. pp. 955–958.

[45] Wenham SR, Willison MR, Narayanan S, Green MA. Efficiency improvement in screen-printed polycrystalline silicon solar cells by plasma treatments. In: 18th IEEE Photovoltaic Specialists Conference; Las Vegas, USA; 1985. pp. 1008–1013.

[46] Szlufcik J, Elgamel HE, Ghannam M, Nijs J, Mertens R. Simple integral screenprinting process for selective emitter polycrystalline silicon solar cells. Applied Physics Letters. 1991; **59**(13): 1583–1584. DOI: 10.1063/1.106291

[47] Blendin G, Faber M, Horzel J, Boubekeur H, Metz A. High throughput inline process for deep phosphorus emitters. In: Proceedings of 26th European Photovoltaic Solar Energy Conference; Hamburg, Germany; 2011. pp. 1372–1376.

[48] Hahn G. Status of selective emitter technology. In: Proceedings of 25th European Photovoltaic Solar Energy Conference and Exhibition/5th World Conference on Photovoltaic Energy Conversion; Valencia, Spain; 2010. pp. 1091–1096.

[49] Gauthier M, Grau M, Nichiporuk O, Madon F, Mong-The Yen V, Le Quang N, Zerga A, Slaoui A, Blanc-Pélissier D, Kaminski A, Lemiti M. Industrial approaches of selective emitter on multicrystalline silicon solar cells. In: Proceedings of 24th European Photovoltaic Solar Energy Conference and Exhibition; Hamburg, Germany; 2009. pp. 1875–1878.

[50] Kim IH, Cho KY, Oh DJ, Shim JM, Lee EJ, Lee HW, Choi JY, Kim JS, Shin JE, Lee SH, Lee HS. Exceeding 19% efficient 6 inch selective emitter crystalline silicon solar cell by screen printed metal contacts: transition of heavy & shallow emitter conditions. In: Proceedings of 25th European Photovoltaic Solar Energy Conference and Exhibition/5th World Conference on Photovoltaic Energy Conversion; Valencia, Spain; 2010. pp. 2694–2697.

[51] Koppes M, Stassen AF. Low cost selective emitter for multicrystalline solar cells with 0.4% point efficiency gain. In: Proceedings of 25th European Photovoltaic Solar Energy Conference and Exhibition/5th World Conference on Photovoltaic Energy Conversion; Valencia, Spain; 2010. pp. 1917–1920.

[52] Book F, Braun S, Herguth A, Dastgheib-Shirazi A, Raabe B, Hahn G. The etch-back selective emitter technology and its application to multicrystalline silicon. In: Proceedings of 35th IEEE Photovoltaic Specialists Conference; Honolulu, Hawaii, USA; 2010. pp. 1309–1314.

[53] Low R J, Gupta A, Bateman N, Ramappa D, Sullivan P, Skinner W, Mullin J, Peters S, Weiss-Wallrath H. High efficiency selective emitter enabled through patterned ion implantation. In: Proceedings of 35th IEEE Photovoltaic Specialists Conference; Honolulu, USA; 2010. pp. 1440–1445.

[54] Poplavskyy D, Scardera G, Abbott M, Meisel A, Chen X, Shah S, Tai E, Terry M, Lemmi F. Silicon ink selective emitter process: Optimization of the short wavelength response. In: Proceedings 25th European Photovoltaic Solar Energy Conference and Exhibition/5th World Conference on Photovoltaic Energy Conversion; Valencia, Spain; 2010. pp. 2416–2419.

[55] Röder TC, Eisele SJ, Grabitz P, Wagner C, Kulushich G, Köhler JR, Werner JH. Add-on laser tailored selective emitter solar cells. Progress in Photovoltaics Research and Applications. 2010;**18**:505–510. DOI: 10.1002/pip.1007

[56] Poulain G, Boulord C, Blanc-Pélissier D, Kaminski A, Gauthier M, Semmache B, Pellegrin Y, Lemiti M. Towards a self-aligned process for selective emitters on silicon solar cells. In: Proceedings of 25th European Photovoltaic Solar Energy Conference and Exhibition/5th World Conference on Photovoltaic Energy Conversion; Valencia, Spain; 2010. pp. 2043–2046.

[57] Rudolph D, Peter K, Meijer A, Doll O, Köhler I. Etch back selective emitter process with single $POCl_3$ diffusion. In: Proceedings of 26th European Photovoltaic Solar Energy Conference and Exhibition; Hamburg, Germany; 2011. pp. 1349–1352.

[58] Lee H, Lee EJ, Kim J, Shin J, Kim J, Choi J, Kim I, Shim J, Oh DJ, Cho KY, Kong J, Seo JK, Lee S, Hallam B, Wenham SR, Lee H. Production of 19.2% efficient large area commercial

grade P-type CZ silicon solar cells based on the laser doped selective emitter technology. In: Proceedings of 26th European Photovoltaic Solar Energy Conference and Exhibition; Hamburg, Germany; 2011. pp. 1390–1393.

[59] Lossen J, Jesswein R, Weiß M, Wütherich T, Dupke M, Katkhouda K, Meyer K, Böscke T, Froitzheim A, Eschrich H, Krokoszinski HJ. The DELFIN concept—a screen printed solar cell with >18.5% efficiency. In: Proceedings of 25th European Photovoltaic Solar Energy Conference and Exhibition/5th World Conference on Photovoltaic Energy Conversion; Valencia, Spain; 2010. pp. 2333–2337.

[60] Horzel J, Szlufcik J, Nijs J. High efficiency industrial screen printed selective emitter solar cells. In: Proceedings of 16th European Photovoltaic Solar Energy Conference and Exhibition; Glasgow, UK; 2000. pp. 1112–1115.

[61] Nijs J, Demesmaeker E, Szlufcik J, Poortmans J, Frisson L, De Clercq K, Ghannam M, Mertens R, Van Overstraeten R. Recent improvements in the screen-printing technology and comparison with the buried-contact technology by 2D-simulation. Solar Energy Materials and Solar Cells. 1996;**41/42**:101–117. DOI: 10.1016/0927-0248(95)00136-0

[62] Stem N, Cid M. Physical limitations for homogeneous and highly doped n-type emitter monocrystalline silicon solar cells. Solid-State Electronics. 2004;**48**:197–205. DOI: 10.1016/j.sse.2003.08.005

[63] Blendin G, Faber M, Horzel J, Boubekeur H, Metz A. High throughput inline process for deep phosphorus emitters. In: Proceedings of 26th European Photovoltaic Solar Energy Conference and Exhibition; Hamburg, Germany; 2011. pp. 1372–1376.

Chapter 6

Self-Adjusting Electrochemical Etching Technique for Producing Nanoporous Silicon Membrane

Norhafizah Burham, Azrul Azlan Hamzah and Burhanuddin Yeop Majlis

Additional information is available at the end of the chapter

http://dx.doi.org/10.5772/67719

Abstract

This chapter presents the technique in producing the nanoporous silicon membrane using electrochemical etching technique. Electrochemical etching technique is a self-adjusting technique due to its ability to control transfer of ion to form pore by manipulating certain parameters. There are several parameters that have been manipulated to study the effect of each parameter to the pore formation by characterizing each component. The project starts with fabrication of silicon membrane and then continues with characterization of HF concentration, current density, doping and also alcohol diluents using field emission scanning electron microscopy (FESEM). The effect of each parameter is discussed in terms of pore size, pore formation and pore structure. Finally, the pore with size less than 100 nm and columnar structure has formed using this technique. The star-shaped structure is also formed through this experimental setup. Improved nanoporous silicon membrane can be applied for filtration and separating particles, especially in an artificial kidney.

Keywords: nanoporous silicon, membrane, electrochemical etching, self-adjusting technique, hydrofluoric acid

1. Introduction

Nanoporous silicon is widely used for separating gas or particle [1–3] and can be applied in biosensors [4], optics [5, 6], tissue engineering [7] and radiotherapy [8]. For those applications, the integration with other components like micropump can produce one complete system. Silicon is chosen for this application because of its physical and chemical stability [9], especially in separating particles due to its biocompatibility and anti-biofouling. Polyethylene glycol (PEG) is used for surface modification for its biocompatibility. The usage of nanoporous silicon membrane is widespread especially in biological filtration. Nanoporous silicon can be combined with micropump, microchannel and microfluidic modules to make a complete device for bioMEMS and LoC applications. There are several methods of pore formation on silicon substrate like electrochemical etching process [10], focused ion beam, electron beam lithography and rapid thermal annealing [2]. Electrochemical etching process is chosen for producing pore on silicon membrane because of its simple experimental setup and easy replication of the pore formation and structure. Recently, the creation of the smallest pore size has been explored to suit various applications. There are various methods that can be explored to vary the pore formation in terms of size and structure of pore silicon [11–17]. The self-adjusting method can create the smallest pore by controlling and manipulating certain parameters during electrochemical etching process, which are current density [18–20], HF concentration [21], time, silicon orientation [22], doping level [23–25], lighting and electrolyte mixture [25]. The characterization of producing nanoporous silicon is discussed in terms of current density, HF concentration, dopant and diluents that influence the formation of pore structure and size of this nanoporous silicon membrane. Field emission scanning electron microscope (FESEM) is used to verify and examine the pore formation due to its capability to visualize the porous silicon structure in nanometer range with higher magnification. The FESEM results also verify the pore formation of porous silicon, whether uniform or non-uniform, and the pore diameters.

2. Theory on electrochemical etching process

Porous silicon can be formed using electrochemical etching technique or anodization process in hydrofluoric solution. Commonly, porous silicon cannot be formed by dipping the silicon in HF solution. But the current flow between two electrodes, which is silicon at the anode and platinum at the cathode in the HF solution, will produce the pores on silicon membrane. Electrochemical etching process is a very simple and economical experimental setup in terms of apparatus and chemical used. **Figure 1** shows different experimental setups used in this electrochemical etching process. **Figure 1** is the simplest experimental setup, which used one Teflon bath to put the HF solution, and two electrodes were dipped in this HF solution, supplying the current to produce pores. Normally, the silicon is placed in the anode and platinum in the cathode. Various other metals can be used at the cathode. But the metal used can hold it in HF solution without eroding it. Teflon cell is used because it has a high level of acid resistance compared to the glass cell. The benefit of this experimental setup is simple and easy to modify. However, the pores are not uniform for both sides of the silicon due to the inhomogeneity resulting from lateral potential drop.

Meanwhile, **Figure 2** shows the second type of electrochemical etching cell in the single-cell approach using a back-side contact. A metal contact is made at the back side of wafer and sealed with o-ring so that only front-side sample will be exposed to anodize electrolyte. This

http://dx.doi.org/10.5772/67719

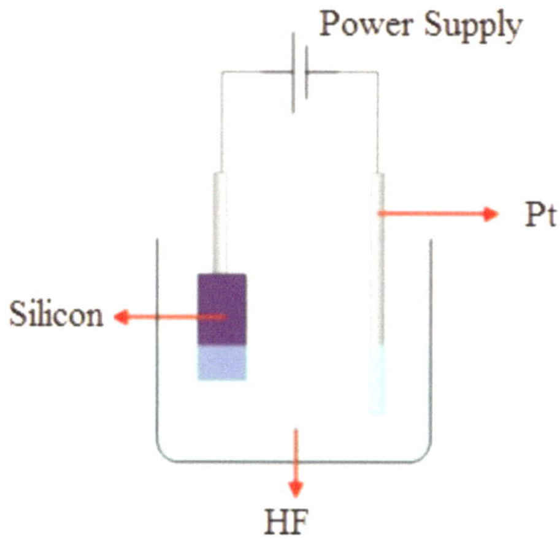

Figure 1. Cross section of a lateral electrochemical etching cell [26].

type of experimental setup is well suited for the front using two cells containing electrolyte, and silicon is placed in the middle of the cell. Furthermore, the cell that leads to good uniformity in porous silicon layers and the simplest interpretation of current-voltage characteristics is most commonly used and offers a good control of thickness and porosity. This cell is also well suited for the front-side illumination of the sample during anodization because of illumination that will affect the pore formation during the process [22].

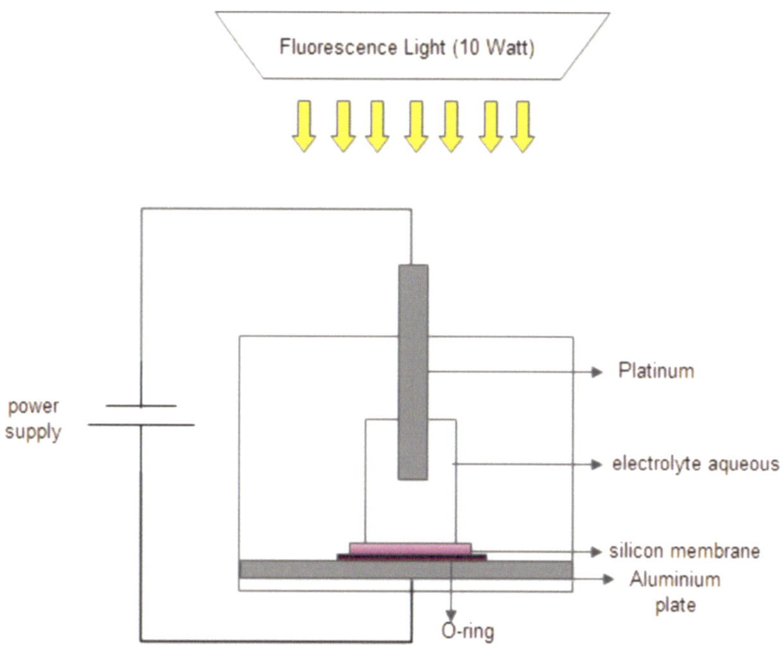

Figure 2. Cross section of a conventional single-tank cell [27].

The third type of cell is the double-tank cell using an electrolytic back-side contact. **Figure 3** shows the equipment used in this process, which consists of two half cells in which platinum electrodes are immersed and the silicon used to separate and isolate the two cells. Both cells used HF solution for electrochemical etching for polished and a back-side contact. The chemical pump is used to circulate the electrolyte solution to remove the gas bubbles generated using anodic reaction and maintain the concentration of HF solution in the cell tank. A good and uniform pore can be obtained using symmetrical and large platinum plates as the cathode and the anode. These two platinum electrodes are connected to power supply, and the current flows from one half to the other through silicon membrane. The back side of silicon membrane acted as a secondary cathode where the proton reduction takes place leading to hydrogen evolution, while the front side of the wafer acted as a secondary anode to form porous silicon.

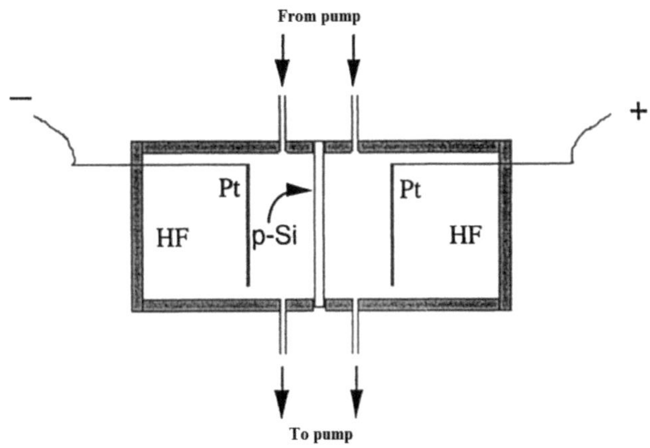

Figure 3. Cross-sectional view of a double-tank cell [22].

2.1. Electrochemical etching mechanism

Silicon is the material that is difficult to dissolve in HF solution except with the aid of the flow. Silicon reacts with HF solution to dissolve the H and F during power supply. In general, some solvents may be used to produce a hole in the surface of the silicon. Nanoporous silicon is formed using ethanoic HF solution that is a mixture between ethanol and HF. However, the organic solvent may also be used to obtain porous silicon like **Dimethylformamide (DMF)** and acetonitrile [28]. The concept of dissolution of silicon ions for the formation of the hole is the same, although any solvent is used. Chemical reaction of the pore formation is expressed in the equation below [19, 29, 30]:

$$Si + 2HF + \lambda h^{+} \rightarrow SiF_2 + 2\,H^{+} + (2 - \lambda)\,e^{-} \quad (1)$$

$$SiF_2 + 2HF \rightarrow SiF_4 + H_2 \quad (2)$$

$$SiF_4 + 2HF \rightarrow H_2\,SiF_6 \quad (3)$$

where h^{+} and e^{-} change between hole and electron and λ is a number of charge change at the first stage.

http://dx.doi.org/10.5772/67719

This equation is also used by some researchers [31–34]. But other equations have been proposed in the process of liquidation in accordance with the method of surface oxidation of silicon, exchange hole and electrons that become a source of divalent silicon on the oxidation level [35]. **Figure 4** shows that model reaction of p-type silicon dissolution in HF solution is used. Ion dissolution measures the current flow caused by the electrochemical etching process that makes the transfer between electrons and holes. The ion is struggling to move in order to break the bonds of Si-O, Si-F and Si-H during the process. Thus, the idea for the formation of pores involves current densities studied to make the resulting hole narrower and straight [19, 36].

Pore formation involving the current density is discussed in **Figure 5**. During electrochemical etching process, the holes will be placed on silicon surface. When a high current density is given, the hole will focus and gather at the boundary layer between silicon crystal and HF solution. So, the electrochemical dissolution of silicon in HF will generate very smooth surface. This process is called electropolishing. On the other hand, if the low current density is applied, a lot of fluoride ions will be placed on the silicon surface rather than the holes. In this situation, etching process is limited because of the lack of hole in silicon surface during ion dissolution. The dissolution rate of fluoride ion to migrate to the electric field is limited because of the lack of hole. So, if the surface is rough, the hole will force out to make a large hole by the uneven pit surface [32].

Briefly, the electrochemical etching process is a process that involves the dissolution of the ion to form pore and electropolishing. Ion dissolution process is dependent on the current density during the electrochemical etching process. The current density also affects the pore structure like spongy or columnar structure [33]. The current density is also among the parameters that can be manipulated to produce the perfect pore structure. Pore formation is influenced by the current density based on graph IV to study either the formation of pores or electropolishing process. Graph IV involving current and voltage used in the formation

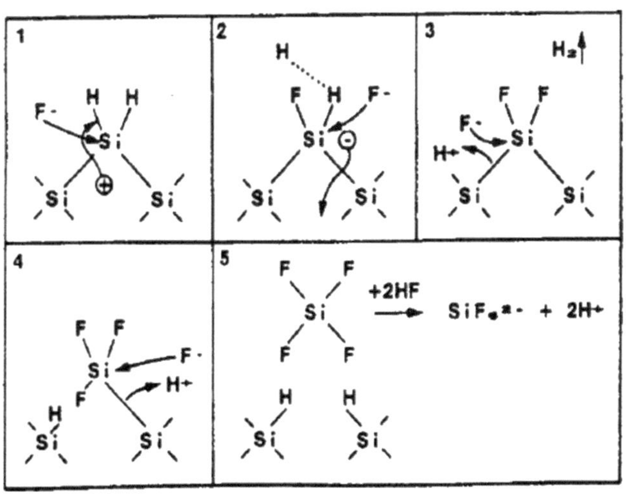

Figure 4. Chemical mechanism of electrochemical etching [31].

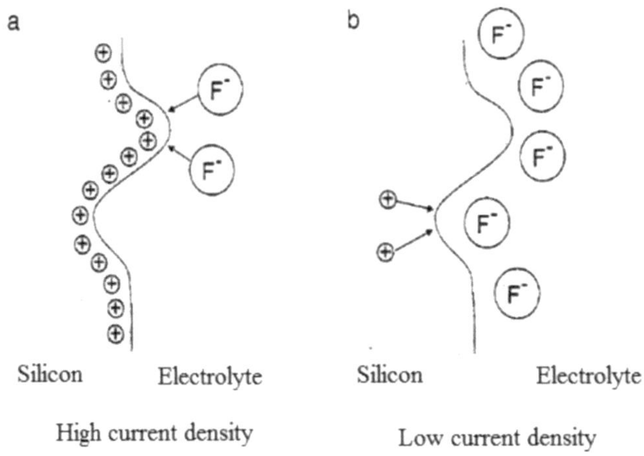

Figure 5. The boundary layer between the silicon crystal and HF solution during ion transfer processes [32].

of pore is shown in **Figure 6**. For each pore formation process and polishing silicon electrochemical reaction has a different equation is divided anode depends on the ability of electrode and HF solution [11]. At low current density, pore silicon can be formed. But the pore was only on the surface of the membrane and structured like a sponge. The current process called simple force of transition forms either pores or electropolishing.

Theoretically, the current density affects the pore formation in silicon. The current will create a line at the end of the pore using space-charge region (SCR) as shown in **Figure 7** [35]. When the electric field is strong enough to build pore with the given current flow, oxide forms on the surface of the pores. The oxide layer will be dissolved by the electrolyte solution and forms pores below the existing holes. When the coating is completely dissolved by the electrolyte, the existing electric field produces a flow line to another, and this process will be repeated.

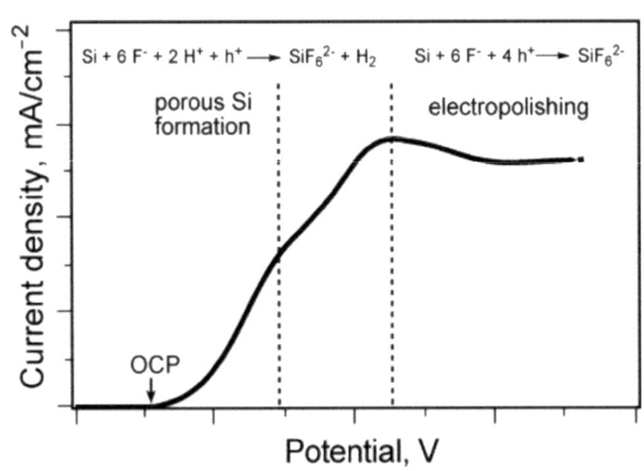

Figure 6. Graph of current density versus different potentials for pore formation and electropolishing [34].

http://dx.doi.org/10.5772/67719

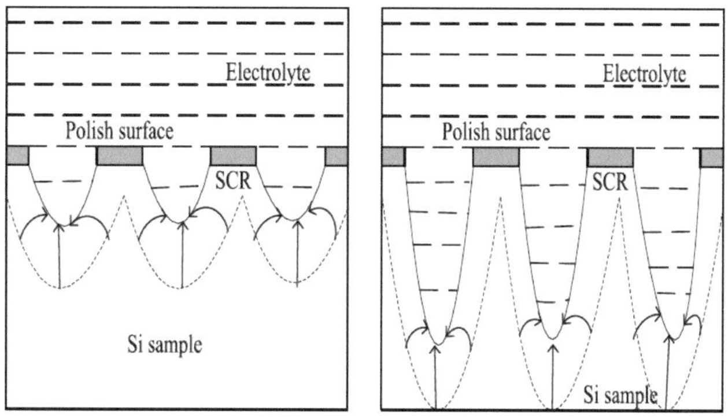

Figure 7. Existence of SCR and current lines bent toward pore tip at the higher electric field strength [37].

Thus, the current flow is capable of dissolving silicon and produces nano- or micro-sized pores depending on the dopant density either p or n. Dissolution will occur two times at the same area as there are currents to break down and encourage the maximum pore formation in the orientation <010> or <001> [35, 36].

However, the straight pores are influenced by other factors such as time and current density etching. The hole formation can be varied by manipulating etching time technique, which is called the current burst model. The current burst is conditioned when the current model is still reacting to the dissolution of the silicon surface. When no current burst model, the formation of a new model of current burst occurred and attacked the area around the edge of the pore caused by inhomogeneity in time and space [38]. When the etching time is increased, current flows in any area of pore tips occur. **Figure 8** shows how the region acts current burst in any possible area during pore formation.

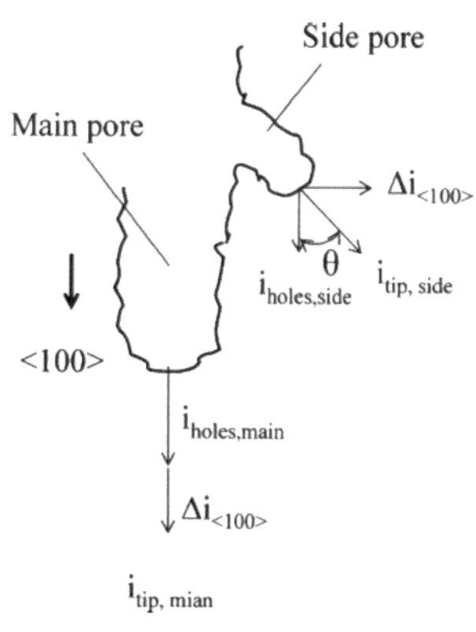

Figure 8. Forming a hole caused by the formation of a new model of current eruption which dissolves holes in different orientations [39].

Electrochemical etching is a self-adjusting technique that manipulates several parameters to get various pore structures and sizes. The dopant-type silicon orientation and photoluminescence factor will affect the pore structure as shown in **Figure 9**. Pore-shaped structure inclined (b) due to the orientation of silicon. Oriented silicons 110 and 111 produce an inclined hole. In order to produce a vertical, straight or columnar pore structure, orientation of silicon 100 can be formed using n-type dopant with back-side illumination. **Figure 9** represents some symbols such as n+, n−, p+ and p− representing the dopant used. The symbols + and − represent the dopant level, which is the amount of charge carriers that exist on the silicon. Symbol + has high density of n/p dopant compared to charge carriers, while the symbol −, a sign of the type of dopant n/p, is less than the charge carriers. Different silicon pore structures can be produced by changing these parameters.

The mechanism of dopant used that affects the formation of the pore structure is discussed. N-type silicon (phosphorus) has the ability to build vertical pore structure, while the p-type silicon (boron) produces uneven pore structure. This is due to the p-type silicon, which does not have a mechanism to control the accumulation of charge carriers to disperse the maximum pore anisotropic silicon and passivate the walls of the hole during the electrochemical

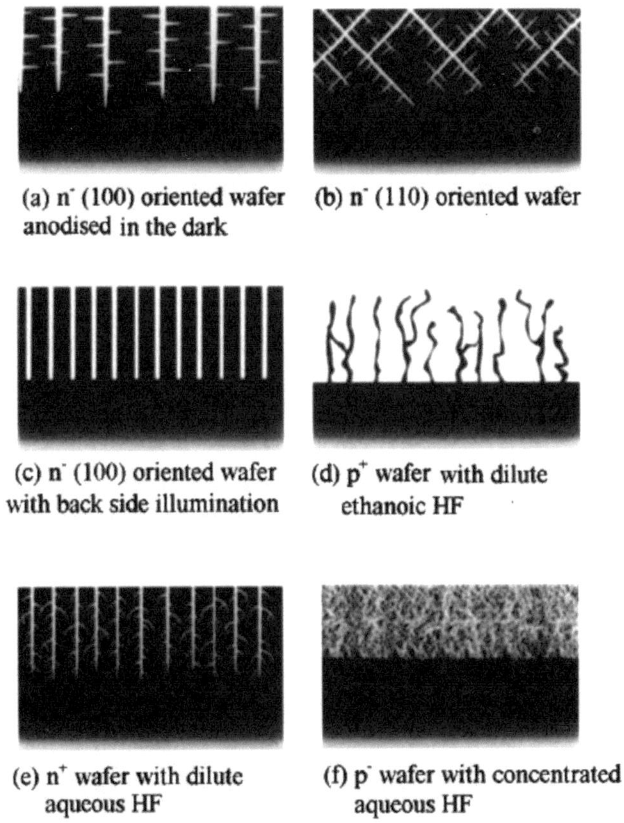

Figure 9. Illustration of pore formation based on dopant factor [22].

etching process. So, pore is difficult to form using this type of dopant [40]. In contrast, the n-type silicon is efficient in collecting minority charge carriers with the help of illumination. Illumination acts as a booster for the formation of a straight channel pore. In addition, the pore wall dissolution spontaneously passivated on dissolution of consequences of the reduction of the hole [41]. Based on the properties of dopants, some chemical reaction model to explain the formation of pores is influenced by the band gap caused by confinement charge during the formation of a small hole. Interstitial concentration decreases, further moving charges in the silicon structure [42]. The charge moves on the surface of the silicon dopant.

Dopant acts as impurities for semiconductor to improve the conductivity of semiconductor materials. The use of doping will facilitate the pore formation during etching process [43]. The difference of the two dopants is having a surplus or one less valence electron. Silicon without dopants also has the same number of electrons and holes. When it comes to the surface conductivity of silicon, the silicon resistance is lower because the conductivity is inversely proportional to the resistance of silicon. The conductivity depends on two parameters, namely, the concentration of charge carriers either electrons or holes and agility carrier. During electrochemical etching process, the reaction of Pt at cathode electrode makes a very small electron movement to balance the charge between silicon and Pt. The dopant will affect pore formation based on valence band on dopant to the electrolyte solution. Valence band is very important to make the charge carriers to the electrode erosion. P-type silicon is a majority carrier. Even without bias, the transport rates of the hole are small to move above the silicon surface. When there is positive bias applied to the silicon electrode, the barrier between the charge transports diminishes, and the valence band of the hole becomes focused onto the surface of the interface. In this situation, this kind of bias terms is used in semiconductors. The majority carriers are electrons, which are n-type silicon.

N-type silicon is exposed to lighting to produce more vent holes formed in the silicon. This is because the light will generate electron-hole pairs near the silicon surface and sweep vent holes on the surface of silicon. Thus, the n-type silicon can produce a straight hole if lighting is given continuously to the silicon surface. Dopant silicon obstacles will affect the formation of the hole because the structure depends on the homogeneity and the size of the hole depends on the current density and time [43]. To produce a sharp, straight or columnar pore structure by using the space-charge region, high lighting and dopant silicon whether n-type or p-type are required [44]. The use of low resistance dopant material will produce a star-shaped hole and a structure, which is not straight even when using the same experimental procedure [37]. Resistance dopants also affect the pore size based on the density of dopant. Silicon with a pore size of 2–4 nm can be formed using silicon-type n and p dopants that have low and medium density, silicon with pore size of 5–50 nm can be generated using a dopant with high density and silicon with pore size of 50 nm to several 10 m can be formed using a dopant with medium density [45]. This is because the mechanism of formation of the vent holes needs to justify the dissolution of the ion [19].

Pore formation depends on the dopant density. **Figure 10** shows the schematic of pore formation mechanism, which starts with the orientation of crystallization of silicon. **Figure 10(a)** shows that the (100) crystallographic face contains strained Si–H bonds, and it tends to be more disposed due to dissolution compared to other faces. In contrast, the (111) face contains Si–H bonds that are more stable and perpendicular to the surface. The differential reactivity of the crystal faces leads

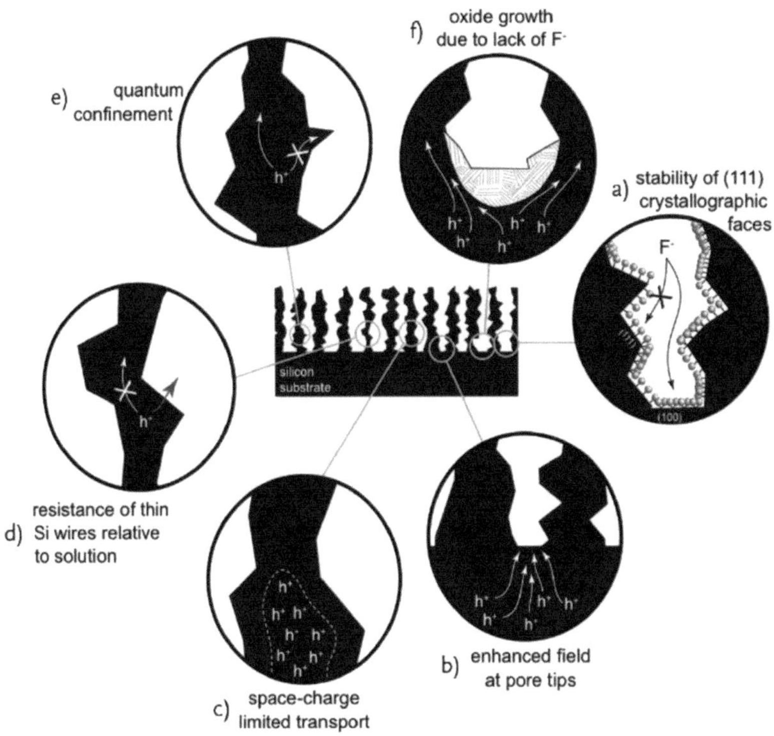

Figure 10. Schematic of pore formation mechanism in porous silicon [46].

to crystallographic pores in order to propagate primarily in the <100> direction. **Figure 10(b)** shows that high radius of curving at the tip, especially at the bottom, of a pore generates a region for enhanced electric field, which can attract valence band holes. **Figure 10(c)** shows that the space-charge region is a region in which carriers are depleted due to band bending while interfacing between silicon and electrolyte. Band bending increases by decreasing the dopant density, so this mechanism is a primary determinant for macropore size formation especially for low-doped n-type Si. In **Figure 10(d)**, as the diameter of a silicon filament decreases, the resistance for transporting the valence band holes increases. At a critical filament diameter (typically a few nm for p-type silicon), injection of the hole into the solution becomes more favorable, and holes do not propagate further down the length of the nanowire. This mechanism is responsible for the lack of electrochemical dissolution of a microporous layer when it has been formed. **Figure 10(e)** shows that the result of increasing the band gap from the quantum confinement excludes the valence band holes from the smallest regions of the porous silicon matrix. (f) If there are no fluoride ions available at the silicon/solution interface, silicon oxide forms at the interface. Valence band holes are then excluded from this region, and they continue to oxidize the silicon/porous silicon interface. This causes pore widening and produces electropolishing porous layer [46].

In short, the pore formation is influenced by many factors for producing holes of various sizes and shapes. The details on pore formation will be discussed further by characterization of these parameters.

http://dx.doi.org/10.5772/67719

3. Fabrication process

Double-sided polished 400 μm thick silicon nitride specimens with <100> orientation were prepared by cutting the substrate to a dimension of 2.54 × 3 cm. The silicon substrates are then cleaned with the standard cleaning procedure. The substrates are dipped in acetone and methanol for 5 minutes in an ultrasonic bath. Then, the samples were dipped into 10% hydrofluoric acid (HF) for 1 minute to remove the stain on the silicon surface. The samples were then rinsed using deionized (DI) water and blasted with nitrogen gas to dry the samples. Finally, the samples were put on a hot plate with the temperature set to 120°C for 15–20 minutes for the hard-bake process to ensure no water remains on the silicon surface.

The silicon samples then undergo the lithography process to pattern the square frame on the substrate. The positive photoresist AZ 4620 was first coated on the silicon substrate using a spin coater with a setting of 500 rpm for 10 seconds and then 2000 rpm for 20 seconds. Next, the substrates were put on a hot plate at 120°C for 1 minute. Only then will the samples be ready for the lithography process. The mask aligner Karl Suss MJB 3 was used to transfer the square pattern on a mask to the silicon substrates. Then, the samples were exposed to UV light for 90 seconds. Afterwards, the substrates were dipped for 4 minutes in an AZ 400K developer to develop the square patterns. After that, the samples went through a hard-bake process for 15 minutes. Next, they were dipped into a buffer oxide etch (BOE) solution to remove unwanted nitrides at the opening frame. The schematic of BOE process is shown in **Figure 11**. The Teflon clamper is used to hold the silicon nitride for this process. The BOE solution is put on the silicon surface for 4 hours in room temperature to produce a mask for KOH etching process.

Then, the substrates were dipped in acetone again to remove the photoresist. For thinning the silicon substrates, 45 g of potassium hydroxide (KOH) pallet was mixed well with 55 ml DI water to make a 45% concentration solution. Next, the KOH solution was put into a beaker containing just DI water. The double-boiling method was used for this etching process as shown in **Figure 12**. This is a repeating process to get the silicon membrane thickness of 5 μm. The KOH process is a very crucial part to get the smoothed surface roughness. The admixture of 45% with isopropyl alcohol (IPA) will improve the surface roughness of the membrane [47].

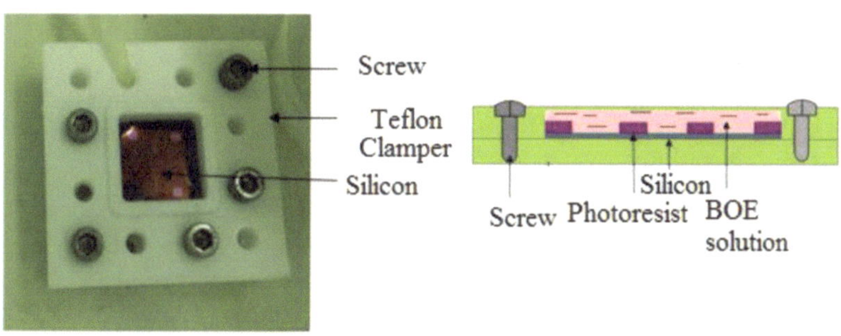

Figure 11. Buffered oxide etchant process to remove unwanted photoresist.

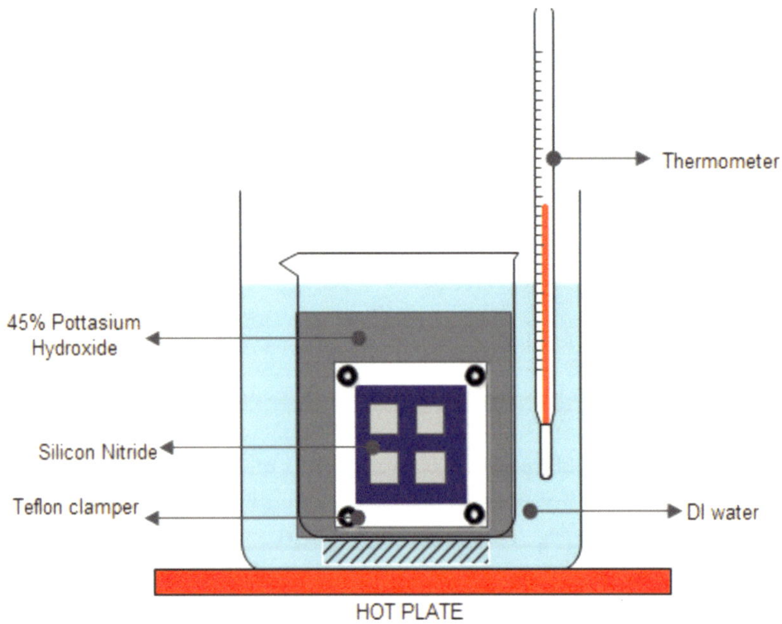

Figure 12. Schematic of double-boiling technique for KOH etching process.

Furthermore, IPA acts as a catalyst to increase the etching rate during bulking away of the silicon substrate. The cross section of silicon membrane is shown in **Figure 13**. The thickness of the membrane is 35 μm after immersing the silicon in KOH solution for 6 hours. So, to obtain a thickness of 5 μm, the silicon will be immersed in KOH solution.

The silicon membrane needs to be handled carefully because the thin membrane is easy to break during the electrochemical etching experiment stage. Electrochemical etching process starts after the membrane is immersed in electrochemical bath in order to remove nitride that

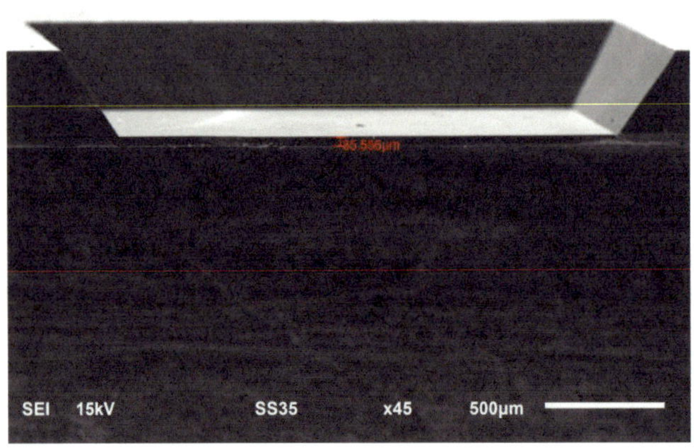

Figure 13. The cross section of silicon membrane after 6 hours of KOH etching process.

remains on the silicon surface using BOE solution. The silicon membrane must be in hydrophobic condition before the process starts. If not, silicon membrane will be dipped again in 10% HF to remove all stains on silicon membrane. Any impurity will affect the pore formation throughout the process. An illustration of the process flow is shown in **Figure 14**.

Electrochemical etching was performed to form pores on the silicon membrane surface using hydrofluoric acid (HF) solution [21, 39, 46, 48]. The experimental setup was shown in **Figure 15** by supplying a constant current between two electrodes immersed in a Teflon cell containing an aqueous solution of HF or diluted HF. An ethanol and HF solution is commonly used among researchers as an electrolyte aqueous solution for electrochemical etch. Ethanol will act as the surfactant in reducing the hydrogen bubble throughout the process [46, 48]. Fluorescence light is put facing the silicon membrane during electrochemical etching process as a catalyst in producing a well-forming pore during pore formation mechanism [15, 17]. Furthermore, photoluminescence is used to assist an electron to attack silicon surface in order to produce the best pore distribution. So, the pore formation will become more uniform toward the end of the process. Finally, FESEM is used to observe and verify the pore structure of the silicon membrane.

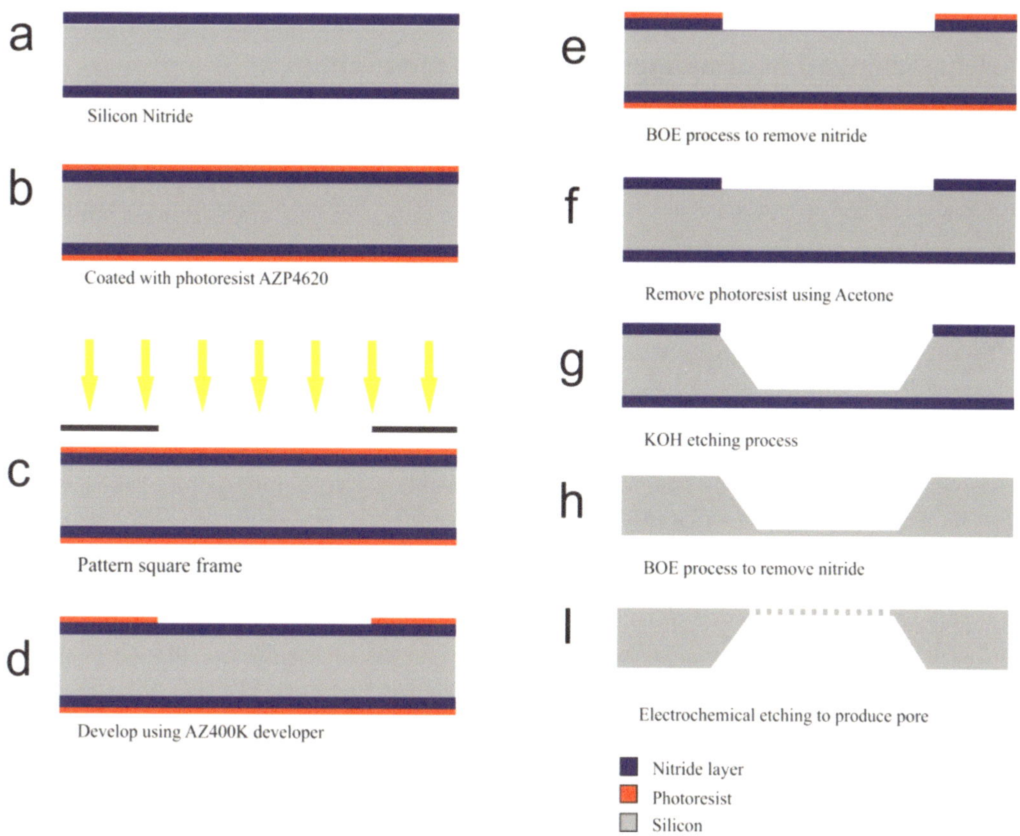

Figure 14. Fabrication process for producing nanoporous silicon membrane.

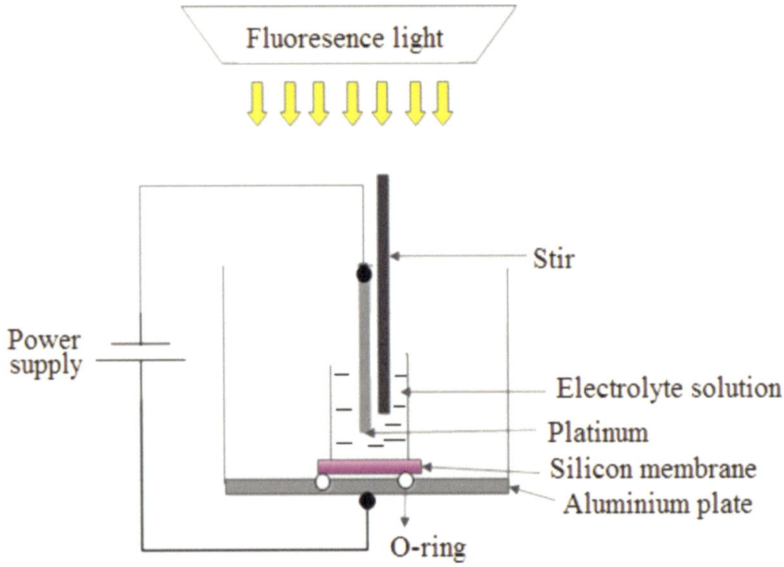

Figure 15. Electrochemical etching process.

4. Characterization of nanoporous silicon membrane

A few experiments have been conducted to study the effect of HF concentration, current density, doping and diluents to the pore formation. The pore formed will be inspected and examined under FESEM Supra VVP5. The pore structure and pore size are examined, and each effect will be characterized in this part.

4.1. HF concentration

The first study is on hydrofluoric acid concentration. The experimental setup used various HF concentrations by differentiating the volume ratio between HF solution and ethanol. This section will discuss the pore formation and size after electrochemical etching process. Forty-nine percent of HF and ethanol are mixed in Teflon beaker with different volume ratios, which are 3:7, 5:5, 7:3 and 9:1. The platinum is placed at the cathode, and undoped silicon is placed at the anode. The current density has been set at 25 mA/cm^2 for 20 minutes. 3 mL of ethanoic HF will be put in the single-tank cell. During the process, the top membrane was faced with photoluminescence of 5 W. After 20 minutes, the silicon has been inspected under FESEM Supra VVP5 to examine the pore size and structure on silicon membrane. **Figure 16** shows that the pore has been formed after electrochemical etching process. FESEM shows that low concentration of HF produces a large pore size compared with high concentration of HF. This is because high-volume HF contains more fluoride ion.

The second experimental setup will study the pore formation by varying time. Other samples were put in this single-tank cell and electrochemical etching process was performed for varying time (30, 40 and 50 minutes). The result will be examined under FESEM to study

http://dx.doi.org/10.5772/67719

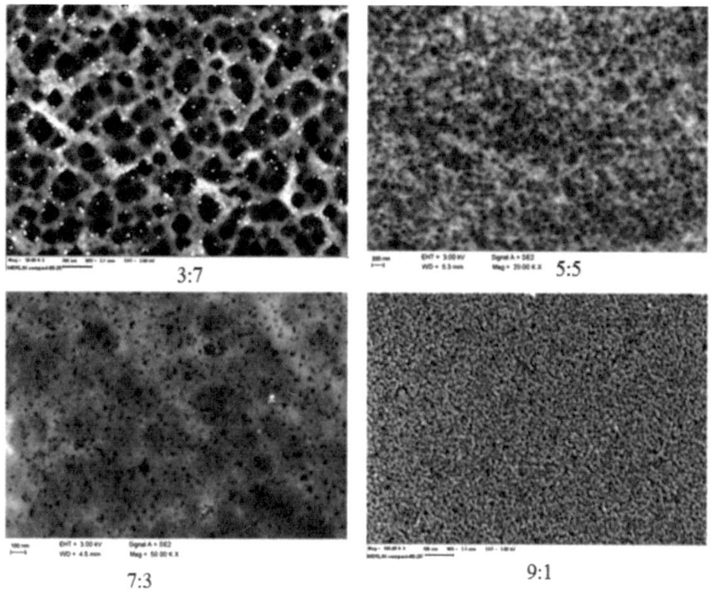

Figure 16. The pore structure of a different volume ratio of HF/ethanol.

the pore size. The average pore sizes are counted by inspecting this sample under Atomic Force Microscope (AFM). The AFM is able to count the pore size by grain/pore analysis. Then, the data will be analysed to study the effect of HF concentration and time to the pore formation. **Figure 17** shows the relation of HF concentration to the pore size at variant time of immersed silicon in electrolyte solution. The graph shows that the pore size becomes wider at low HF concentration. Furthermore, time taken for electrochemical etching process

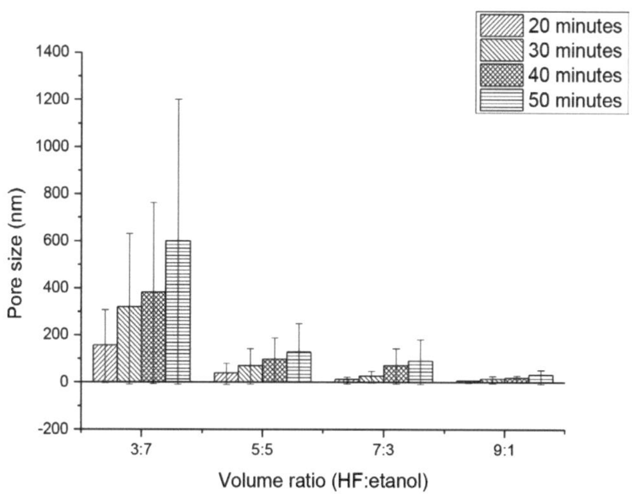

Figure 17. The effect of HF concentration in terms of pore size at different times.

is one of the factors that make the pore become wider. Low HF concentration produces non-uniform pore structure. This pore formation mechanism starts with migration between electron and holes in order to attract charge carrier when applied in electric field. The migration of electron and holes happens in competition between Si–O, Si–F and Si–H bond formation. The small amount or volume ratio of HF in electrolyte solution will generate the oxide. The oxide layer is generated on the silicon surface. F^- ions are trying to attack rapidly to avoid oxide layer to grow and to avoid the water molecules to take over the role of nucleophile. The lack of F ions makes the oxide unremovable from the silicon surface. It is because the insulated oxide terminates the pore propagation [46].

Table 1 shows that the pore size range was measured under FESEM. The effect of volume ratio of HF and ethanol during electrochemical etching process is studied by measuring the pore size. It has been found that the HF volume is a key in order to form nanosized pores. The highest HF amount will reduce the thickness of space-charge layer and increase the pore-tip current density [39]. The high volume of HF can prevent the native oxide growth throughout the process because dissolution silicon oxide rate is increased by increasing the HF volume, which automatically increases the critical current density at the covered oxide layer on silicon. In this case, the small pore structure is formed by increasing the HF concentration.

The immersed time also affects the pore formation. **Table 1** shows the longest immersed time that makes the pore diameter wider. 20 and 50 minutes as immersed time are compared. The result shows that 20 minutes of the uniform pore formation produced is compared to 50 minutes. It is because the longer immersed time makes the pore become wide due to the breakage of the pore wall.

The pore size formed in different HF concentrations using dopant boron has been studied [21]. For the dopant silicon, pore size depends on dopant concentration to produce variant size of pore. For example, mesoporous (5–50 nm) is produced using highly doped level. Meanwhile, a macroporous (50 nm–10 μm) can be produced using medium-doped level and microporous (2–4 nm) using low dopant. **Figure 18** shows that the finding of pore formation using highly doped boron has been inspected under FESEM and the visual color observation on silicon surface. A volume ratio of 3:7 produces a pore size of 10 nm; 4:6, 8 nm; and 6:4, 3 nm. From the result, the HF concentration can affect pore formation for various silicon substrates, regardless of their being doped or undoped.

Time (minutes)	**Pore diameter**			
	3:7	**5:5**	**7:3**	**9:1**
20	60–200 nm	30–50 nm	9–19 nm	5–10 nm
30	220–360 nm	40–100 nm	16–34 nm	13–25 nm
40	200–500 nm	50–150 nm	57–101 nm	17–30 nm
50	600 nm–1 μm	50–300 nm	58–120 nm	30–50 nm

Table 1. The range of pore size with different amounts of HF and ethanol.

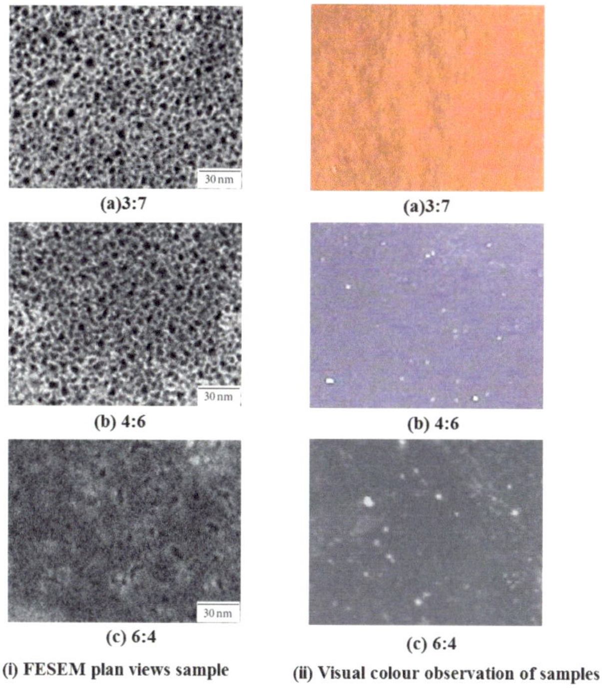

Figure 18. Effect of pore formation for variant HF concentration using p-type silicon [21].

4.2. Current density

Current density is one of the factors that affect the pore formation. The previous experimental setup demonstrates that the HF concentration affects the pore formation in terms of pore size. This experimental setup examines the current density affecting the pore structure, etching rate and also pore size. This study used the aforementioned experimental setup using p-type silicon membrane with resistivity of 1–100 Ωcm. The first set of experiments is to gauge the etching rate of porous layer by manipulating the current density. The current density is varied from 5 to 30 mA/cm^2 using ethanoic HF as electrolyte solution. The porous layer is measured using FESEM by cutting the silicon membrane to get the cross section of porous layer. The silicon membrane is immersed in ethanoic HF solution for 1 hour. A 5 mA/cm^2 current density gives an etch rate of 12.6 nm/h, while a 30 mA/cm^2 current density gives an etch rate of 1.3 μm/h. It was observed that etch rate increases linearly with the applied current, as illustrated in **Figure 19**. In this experimental setup, there are certain parameters that have been fixed in order to gauge the sole effect of current density on the etch rate, which include photoillumination, HF concentration, distance between anode and cathode and stirring time.

Ethanoic HF electrolyte solution etching rate is very slow. In contrast, the HF solution has a quick etching rate. The etch rate of μm/h is observed for current density changing from 5 to 20 mA/cm^2 for 5% HF electrolyte solution [19]. The ethanoic HF shows the same trend graph between etching rate and current density. However, the ethanoic HF etch rate became slower due to the effect of ethanol which acts as surfactant in decreasing the hydrogen bubble [20, 21].

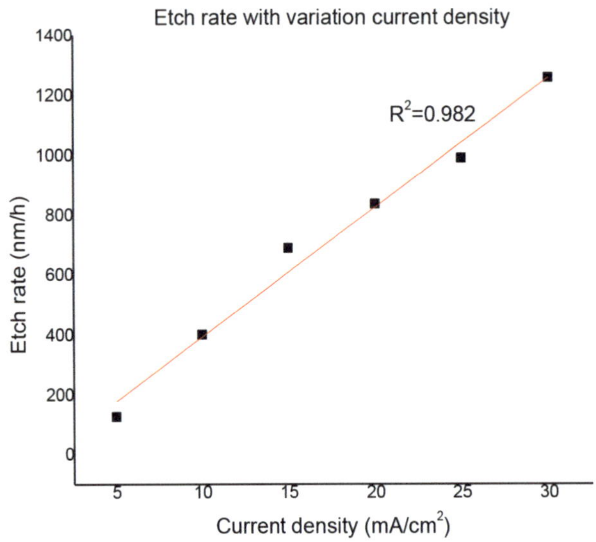

Figure 19. Correlation between the porous layer etching rate and current density in ethanoic HF solution.

Next, the effect of current density to the pore size is studied by grouping current density by three, which is the low current density (5–30 mA/cm²), medium current density (60–100 mA/cm²) and high current density which is set at 200 mA/cm² upwards. Electrochemical formation of silicon membrane is observed within this current level.

The pore size for various current densities has been plotted in **Figure 20** for different current density levels used for electrochemical etching process. The pore size is less than 30 nm for

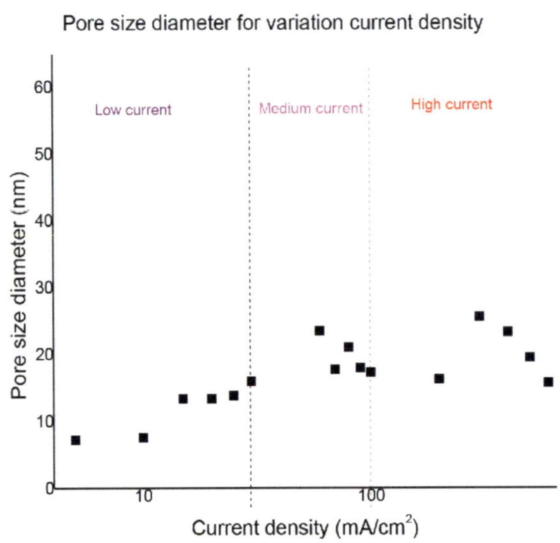

Figure 20. The size of pore by variation of current density.

http://dx.doi.org/10.5772/67719

different current densities applied. The current density that has less effect on the pore size has been observed due to the doping level effect. The p-doped silicon substrate has been used to perform this electrochemical etching process. From previous studies, pore size was affected by concentration of HF solution [14] and doping level [22, 23].

Furthermore, the effect of current density to the pore structure is studied by varying the current density. This experimental setup is used to study the cross section of porous layer. The low and medium current densities that produce spongy porous silicon layers have been observed as shown in **Figures 21** and **22**. For filtering particle use, the columnar silicon structure is the most appropriate structure to confirm that all particles can be penetrated through the silicon membrane without any clogging in the middle of separating process.

The dissolution process is affected by current density during electrochemical etch. Usually, pore formation mechanism begins with the migration of electron and holes in order to attract charge carriers by applying electric field. The chemistry reaction that occurs on the silicon surface involves the competition between Si–O, Si–F and Si–H bond formations [24].

Pore formation is formed by the number of holes on the silicon substrate. It enables to align themselves toward the chemical reaction to follow the trend of current line according to the high current density [25, 26]. The strong electric field can be gained by high current density that tries to utilize a polarizing effect on the substrate. Holes intend to gather at defective surface site. Based on an electron excitation, silicon or dopant atom is able to move into the lattice or non-lattice sites of the silicon crystallites.

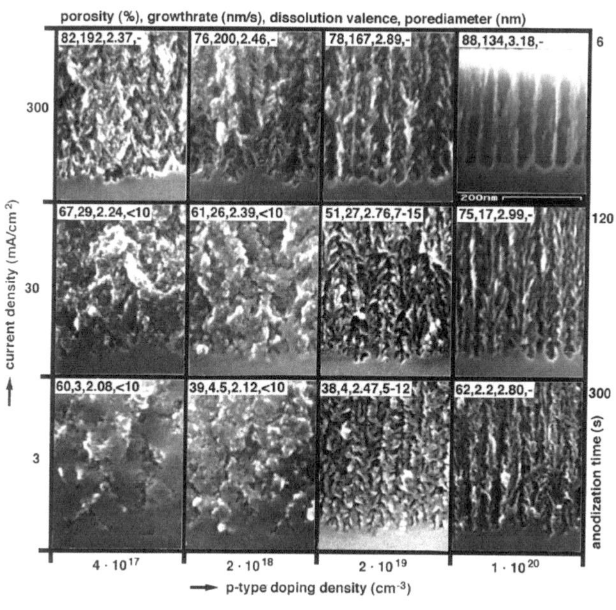

Figure 21. Scanning electron micrographs (SEM) of porous silicon for p-type doped <100> silicon [19].

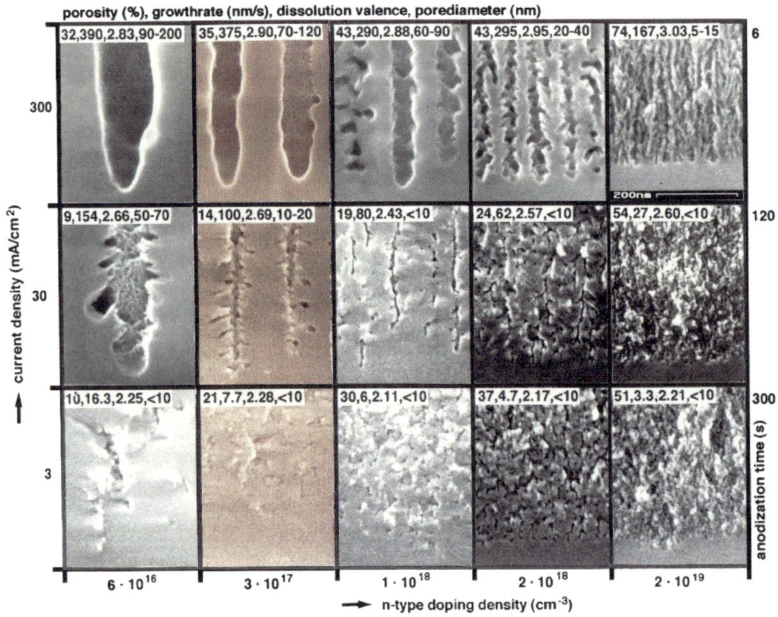

Figure 22. Scanning electron micrographs of porous silicon for n-type doped <100> silicon [19].

The F^- ions in the electrolyte aqueous solution are encouraged to transfer to the silicon substrate, with the succeeding reaction resulting in dissolution. These holes prefer to gather ions at the pore-tip border in silicon bulk because of the low potential energy as compared to the wall area [8, 27]. **Figure 21** shows the porous layer under different current densities and time immersed to HF solution by varying the dopant density. Highly doped p-type silicon is able to produce the columnar porous layer. Meanwhile, n-type silicon is capable to produce columnar structure using high current density at low to medium dopant density.

4.3. Doping

Doping can be affected by pore structure of silicon, changing from being undoped to being doped to either phosphorus or boron, as shown in **Figure 23**. The pore structure is observed by varying the silicon substrate which is undoped, phosphorus doped and boron doped. Undoped silicon membrane is used as an indicator to verify the doping effect. It shows that the square pore structure can be formed on silicon membrane surface. Undoped silicon formed a scattered pore with different sizes. But phosphorus-doped silicon observed the irregular pore formation. The circular pore shape can be formed when irregular shape breaks the existing pore structure and finally forms a larger pore. For this matter, it was observed that the immersed time affects the pore formation. However, the immersed time will not affect to the doped silicon bulk because the doped level can determine the range of pore size [14, 18, 19].

The effect of time on pore structure is studied during immersion of silicon substrate in HF electrolyte. The pore size increases consistently when increasing immersed time for undoped silicon substrate as shown in **Figure 24**. In order to gauge the sole effect on doping type, the

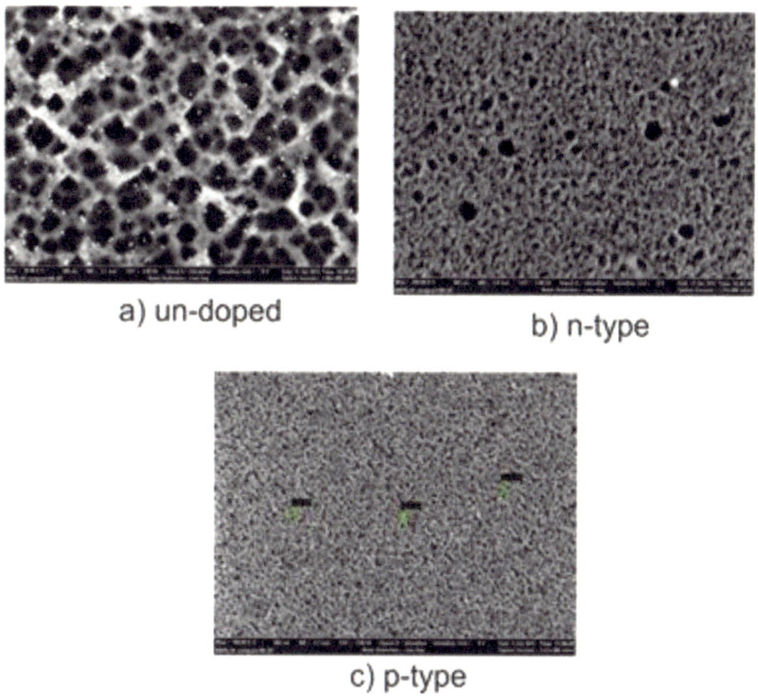

Figure 23. Shape of porous on different doping.

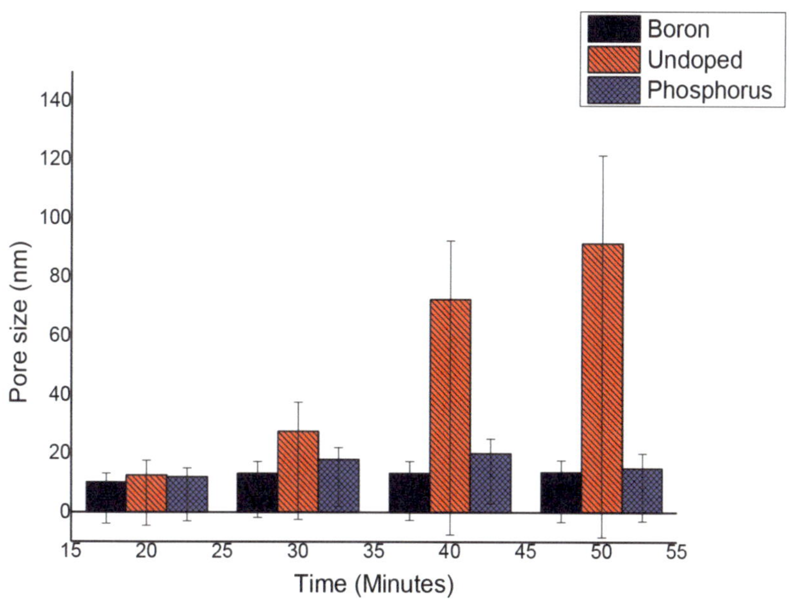

Figure 24. Pore formation by varying immersed time during electrochemical etching process.

amount of HF, distance between anode and cathode and stirring rpm were fixed. For this setup, pore formation for boron- and phosphorus-doped silicon substrate is not affected by time. The pore size with range of 12–20 nm has been formed for doped silicon substrate. The pore size diameter has been affected by doping density. As aforementioned, the doping density can be used to categorize the pore size diameter which is mesoporous, microporous and macroporous that has been classified by International Union of Pure and Applied Chemistry (IUPAC).

4.4. Diluents

The surface morphology of silicon membrane is discussed in this part to examine the effect of alcohol to the pore formation. Methanol, ethanol and propanol are used in this experimental setup. The pore formation after electrochemical etching process is observed when a current density of 25 mA/cm^2 is supplied for 30 minutes. The quantity of holes in electrode surface and the diffusion of fluoride ion will control the mechanism of pore formation. The passivation of pore wall can be boosted by electric field supplied during the pore formation [12, 14, 49]. In this experimental setup, three types of silicon substrate are used, namely, undoped (>80 Ωcm), n-type (resistivity 0–100 Ωcm) and p-type (resistivity 0–100 Ωcm).

Figures 25–27 show the effect of alcohol diluents for variant silicon substrate. It indicates that pore formation not solely depends on the alcohol diluent, but the dopant also gives the effect too. It has been examined that p-type silicon substrate is hard for pore creation

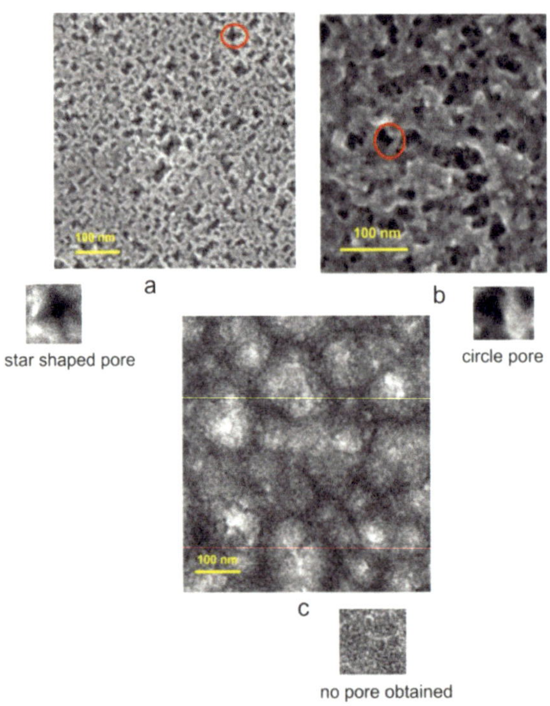

Figure 25. The pore structure for the mixture of HF + methanol as electrolyte solution: (a) undoped, (b) n-type (c) p-type.

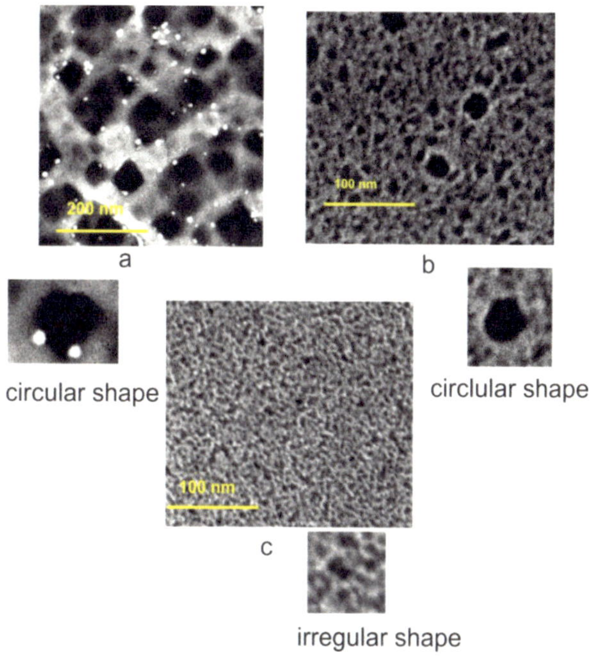

Figure 26. The pore structure for the mixture of HF + ethanol as electrolyte solution: (a) undoped, (b) n-type and (c) p-type.

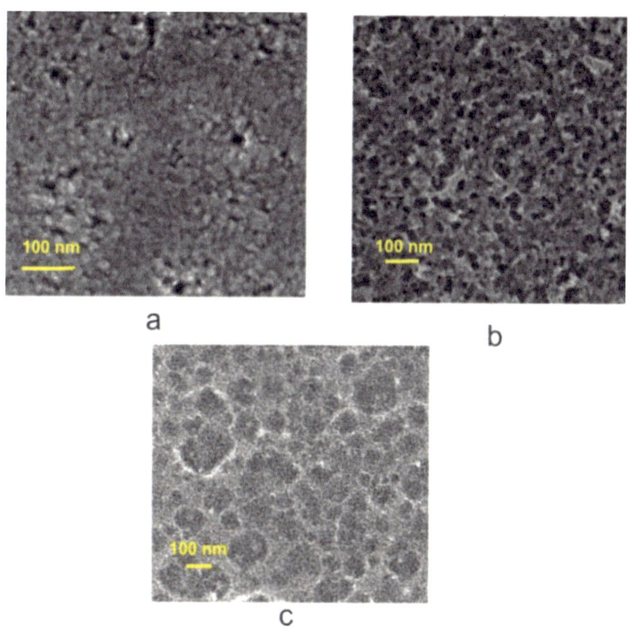

Figure 27. The pore structure for the mixture of HF + propanol as electrolyte solution: (a) undoped, (b) n-type and (c) p-type.

compared to others which are undoped and n-type silicon substrate. This is because of the disability of p-type substrate to control the charge carrier collection at the pore tips for anisotropic dissolution of silicon. So, ion can hardly attack the silicon during the pore formation [40]. Besides that, n-type silicon substrate is more efficient to collect the charge carrier with the assistance of photoluminescence to promote columnar pore structure due to the depletion of holes during pore passivity against dissolution [41]. P-type silicon substrate is suggested to use high current density to get the columnar and well-ordered pore structure [33].

In the first part, methanol and 49% hydrofluoric acid that are mixed become electrolyte solution. Under FESEM inspection shown in **Figure 25**, it is found that star-shaped pore is observed using undoped silicon substrate with less than 40 nm pore size. Meanwhile, non-uniform pore is examined on n-type substrate with less than 40 nm pore size. The pore is formed on the undoped and n-type silicon surface. However, pore is not observed on p-type silicon substrate due to its mechanism, which is hard to control and collect the charge carrier as mentioned before [50].

In the second part, electrolyte solution is changed to the mixture of hydrofluoric acid and ethanol with the aforesaid parameters. As shown in **Figure 26**, pores are formed on all silicon surfaces with variant pore structure and uniformity. An irregular-shaped pore with 150 nm size is observed on undoped silicon substrate. Then, the irregular-shaped pore with circular pore structure is examined on n-type silicon substrate. According to FESEM, the circular pore has larger size compared to irregular-shaped pore due to its breakage during the process. Besides that, an irregular shape is observed with well-ordered pore on the p-type silicon membrane surface.

In the third set of experiment, the mixture of propanol and hydrofluoric acid as electrolyte solution is used. The current density is set at 25 mA/cm^2 for 30 minutes for each silicon substrate. The pore formation becomes unstable for the three silicon substrates. The undoped silicon substrate gives non-uniform pore structure. It is the same with n-type substrates which observed the irregular shape plus the bad pore structure. Then, for p-type substrate it is observed that pore cannot be formed well in electrolyte aqueous solution. **Figure 27** shows a micrograph of the structure of pore when using this solution.

Based on a few experiments shown, alcohol diluents affect the pore formation. By altering this electrolyte aqueous solution, the variation of pore-like circular and star-shaped pore structures has been explored in this setup. Besides, in a mixture of HF and propanol, it is difficult to identify clearly the pore structure using this electrolyte solution. Normally, alcohol diluents are called as amphiphilic surfactants. The amphiphilic surfactants are hydrocarbon-based surfactants that are absorbed on silicon surface with their non-polar tails attached on hydrophobic silicon surface [40]. The shortest hydrocarbon chain is methanol, which results nearly in well-ordered pore structure. Furthermore, variant diluents, whether organic or nonorganic, can alter the pore structure. An organic surfactant like DMF and Tetramethylammonium hydroxide (TMAH) also gives a different shape [51].

The combination of HF and ethanol is the most stable due to the sharp pore structure and formation [52]. On the other hand, the worst pore formation is observed using the mixture of propanol and hydrofluoric acid. The highest etching rate can be observed using propanol due to the fastest etched rate which is 20 nm/minute by inspecting the porous cross section. Even though faster time produces pore structure, the pore is difficult to observe using these diluents on the silicon substrate surface. The surface roughness also can be inspected to observe the pore structure on the silicon surface [25].

A star-shaped pore structure is studied to find out whether this pore structure can be replicated again. This process is conducted to optimize this parameter whether star pore can be formed. The aforementioned parameter is used for this experimental setup. The immersed time is varied for each 10-minute start, from 20 to 50 minutes. In **Figure 25**, star-shaped pore structure was formed in this electrolyte aqueous solution with aforementioned parameter. The well-ordered pore is formed in immersed time between 30 and 50 minutes. However, wide star shape is formed by increasing immersed time due to pore breakage. The pore structure becomes wider as shown in **Figure 28** by increasing immersed time.

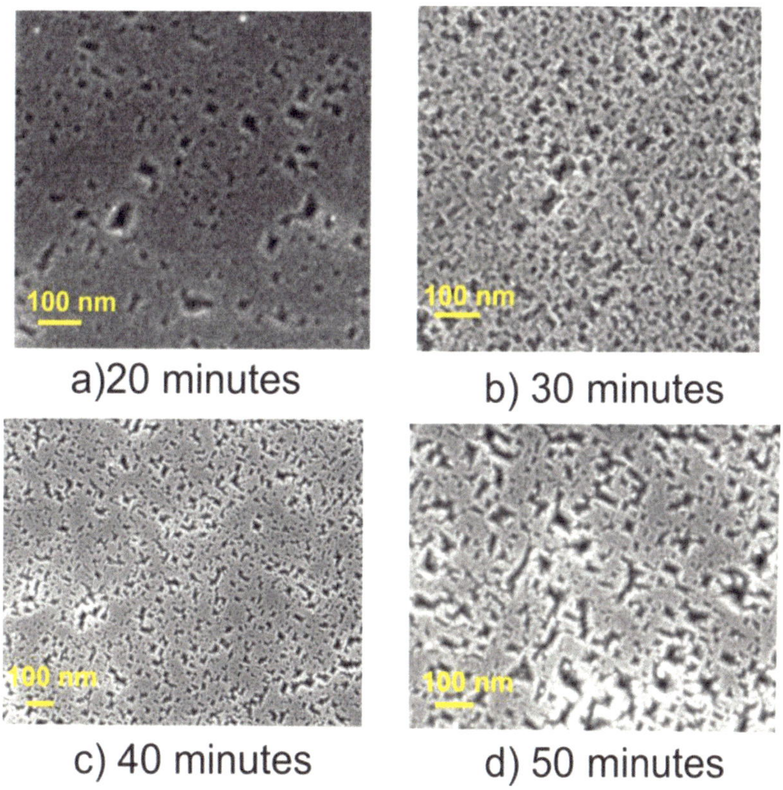

Figure 28. The pore formation by varied immersed time for ethanol HF + methanol solution.

5. Conclusion

Pore formation that can be varied by manipulating parameters like dopant, current density, HF concentration and diluent is discussed in this topic. This technique is called a self-adjusting technique because the ion will move by itself to produce various pore structures. The pore can be used in various applications like filtration system, biosensor and microfluidic modules. For filtration system, the columnar pore structure gives the most advantages to ensure the particle can be separated efficiently. The pore size can be varied depending on the size of particles to be sorted out. According to this study, the electrochemical etching process is a very easy technique in producing pore due to its simple experimental setup and the chemicals used.

Author details

Norhafizah Burham[1,2*], Azrul Azlan Hamzah[1] and Burhanuddin Yeop Majlis[1]

*Address all correspondence to: fiezah@msn.com

1 Institute of Microengineering and Nanoelectronics (IMEN), Universiti Kebangsaan Malaysia, Bangi, Selangor, Malaysia

2 Faculty of Electrical Engineering, Universiti Teknologi Mara (UiTM), Shah Alam, Selangor, Malaysia

References

[1] H. D. Tong, H. V. Jansen, V. J. Gadgil, C. G. Bostan, E. Berenschot, C. J. M. van Rijn, et al, "Silicon nitride nanosieve membrane," *Nano Letters*, vol. 4, pp. 283–287, 2004.

[2] C. C. Striemer, T. R. Gaborski, J. L. McGrath, and P. M. Fauchet, "Charge- and size-based separation of macromolecules using ultrathin silicon membranes," *Nature Publishing Group*, vol. 445, pp. 749–753, 2007.

[3] W. H. Fissell, A. Dubnishev, A. N. Eldridgeb, A. J. Fleischman, A. L. Zydneyc and S. Roy, "High-performance silicon nanopore hemofiltration membranes," *Journal of Membrane Science*, vol. 326, pp. 58–63, 2009.

[4] M. R. Lee and P. M. Fauchet, "Two-dimensional silicon photonic crystal based biosensing platform for protein detection," *Optics Express*, vol. 15, pp. 4530–4535, 2007.

[5] A. V. Pavlikov, A. V. Lartsev, I. A. Gayduchenko and V. Y Timoshenko, "Optical properties of materials based on oxidized porous silicon and their applications for UV protection," *Microelectronic Engineering*, vol. 90, pp. 96–98, 2012.

[6] D. Lockwood, A. Wang and B. Bryskiewicz, "Optical absorption evidence for quantum confinement effects in porous silicon," *Solid State Communications*, vol. 89, pp. 587–589, 1994.

[7] S. D. Alvarez, A. M. Derfus, M. P. Schwartz, S. N. Bhatia and M. J. Sailor, "The compatibility of hepatocytes with chemically modified porous silicon with reference to in vitro biosensors," *Biomaterials*, vol. 30, pp. 26–34, 2009.

[8] C. Talamonti, M. Bruzzi, L. Marrazzo, D. Menichelli, M. Scaringella and M. Bucciolini, "Bidimensional silicon dosimeter: development and characterization," *Nuclear Instruments and Methods in Physics Research Section A: Accelerators, Spectrometers, Detectors and Associated Equipment*, vol. 658, pp. 84–89, 2011.

[9] S. P. Adiga, C. Jin, L. A. Curtiss, N. A. Monteiro-Riviere and R. J. Narayan, "Nanoporous membranes for medical and biological applications," *Wiley Interdisciplinary Reviews: Nanomedicine and Nanobiotechnology*, vol. 1, pp. 568–581, 2009.

[10] A. Hamzah, H. Z. Abidin, B. Y. Majlis, M. M. Nor, A. Ismardi, G. Sugandi, et al, "Electrochemically deposited and etched membranes with precisely sized micropores for biological fluids microfiltration," *Journal of Micromechanics and Microengineering*, vol. 23, p. 074007, 2013.

[11] D. R. Turner, "Electropolishing silicon in hydrofluoric acid solutions," *Journal of the Electrochemical Society*, vol. 105, pp. 402–408, 1958.

[12] M. J. J. Theunissen, "Etch channel formation during anodic dissolution of n-type silicon in aqueous hydrofluoric acid," *Journal of the Electrochemical Society*, vol. 119, pp. 351–360, 1972.

[13] M. I. J. Beale, J. D. Benjamin, M. J. Uren, N. G. Chew and A. G. Cullis, "An experimental and theoretical study of the formation and microstructure of porous silicon," *Journal of Crystal Growth*, vol. 73, pp. 622–636, 1985.

[14] V. Lehmann and H. Föll, "Formation mechanism and properties of electrochemically etched trenches in n-type silicon," *Journal of the Electrochemical Society*, vol. 137, pp. 653–659, 1990.

[15] V. Lehmann, "Porous silicon formation and other photoelectrochemical effects at silicon electrodes anodized in hydrofluoric acid," *Applied Surface Science*, vol. 106, pp. 402–405, 1995.

[16] V. Lehmann, "The physics of macroporous silicon formation," *Thin Solid Films*, vol. 255, pp. 1–4, 1995.

[17] V. Lehmann, "The physics of macropore formation in low doped n-type silicon," *Journal of the Electrochemical Society*, vol. 140, pp. 2836–2843, 1993.

[18] S. Yaakob, M. A. B. Ismail, N. H. H. A. Bakar and K. Ibrahim, "The formation and morphology of highly doped n-type porous silicon: Effect of short etching time at high current density and evidence of simultaneous chemical and electrochemical dissolutions," *Journal of Physical Science*, vol. 23, pp. 17–31, 2012.

[19] V. Lehmann, R. Stengl and A. Luigart, "On the morphology and the electrochemical formation mechanism of mesoporous silicon," *Materials Science and Engineering*, vol. B69–70, pp. 11–22, 2000.

[20] R. M. Mehra, V. Agarwal, V. K. Jain and P. C. Mathur, "Influence of anodisation time, current density and electrolyte concentration on the photoconductivity spectra of porous silicon," *Thin Solid Films*, vol. 315, pp. 281–285, 1998.

[21] P. Kumar, P. Lemmens, M. Ghosh, F. Ludwig and M. Schilling, "Effect of HF concentration on physical and electronic properties of electrochemically formed nanoporous silicon," *Journal of Nanomaterials*, vol. 7, pp. 1–7, 2009.

[22] L. T. Canham, *Properties of Porous Silicon*. United Kingdom: INSPEC, The Institution of Electrical Engineers, 1997.

[23] Z. Li, L. Zhao, H. Diao, C. Zhou, H. Li and W. Wang, "Macroporous silicon formation on low resistivity p-type c-Si substrate by metal catalyzed electrochemical etching," *International Journal of Electrochemical Science*, vol. 8, pp. 1163–1169, 2013.

[24] F. Gaspard, A. Bsiesy, M. Ligeon, F. Muller and R. Herino, "Charge exchange mechanism responsible for p-type silicon dissolution during porous silicon formation," *Journal of the Electrochemical Society*, vol. 136, pp. 3043–3046, 1989.

[25] C.-G. Kang, M.-S. Kang, J.-H. Yang, J.-H. Jin, S.-I. Hong and N.-K. Min, "Comparison of nano-porous silicon prepared by photoelectrochemical etching in HF-ethanol and HF-acetonitrile solutions," *Journal of the Korean Physical Society*, vol. 42, pp. S693–S697, 2003.

[26] N. Burham, A. A. Hamzah and B. Y. Majlis, "Effect of hydrofluoric acid (HF) concentration to pores size diameter of silicon membrane," *Bio-Medical Materials and Engineering*, 24, pp. 2203–2209, 2014.

[27] N. Burham, A. A. Hamzah and B. Y. Majlis, "Effects of alcohol diluents on nanopore structure of electrochemically etched silicon membrane," *Microelectronic Engineering*, vol. 141, pp. 160–167, 2015.

[28] M. M. Rieger and P. A. Kohl, "Mechanism of (111) silicon etching in HF-acetonitrile," *Journal of the Electrochemical Society*, vol. 142, pp. 1490–1495, 1995.

[29] P. Steiner and W. Lang, "Micromachining applications of porous silicon," *Thin Solid Films*, vol. 255, pp. 52–58, 1995.

[30] R. Memming and G. Schwandt, "Anodic dissolution of silicon in hydrofluoric acid solutions," *Surface Science*, vol. 4, pp. 109–124, 1966.

[31] V. Lehmann and U. Gösele, "Porous silicon formation: a quantum wire effect," *Applied Physics Letters*, vol. 58, pp. 856–858, 1991.

[32] W. Lang, P. Steiner, and H. Sandmaier, "Porous silicon: A novel material for microsystems," *Sensors and Actuators A Physical*, vol. 51, pp. 31–36, 1995.

[33] N. Burham, A. A. Hamzah and B. Y. Majlis, "Effect of current density on silicon surface in electrochemical etching," *Micro & Nano Letters*, vol. 9, pp. 850–853, 2014.

[34] X. G. Zhang, "Mechanism of pore formation on n-type silicon," *Journal of the Electrochemical Society*, vol. 138, pp. 3750–3756, 1991.

[35] J. Carstensen, M. Christophersen, and H. Föll, "Pore formation mechanisms for the Si-HF system," *Materials Science and Engineering*, vol. B69–70, pp. 23–28, 2000.

[36] N. I. Rusli, M. S. Z. Abidin, B. Astuti, N. K. Ali and A. M. Hashim, "Formation of porous silicon: Mechanism of macropores formation in n-type Si," *Sains Malaysiana*, vol. 42, pp. 643–648, 2013.

[37] S. Nadia, N. K. Ali, M. R. Ahmad and S. M. Haidary, "Effect of dopant concentration on the pore formation of porous silicon on n-type silicon," in *Photonics (ICP), 2014 IEEE 5th International Conference* on, 2014, pp. 53–55.

[38] H. Föll, M. Christophersen, J. Carstensen and G. Hasse, "Formation and application of porous silicon," *Materials Science and Engineering*, vol. R 39, pp. 93–141, 2002.

[39] X. G. Zhang, "Morphology and formation mechanisms of porous silicon," *Journal of the Electrochemical Society*, vol. 151, pp. 69–80, 2004.

[40] K. Chao, S. Kao, C. Yang, M. Hseu and T. Tsai, "Formation of high aspect ratio macropore array on p-type silicon," *Electrochemical and Solid-State Letters*, vol. 3, pp. 489–492, 2000.

[41] V. Lehmann, "Developments in porous silicon research," *Materials Letters*, vol. 28, pp. 245–249, 1996.

[42] R. L. Smith and S. D. Collins, "Porous silicon formation mechanisms," *Journal of Applied Physics*, vol. 71, p. R1, 1992.

[43] C. R. Miranda, M. R. Baldan, A. F. Beloto and N. G. Ferreira, "Morphological and optical characteristics of porous silicon produced by anodization process in HF-acetonitrile and HF-ethanol solutions," *Journal of the Brazilian Chemical Society*, vol. 19, pp. 769–774, 2008.

[44] O. Bisi, S. Ossicini and L. Pavesi, "Porous silicon: A quantum sponge structure for silicon based optoelectronics," *Surface Science Reports*, vol. 38, pp. 1–126, 2000.

[45] P. Granitzer and K. Rumpf, "Porous silicon: A versatile host material," *Materials*, vol. 3, pp. 943–998, 2010.

[46] M. J. Sailor, "Fundamentals of porous silicon preparation," in *Porous Silicon in Practice: Preparation, Characterization and Applications*, ed: Wiley-VCH Verlag GmbH & Co. KGaA, pp. 1–42, 2012.

[47] N. Burham, A. A. Hamzah and B. Y. Majlis, "Effect of isopropyl alcohol (IPA) on etching rate and surface roughness of silicon etched in KOH solution," in *Micro and Nanoelectronics (RSM), 2015 IEEE Regional Symposium* on, 2015, pp. 1–4.

[48] J. Salonen and V. P. Lehto, "Fabrication and chemical surface modification of mesoporous silicon for biomedical applications," *Chemical Engineering Journal*, vol. 137, pp. 162–172, 2008.

[49] V. Lehmann and S. Rönnebeck, "The physics of macropore formation in low-doped p-type silicon," *Journal of the Electrochemical Society*, vol. 146, pp. 2968–2975, 1999.

[50] T. Urata, K. Fukami, T. Sakka and Y. H. Ogata, "Pore formation in p-type silicon in solutions containing different types of alcohol," *Nanoscale Research Letters*, vol. 7, pp. 329–333, 2012.

[51] X. Q. Bao, J. W. Jiao, Y. L. Wang, W. N. Kyoung and H. Choi, "Macropore formation without illumination on low doped n-type silicon," *Journal of the Electrochemical Society*, vol. 154, pp. D175–D181, 2007.

[52] P. Stroeve and N. Ileri, "Biotechnical and other applications of nanoporous membranes," *Cell Press*, vol. 29, pp. 259–266, 2011.

Chapter 7

Macroporous Silicon: Technology and Applications

Didac Vega Bru and Ángel Rodríguez Martínez

Additional information is available at the end of the chapter

http://dx.doi.org/10.5772/67698

Abstract

Macroporous silicon (MPS) is a versatile material that since its origin in the early 1990s has seen intense research and has found applications in many fields. MPS is a key technology in photonic crystals research, and optic and photonic applications are its main applications. However, this chapter is devoted to several of the non-photonic uses of MPS. In particular, new electronic and MEMS devices and applications will be described. Furthermore, in this chapter, the technology of MPS fabrication will be presented.

Keywords: macroporous silicon, electrochemical etching, field effect transistors, supercapacitors, microneedles, MEMS

1. Introduction

Macroporous silicon (MPS) is a novel material, which was described as such in the early 1990s by Lehmann in his pioneering works [1–3] and by other researchers [4, 5]. Since its first description, MPS has attracted great interest, and it has been suggested for many applications in several fields such as electronics [6], optics [7], photonics [8], solar cells [9], and even energy storage [10], fuel cells [11] or catalysis [12].

One particular form of MPS consists on an ordered arrangement of the pores that are etched on the silicon bulk. In such form, the primary interest of MPS is as photonic crystals (PCs). However, as it will be revealed throughout the chapter, many other applications have been proposed for MPS, which does not make use of that interesting optical properties. For such cases, the use of an ordered array of pores may not be necessary, but in many cases, this can give certain advantages.

In the present chapter, a brief review on the technology of macroprorous silicon, its fabrication techniques and its applications will be presented. In this chapter, special emphasis is made regarding the *non-photonic* applications of MPS.

2. Brief history of macroporous silicon

Electrochemical etching (EE) of silicon (and germanium) was known from the late 1950s. The works of Uhlir [13], Turner [14] and others [15] were the first to study the semiconductor dissolution in an electrolyte under anodic conditions. These initial works dealt with electropolishing of the semiconductor in order to obtain flat, defect-free surfaces or for bulk thinning. However, it was noticed that under certain conditions, pits were formed instead of a uniform etching over the treated surface [13, 14]. In particular, low voltages and currents, or the pre-existence of surface defects, were found to promote surface pitting, preferentially at surface irregularities or defects. Therefore, this allows removing material locally in a more or less controlled manner. The processing of surfaces done by this technique in the described way gives rise to *porous materials*, in which a portion or the whole of it is *porosified*, resulting in a sponge-like structure. An example of this process is schematically shown in **Figure 1**.

Hydrofluoric acid (HF) is one of the electrolytes used for EE. Its interest is that, in ambient conditions, it reacts very slowly with silicon[1] but readily etches the semiconductor when an electric current flows through the electrochemical cell. The anodic oxidation of silicon in hydrofluoric acid with the correct potential and current produces *porous silicon*[2] (PS). The first specimens of PS were found to be *microporous* with a random distribution of the etched voids. However, this first form of PS was not initially distinguished as microporous, and, nowadays, the term PS is still used to refer to its microporous form.

After the discovery, the interest of this new material in fields such as electronics and photonics became clear. Later on, the macroporous form of PS was developed, and it has also shown its potential in fields as different as catalysis [16], drug delivery [17], energy storage [18] or optical

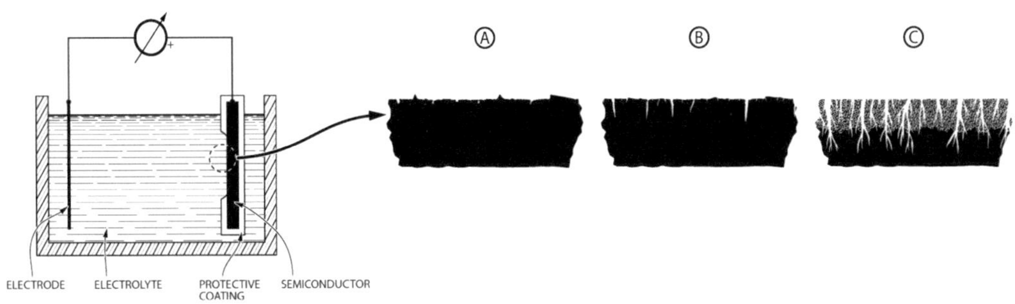

Figure 1. Electrochemical set-up for semiconductor etching and electropolishing. Details of the surface during several steps of the etching process are shown in panels (A), (B) and (C). Initially, the surface has a low content of defects such as pits, dislocations or protrusions. After the anodic treatment is started with the right conditions, the surface (B) shows a certain degree of electropolishing but existing pits start growing, and new pits evolve. After a prolonged treatment, the surface and bulk (C) show noticeable change, and pores grow deep into the sample creating a sponge-like material.

---

[1]Silicon dissolution in HF is a thermodynamically favourable process; however, the dissolution rate is very small: $R_{Si} = 2.5 \times 10^{23}$ cm^{-2} s^{-1} (about 0.3 Å/min) at room temperature [33].

[2]It must be noted that electrochemical etching is one possible method to obtain porous silicon. A different technique is stain etching. This is a purely chemical method based on the chemistry of silicon in HF – HNO_3 mixtures, which also produces PS. See for example [24].

sensors [7]. The first instances of MPS can be found in the early 1990s in Lehmann's pioneering works on fabrication [1, 2] and application of macroporous silicon [19]. In his works, he focused primarily on the formation of ordered MPS in a controlled manner. The fabrication technique used in his works was the *photo electrochemical etching* (PECE) of silicon in HF and was thoroughly characterized [3]. Nevertheless, PECE is not the only method available for MPS fabrication, the other main one being *(deep) reactive ion etching* ([D]RIE). MPS will be covered in more detail in the next section. In this chapter, particular emphasis will be made on the photo electrochemical fabrication process as described by Lehmann.

2.1. What is macroporous silicon?

Macroporous silicon is a particular form of porous silicon. One way of classifying porous silicon is looking at the 'pore size'. According to the IUPAC [20], the material can be designated as *microporous* for pore sizes below 2 nm, then *mesoporous* for sizes under 50 nm, and *macroporous* for pore diameters larger that 50 nm. However, this simple classification does not give enough information about the material, as the pore morphology can vary greatly, and therefore, the 'pore size' cannot be precisely defined. In spite of these limitations, the average pore diameter d_{pore} is often regarded as the 'pore size'. MPS has an average pore diameter greater than 50 nm, and in practical realizations of the material, typical pores are larger than 100 nm.

Besides pore size, MPS can also be described either as ordered or random regarding the surface pore growth site distribution; this is shown in **Figure 2**. Nevertheless, the individual pore diameter distribution is not generally accounted in the classification of the material. Other morphological characteristics of macroporous silicon such as pore shape, branching, depth or depth profile are not normally considered when classifying MPS.

MPS is a versatile material that has some very interesting properties. To begin with, silicon is the material of choice in the semiconductor and MEMS industry. This opens the possibility to use many of the existing production methods, techniques, experience and equipment available in VLSI foundries and research labs. As a result, important costs savings can be achieved for the development and fabrication of devices based on MPS. Furthermore, it has been shown that by employing standard CMOS fabrication methods, it is possible to integrate MPS (and PS in general) into VLSI devices [21]. The possibility of monolithic integration is a clear advantage

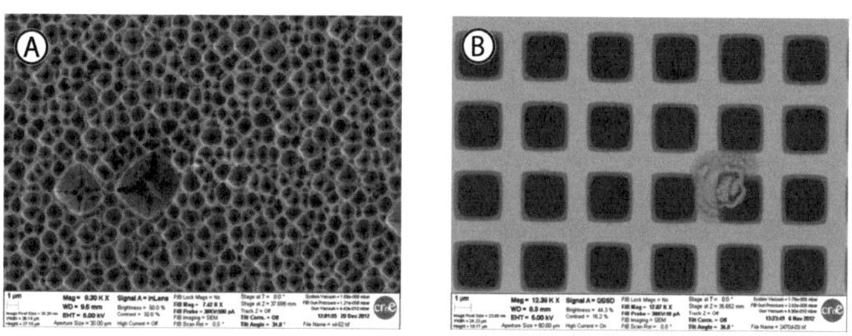

Figure 2. Top view of random (A) and ordered (B) macroporous silicon. Pore radius may not be uniform, as shown in the random sample (A).

of MPS for next-generation device fabrication. Furthermore, MPS can be processed to create complex structures, as those of **Figure 3** from Ref. [22]. The as-etched structures can be further processed to alter some morphological properties [23], to passivate the surface or to functionalize the MPS device with coatings such as catalysts [12], nanoparticles and others.

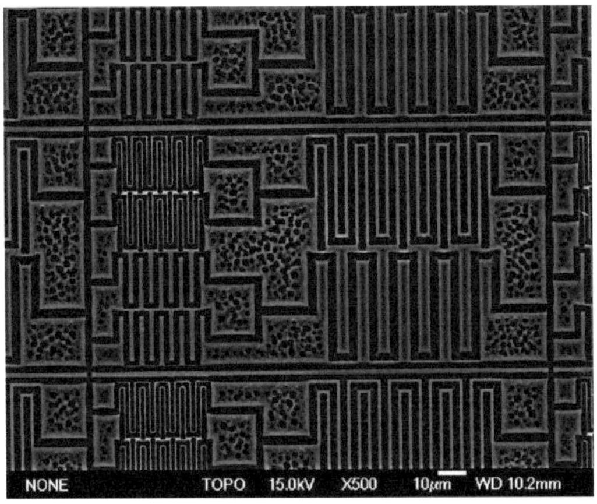

Figure 3. Complex MPS structure fabricated with the PECE of silicon in HF. Reprinted from [22], Copyright 2002, with permission from Elsevier.

3. Fabrication of macroporous silicon

There exist several microfabrication and micromachining techniques that allow the production of MPS devices. Due to its historical importance, and because it allows micromachining in three dimensions, the *electrochemical etching* of silicon remains one of the most important and versatile fabrication methods, as will be shown later. The next method in importance is *reactive ion etching*. Other remarkable methods[3] include electroless etching of silicon[4] (stain etching [24], metal-assisted etching [25], etc.), direct laser writing (DLW) [26], lithographic methods (layer-by-layer [27], interference lithography [28], block copolymer [29] ...) and glancing angle deposition [30].

3.1. Reactive ion etching

RIE is a well-known method for shaping silicon in microelectronics and MEMS device fabrication [31]. A succinct description is as follows: this is a dry etching process where a plasma of a specific mixture of gases is used to selectively remove material from the stock. This process is done in a low-pressure chamber. A mask is necessary to define the affected regions, and the etch proceeds anisotropically in the unmasked areas. The material is eroded at a speed that depends on the gas mixture, temperature, chamber pressure and power of the plasma. As described, RIE is simply a necessary step in the fabrication of MPS, as the mask can be defined

[3]An exposition of the different methods for PC fabrication and their possibilities can be seen in Ref. [89].

[4]For a general treatment on the subject, see Ref. [33].

by several means. For instance, layer by layer or DLW may or may not require the use of RIE for the fabrication of MPS and PCs.

The most interesting aspects of RIE for the processing of MPS are that it allows defining very small features, and that it is completely integrated into the VLSI fabrication process flow. Using RIE to create the pores gives great freedom in the mask layout as both processes are independent. This allows adding point defects, linear defects or other lattice disordering elements. This ability is of great interest for light processing and optical devices, such as waveguides, interferometers or resonators. It is, however, not limited to the photonics field, as it has also been demonstrated for electronic applications such as capacitors.

In spite of these advantages, the fabrication of MPS with RIE has some limitations. Firstly, etching high aspect ratio[5] (AR) pores into silicon using plain RIE can be tricky. Effects such as masking, microloading, footing or redeposition must be taken into account, otherwise the resulting pores will not have a uniform profile in depth. These effects are particularly serious when trying to develop high AR trenches, as pores may close after a certain depth and grow no further. For deeper pores, a DRIE process must be adopted. Nevertheless, DRIE leaves 'large' scallops on the sidewalls that are generally undesired. Besides this, MPS fabricated by RIE is limited to 2-d structures, as the etched features depend on a planar mask[6] and material etching is essentially in the direction of the plasma electric field.

3.2. Electrochemical etching

Electrochemical etching of silicon is a very versatile method for the fabrication of MPS. As was found from the initial works on silicon etching, silicon can be selectively dissolved when immersed in an HF bath, and an electric current flows through. Therefore, this in an *electro-chemical* method as a current is needed to activate the chemical reactions that dissolve silicon. In the absence of such current, the silicon substrate remains practically unaltered. Furthermore, to perform the etching, the silicon must be the *anode* in the electrochemical system. Otherwise, there will be no reaction at the surface, as for the reactions to occur holes must be exchanged. Both n- and p-type silicon can be dissolved by this technique.

The pore evolution and morphology using the EE method depends on several factors. On one hand, the current density and electrode potential determine de amount of etched silicon but have little influence in the advance of the 'pore front', that is the etch velocity. On the other hand, the electrolyte concentration, temperature, and optional additives affect the etch velocity and pore shape, and limit the practical operation area of the electrical parameters [1, 32]. Furthermore, the Si bulk characteristics such as doping type and density, crystal orientation, and carrier lifetime strongly determine the morphology of the pores, and the obtainable pore dimensions [32].

In the porous layer formation regime, it has been found that currents through the electrochemical cell have a similar shape to that shown in **Figure 4** (for n-type silicon). In particular, PS formation

---

[5]'Aspect ratio' is defined as the ratio between the pore width over the pore length.

[6]Some attempts have been made into creation of 3-d PCs exploiting the scalloping effect of RIE; however, progress has been slow [90].

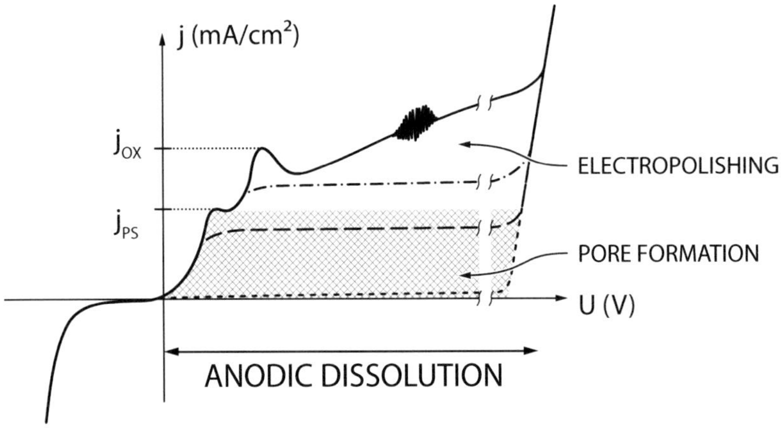

Figure 4. Currents through an H/Si electrochemical system for different applied potential and illumination. The shown $j(v, \Phi_e)$ is for an n-type silicon bulk. Solid line (—) is for 'infinite' illumination, dash-dot (— •) is for high illumination intensity, dash (—□—) is for medium IR intensity, and short dash (-□-) is for no illumination condition.

occurs if the current density is kept below J_{PS}. Larger currents will start to electropolish the surface of the semiconductor. For very large currents, $j > J_{ox}$, only electropolishing will take place. The chemical processes involved in the dissolution of silicon are quite complex and depend on the local current intensity and fluoride ion concentration. Two competing mechanisms act on the silicon surface: formation of SiO_2 and subsequent dissolution by HF, and direct dissolution of silicon [4, 33]. Oxide formation and dissolution consume four holes, and thus have valence 4 ($\lambda = 4$). On the other hand, direct dissolution of silicon requires two to three holes depending on light intensity: low intensity favours $\lambda = 3$, while high intensity favours $\lambda = 2$. More insight on the chemicophysical aspects of silicon dissolution can be found in the works of Lehmann [2, 3], Zhang [4, 32] and Kolasinski [33].

The electrical characteristics of the silicon wafer substrate are determinant for the morphology of the etched pores. The size of the resulting pores is strongly dependent on the wafer resistivity (as seen in **Figure 5**) and the use of a mask to define the nucleation centres. From Lehmann's [3] and Zhang's [4] work, in the MPS etching regime, the pore morphology and growth are basically dependent on the space charge region (SCR) formed in the pore region in contact with the electrolyte. It was proposed that etching occurs at localized sites with a current density of J_{PS} whose value is just a function of the weight concentration of hydrofluoric acid: $J_{PS} = Kc^{3/2} \exp(-E_a/kT)$. The pore growth speed depends only on the current passing through the etched areas; thus, it can be calculated by the removed charge and valence of the reaction: $v = J_{PS}/nq_e N_{Si}$. Here, $K = 3300$ mA cm^{-2} is an empirically determined constant, c (wt. % HF) is the electrolyte concentration, $E_a = 343$ meV is the empirically found activation energy, k is the Boltzmann constant, T is the temperature in Kelvin, n is the reaction valence, q_e is the electron charge (1.602×10^{-19} C), and N_{Si} is the atomic density of silicon (5×10^{22} cm^{-3}) [2]. Typical concentration of HF used for MPS etching is around 2.4–10%.

In the case of ordered MPS, where a mask is used to define the pore growth sites (nucleation centres), etching current at the tips will be such that $j_{tip} = J_{PS}$; thus, the pore will advance at the

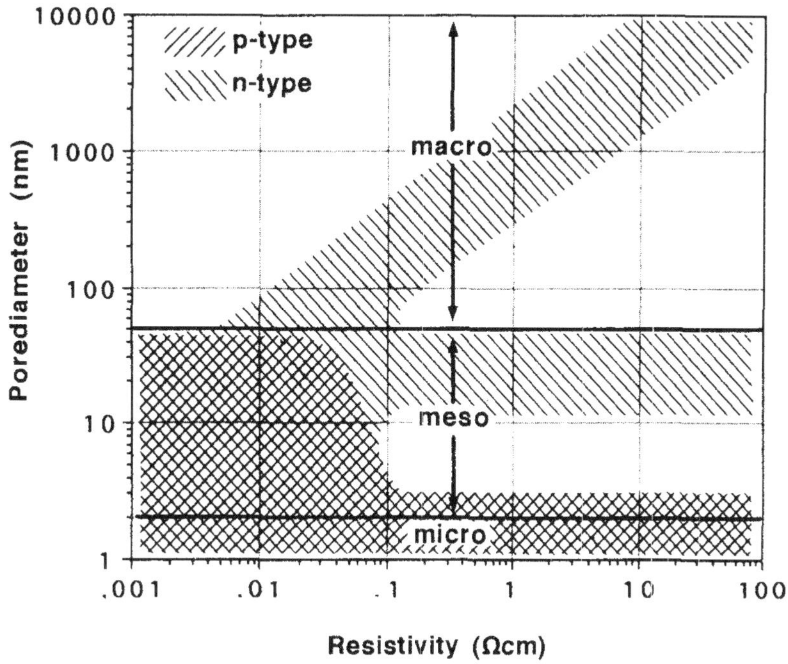

Figure 5. Etched pore diameter for different silicon doping type and concentration. © 1996 IEEE. Reprinted, with permission, from [19].

same speed irrespective of pore size. However, by adjusting the total current through the Si substrate, it is possible to change the actual etched area (and thus pore diameter) since porosity $p = A_{total}/A_{etched} = j/J_{PS}$. The pattern for the pore distribution can be fabricated using any lithographic means like UV or NIL. The mask can be a corrosion-resistant material such as silicon nitride, but, given that MPS etching is highly anisotropic and selective under the correct parameters, it is possible to transfer the pattern directly to the silicon and create *etch pits* at the desired sites that will act as nucleation points. This will promote the pore formation at *only* these sites. The unreacted silicon 'left behind' the pore tip is further passivated by hydrogen adsorption reducing the already low Si etch rate in HF.

As holes are needed for the porosification, for n-type silicon, these need to be generated somehow. One of the most extended methods is PECE, after the work by Lehmann [2]. This method requires illuminating the work silicon with suitable IR light; thus, the etching cell and controlling system must be similar to the one depicted in **Figure 6**. In such system, the cell current is proportional to the illumination intensity. Knowing the current limits from **Figure 4**, the computer control can be programmed to adjust the photogenerated current and cell potential to keep the pore growth stable. Electrolyte temperature is also controlled, as variations of it will change the J_{PS} and thus the etching speed. HF is pumped through the cell to remove any gas bubbles that may form from the release of hydrogen during etching and to provide fresh electrolyte. The electrical contact on the Si substrate may be done by coating with a thin transparent conductor or by physical contact with a different electrolyte on the backside. The simple configuration shown in **Figure 6** uses a high-doped layer on the backside with a probe to contact the backside. Oxygen

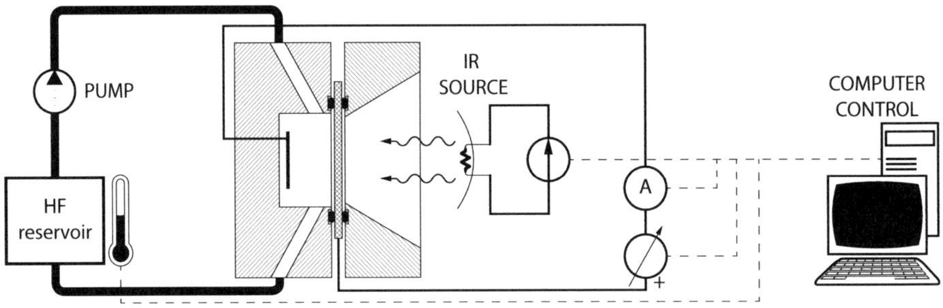

Figure 6. Etching system used in the MNT group for the fabrication of MPS.

dissolved in the electrolyte can also affect the pore shape, particularly if these are very deep. To reduce the unwanted oxidation, purging the electrolyte with bubbling N_2 is recommended.

It was seen that pore diameter (porosity) can be controlled in MPS by adjusting the photogenerated etching current. This is exploited to create 3-d structure by freely changing the IR illumination intensity as the pores progress. In spite of the limits to the achievable shapes imposed by the electrochemical system, it is possible to create very complex structures like chirped modulations [34], embedded defects [35], etc. MPS with pores with diameters down to 100 nm has been reported [36]; however, only straight pores have been shown, and surface quality is still low. Better pores are obtained for $d_{pore} = 275$ μm [37]; however, those are still straight. The smallest reported 3-d MPS is about 500 nm pitch [38].

3.3. Surface treatment and functionalization

After the MPS layer is etched, several post-processing steps can be applied to the device. This may be of particular importance for MPS fabricated using EE. This is because the electrochemical process leaves a thin microporous silicon layer on the pores' walls. The thickness and roughness of this microporous layer will depend on the electrolyte composition and concentration, as well as on the substrate resistivity and applied potential [32] (see **Figure 7a**). A simple way to remove this layer is to perform a short thermal oxidation of the MPS structure. The microporous layer will readily oxidize [39]. After the oxidation step, the oxidized PS layer can be removed by dipping the MPS structure in an HF solution, leaving a very smooth and defect free surface on the walls of the pores. The so-prepared surface can be further processed, for instance to create high-quality oxide layers, as seen in **Figure 7b**.

This oxidation-oxide removal cycle can be repeated several times in order to create enlarged, cylindrical pores.[7] This is useful when creating 3-d MPS structures and strong modulation is desired, such as pores interconnecting with each other at specific points and resulting in an opal-like structure (see **Figure 8**). The pore shape can be further altered by etching the freshly fabricated MPS structure in diluted wet chemical etchants such as KOH or TMAH. This allows obtaining flat faces that are not possible from PECE alone [40].

---

[7]The actual pore cross-sectional shape obtained with EE is not perfectly cylindrical and depends on potential applied and other factors [19].

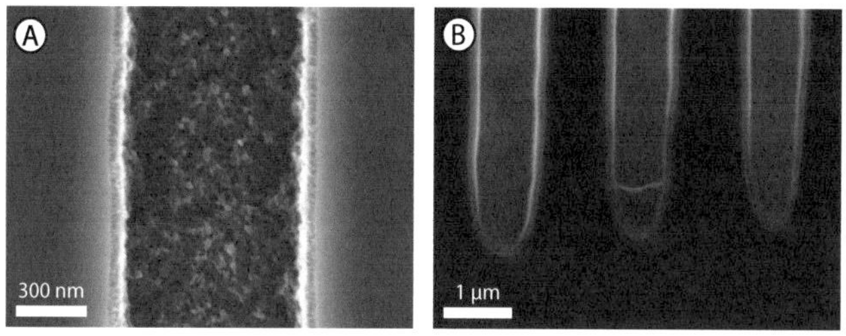

Figure 7. Detail of the pore surface of a MPS structure fabricated by EE. (A) shows a freshly etched pore revealing a layer of microporous silicon in the walls. After oxidizing, stripping and further oxidizing, a much smoother, microporous-free surface is seen in (B).

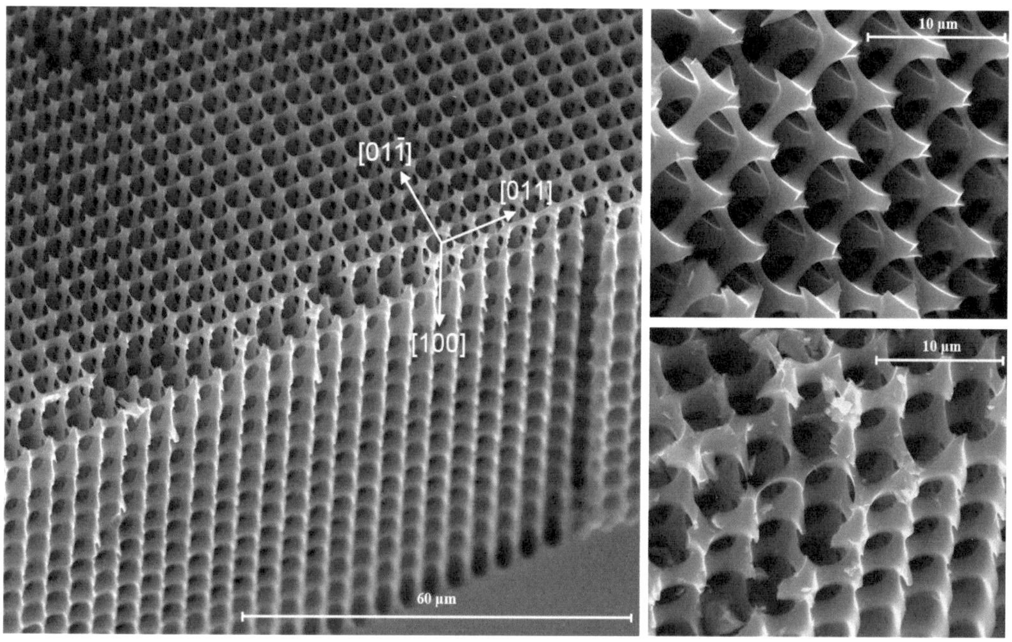

Figure 8. Bird's-eye views of the 3-d MPS modulated pores and subsequent widening of the pores. Reprinted from [23], Copyright 2008, with permission from Elsevier.

3.3.1. Functionalization

The increased available surface in MPS has been of interest for sensors. To improve the adsorption of species on the surface or the sensitivity of the device, several surface functionalization approaches have been proposed. For instance, carbonization of the PS structure helps in passivating and stabilizing the porous layer better than oxidation [41] or to avoid oxidation altogether. This is of particular interest for photoluminescence applications [41] and humidity sensors [42]. Other surface treatments have been proposed depending on the application. For example, using the atomic-layer deposition (ALD) method, it is possible to cover the walls of a MPS structure conformally with materials such as alumina (Al_2O_3), titania (TiO_2) or tantalum oxide (Ta_2O_5), which have applications in sensing devices [43] and electronic devices [44]. Other surface treatments of great interest for energy applications or chemical reactors are the deposition of nanoporous materials like zeolites [45] and catalysts like cobalt, platinum or palladium [12, 16].

3.4. Porous membranes

Certain applications require the passage or filtering of a fluid. Other applications seek that the passing fluids undergo certain chemical reactions. In either case, having a very large surface is beneficial. The main advantages are the fast mass and heat transfer, the short diffusion length and pressure drop, and the precise control of the process, resulting in higher product yields (for chemical reactors) and increased efficiency [16]. For this application, MPS has been proposed as a better alternative to common monoliths used in catalysis [12]. MPS membranes have also been suggested as ratchets [46] for physically selecting particles. Membranes have also been found useful for microfluidics [17] and for electrodeposition inside the pores.

The fabrication of the MPS membranes can be done by dissolving the remaining silicon bulk on the backside after the electrochemical etching of the pores. A common technique is to use alkaline anisotropic etchants typical in MEMS processing [47]: potassium hydroxide (KOH) and tetramethylammonium hydroxide (TMAH). The membrane is formed when the removed silicon reaches the pore tips. When performing such process, it is necessary to protect the porous volume, as it will etch much faster than the silicon bulk. Furthermore, it may be interesting to selectively etch the MPS structure to reveal only part of the pores. For this reason, first a protective layer must be prepared on the structure surface. This can be accomplished by oxidation of the MPS sample. This protective layer must have a minimum thickness: the dissolution process is not perfectly uniform, leaving a roughened surface [48]. Therefore, in some places, the pore tips will be uncovered earlier. The protective layer must be thick enough to withstand until the full membrane is finished, otherwise the etchant will leak and rapidly dissolve the porous volume.

This process has to be carefully monitored to avoid overetching and the destruction of the protective layer. In this regard, TMAH has a greater Si-SiO_2 selectivity [47], therefore making this process more robust. The actual method employed for the membrane fabrication will ultimately depend on any further processing and application of the MPS device. In particular, KOH may not be suitable if there are posterior high-temperature steps or the process flow has to be CMOS compatible.

3.4.1. *In situ*

As an alternative to alkaline etching of silicon, there is the possibility to create the membrane during the electrochemical pore growth phase. Two approaches are possible: entering the electropolishing regime during EE [49], and creating a sacrificial PS layer at the bottom [50]. The latter approach has been successfully applied for MEMS devices where the sacrificial micro-PS is formed underneath the actual device, such as cantilevers [51], thermal actuators [52] or sensing devices [53]. After the micro-PS is formed, the devices are released by dissolving this porous volume [50]. This can be done with a sufficiently diluted etchant or with plasma. Alternatively, one can oxidize the PS and then remove the oxide by an HF dip [50, 54]. This method can be also applied to MPS structures by a double-layer technique: first etching the macroporous volume and afterwards evolving the micro-PS beneath the macropores [54]. This may require changing the HF solution between etching phases.

The fabrication of an in situ membrane using the electropolishing regime is very similar to the sacrificial method. As in the previous method, it can be applied for both micro-PS and MPS. In this case, only the current density (and potential if necessary) is increased above the j_{PS} threshold [55]. This will promote the electropolishing of silicon at the pore tip depth and can be performed for an extended period of time to completely detach the MPS membrane from the Si substrate [49] (see **Figure 9**). For very high AR structures, it is important to take into account the acid concentration and the diffusion of reactants at the pore ends [2].

3.5. Pore filling

Conformal coating of the high AR structures present in MPS is often straightforward. Several techniques common in microfabrication are available for the task (e.g. ALD) [56]. However, the complete filling of such structures can be more difficult. Such methods can be very slow to fully fill the pore volume or have conformality issues for very thick depositions. An alternative method to fill the pores is to use either *electroless deposition* or *electrodeposition* (also *electroplating*). These methods are also common in MEMS [56] and CMOS microfabrication for the metallization layers of VLSI circuits [57].

Electroless deposition and electrodeposition are commonly used to deposit metals, though these processes are not limited to those materials. Copper is typically used in VLSI

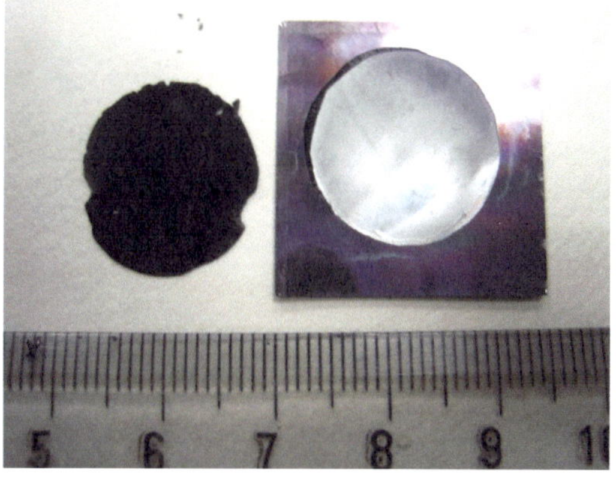

Figure 9. MPS membrane fabricated with in situ electropolishing. The black surface corresponds to the backside of the detached membrane; at the right is the supporting silicon substrate where the MPS was etched. © 2008 IEEE. Reprinted, with permission, from [49].

microfabrication. However, Cu easily diffuses through SiO_2 and Si [58], and it has poor adhesion to dielectrics like SiO_2. Noble metals can also be deposited by these means, for example gold; however, this would result in exceptionally expensive devices as the total volume of a MPS structure is large. Some good candidates for pore filling are nickel and aluminium. Aluminium plating would be the preferred material for an industrial process; however, Al plating is more complex (non-aqueous solution [59]) and has higher safety risks (inflammable or explosive products). On the other hand, nickel plating is simple and well known in industry, both electroless and electrodeposition. A MPS structure can be directly plated without further processing: Ni will deposit along the walls of the pores and exposed surface. However, for high AR pores, the deposition can be non-uniform, resulting in early pore closing and leaving voids inside the filling [57]. A way to avoid such issues is to use MPS membranes. Open membranes allow the passage of the electrolyte, thus reducing concentration gradients. Electrode size and electrolyte are also important for uniformity of the deposition.

Focusing in Ni electroplating, a typical plating bath is a dissolution of nickel sulphamate, nickel chloride and boric acid for pH correction. This plating solution is used at about 55°C. The resulting surface is a smooth, matte finish deposit. In particular, in the work of Vega et al. [18], this technique was used to fill MPS membranes as seen in **Figure 10**. The membranes were initially oxidized, and a metal evaporation was performed on one of the sides (the front face was used due to the better, smooth, surface). The so-prepared membrane was contacted on the metal face and afterwards insulated to avoid deposition on this face. The sample was submerged in the plating solution with the free face looking the counterelectrode, and the plating was performed at a mean current of 30 mA/cm^2. Deposition speed varied linearly with current, but the open surface (total pore area) determined the actual speed and had to be calibrated for each different MPS structure. The electrodeposits can be performed both with DC current and with pulsed current. As the silicon structure is protected by an insulating layer of silicon dioxide, Ni only deposits on the advancing Ni front inside the pores. This results in a smooth and void-free pore filling as seen in **Figure 10**.

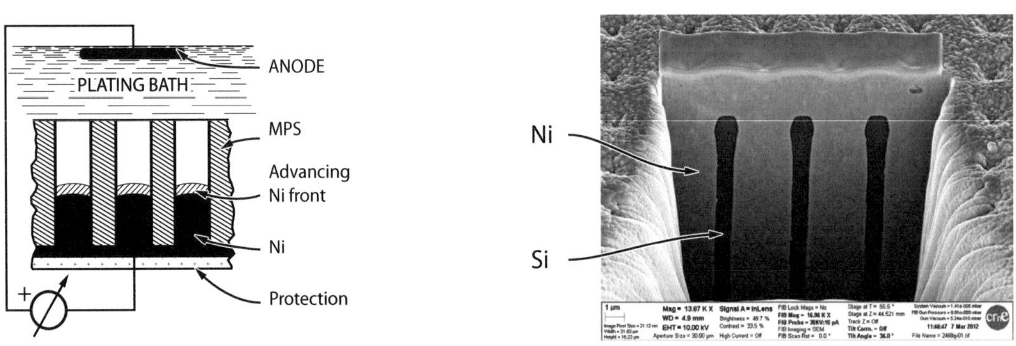

Figure 10. SEM micrograph of a macroporous silicon membrane filled with nickel. The left panel shows schematically the deposition on Ni inside the pores. To the right, a SEM image of a filled MPS structure is cut using FIB milling to expose the interior of the membrane; a platinum layer is used to aid the milling procedure. It can be seen that nickel fully covers the pore interior as well as the surface of the sample.

4. Applications of macroporous silicon

In this section, a brief description of a selection of the non-photonic applications of MPS will be presented. In particular, applications in the electronic devices and MEMS fields will be examined. However, bear in mind that MPS can be applied to a very broad range of fields, only some applications can be introduced here.

4.1. Field effect transistors

Power applications either demand operating at high voltages, high currents or both. In power MOS-FET devices, the gate oxide thickness determines the maximum potential the transistor can withstand (in the absence of a drift region). However, thick oxides will increase the threshold voltage, V_{th}. Furthermore, MOS transistors have a negative thermal coefficient, therefore for high currents, on-resistance, R_{ON}, will increase and thus limit the maximum current. In this situation, a large number of devices connected in parallel may be necessary to handle the demanded current density.

In these circumstances, power MOS structures evolved to *trench transistors* in the early 1990s and to the *superjunction* (SJ) *devices* [60] in mid-1990s. The outcome is a device with a much reduced R_{ON}. A summary of the vertical trench device structures is shown in **Figure 11**. Having millions of 'micro transistors', each one will handle a tiny amount of the requested current. As seen from **Figure 11**, the construction of these devices is fairly complex, requiring many steps and masks.

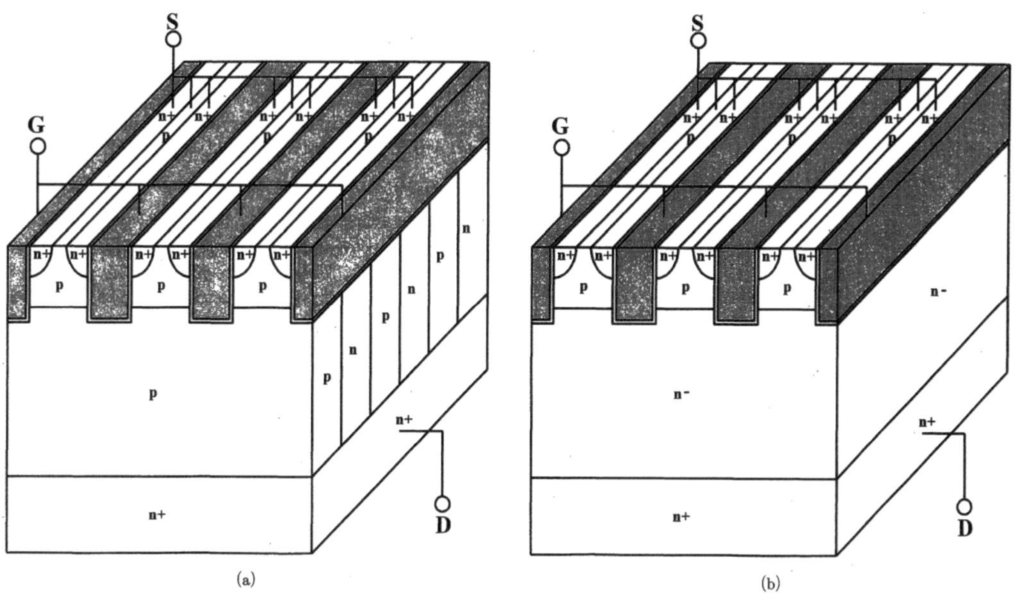

Figure 11. Structure of vertical MOS-FETs, (a) n-channel, trench gate and (b) conventional. Copyright 1997 The Japan Society of Applied Physics, from [60].

MPS provides an alternative to simplify the fabrication of trench FET transistors. Basically, the aim is to take advantage of the ease to obtain a large number of pores or trenches. EE of silicon provides a material with a controllable density of pores, which can be arranged freely (under certain constraints), and with a profile, which can be further adjusted to meet the device performance goals. The pores can subsequently be coated or filled by appropriate means (e.g. chemical vapor deposition (CVD) or thermal oxidation) to make either the gate or the channel of the device. Overall, the obtained structure is simpler and easier to fabricate, thanks to the reduced number of processing steps required. In Ref. [6], the suitability of such MPS structures was studied theoretically.

According to the definition of the channel region, two kinds of devices can be distinguished: vertical structures are those whose channel corresponds to the pore volume, while the gate is formed by the MPS surrounding material; and horizontal devices are those with the roles reversed, that is the channel is the MPS bulk material, while the pore volume constitutes the gate. Thus, carrier flow takes place in the pores for the vertical devices and through the MPS substrate for the horizontal devices.

Using ordered MPS, some of the proposed FET have the configurations shown in **Figure 12**: vertical J-FET and MOS-FET, and horizontal J-FET and insulated gate (IG) J-FET.[8] Using a square lattice with a pitch of 4 μm, a density of over 10million unit devices per cm^2 is achieved. MPS of the given feature size fabricated using the PECE technique employs low-doped silicon substrates. In particular, n-type, phosphorous doped, $\rho = 3\ \Omega$ cm, ⟨100⟩ oriented wafers are suitable for the above case. For the vertical devices, where the MPS substrate acts as the gate, this is not a limiting factor. However, for the horizontal operating devices, this is indeed an issue. To overcome this drawback, a doping of the as-etched MPS must be performed. Furthermore, removing the microporous layer after EE to get a smooth, defect-free, surface will slightly widen the pores. This must be taken into account when designing the

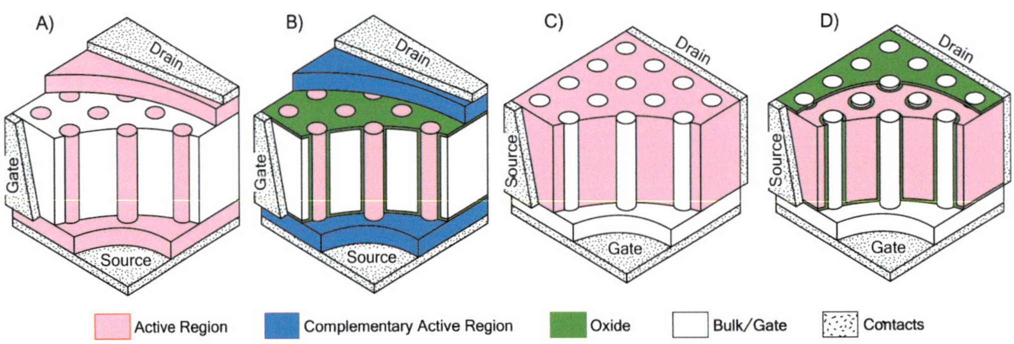

Figure 12. Breakout view of the proposed devices. For vertical structures (J-FET [a] and MOS-FET [b]), the channel corresponds to the pores, while for horizontal structures (J-FET [C] and IG J-FET [D]), the channel is the bulk porous silicon. © 2011 IEEE. Reprinted, with permission, from [6].

[8]As will be shown, it can operate with both positive and negative V_{GS}, and thus is also referred as *enhanced J-FET*.

http://dx.doi.org/10.5772/67698

devices. For the MOS or IG devices, a thin layer of insulating material, such as silicon oxide, can be deposited or thermally grown on the pores' walls.

After preparing the MPS supporting structure, the pores are filled. For the vertical structures, a CVD process may be performed to fill the pores with a semiconductor, such as silicon, of the desired doping. In contrast, for the horizontal ones, this step is not as critical. To form the gate, one can either choose a CVD process to coat with a relatively thin film of conducting or semiconducting material, or, alternatively, use an electrodeposition method, such as the one described in section *pore filling* to fill the pores with a metal.

From the work in Ref. [6], it can be seen that the overall specific on-resistance, R_{ON}, is comparable to that of more complex trench and SJ devices. It can be noticed that J-FET devices have larger R_{ON}, but this is to be expected as it solely depends on the channel resistivity, whereas MOS devices operate in strong inversion. Reducing the drift region (particularly for J-FET devices) and exploring alternative pore profiles may help in reducing R_{ON}. Regarding the breakdown voltage V_{br} of the devices, it was found that for all of them it was around 40 V. The electric field distribution revealed that this was caused mainly by the sharp edges found near the gate-drain region. The obtained V_{br} results are similar to those other found in literature (with no drift region for high-voltage operation). Nevertheless, a higher breakdown voltage can be obtained by adding a drift region and physically separating the gate from the drain, at the cost of a more complex fabrication process.

A summary of the device characteristics from Ref. [6] is given in **Table 1**. One of the most notable characteristics is the large current densities that these devices can handle. For instance,[9] the vertical J-FET from Ref. [6] is able to handle near 600 A/cm^2. In general, the performance of the proposed devices is comparable to that of more complex trench FETs. Finally, for the IG device, two things stand out: when operating in *enhancement mode* ($V_{GS} > 0\ V$), the transfer function is surprisingly linear, and that achieving complete turn-off of the device may require large potentials.

Regarding the small signal behaviour of the devices, the obtained results show that individual cell devices have moderate capacitance. Though these devices may be unsuitable for their use in signal applications, they are satisfactory for fast switching of loads in power applications.

4.2. Supercapacitors

MPS has also been shown to be applicable for energy storage since the beginnings of its research, see, for example, the work by Lehmann [61] or IBM [62]. The purpose of using MPS for capacitors is twofold: on one hand, there is the need of large capacitance devices for energy storage [10]; on the other hand, there is a drive towards higher integration, be it for cost and space savings or for efficiency reasons [63].

To fabricate high-capacity devices using MPS, one takes advantage of the greatly enlarged available surface area. Basically, there are few ways to increase capacity: electrochemical means (Helmholtz double layer or chemical energy storage[10]), decrease plate separation,

---

[9]The current density is calculated over the cross section of the channel.

[10]The former case would apply to capacitors, while the latter case would be for batteries.

		Vertical		Horizontal		
		J-FET	MOS-FET	J-FET	IG J-FET	
Specific on-resistance	R_{ON}	5.1	0.67	0.74	0.074	mΩ cm^2
Gate bias*	V_{GS}	0	10	0	10	V
Transconductance gain	g_m	16.6	70.6	616.7	875.0	μS
Pinch-off/threshold Voltage	V_{po} or V_{th}	−3.21	0.45	−3.8	−3.5	V
Input gate capacitance	C_G	11.6	40.0	17.4†	70.0†	fF
Gate oxide thickness	t_{ox}	–	100	–	100	nm
Pore diameter	d_{pore}	2	2	3	3.6	μm

These results correspond to: channel doping is 10^{-18} cm^{-3}.
Gate doping is 10^{-16} cm^{-3}.
Pore length $l_{pore} = 20$ μm.
Drain-source potential was set to $V_{DS} = 8$ V for parameter determination.
*Value used for parameter determination.
†The given values are for a single cell and may not scale linearly.

Table 1. Transistor characteristics summary for a single pore device, from Ref. [6].

increase the plate surface, increase the dielectric constant or a combination of the former. *Electrolytic capacitors* (ELCs) and EDLCs are based on the double layer effect,[11] while *electrostatic capacitors* (ESCs) rely only on charge accumulation. In particular, ELCs are capable of offering moderate-to-large capacity densities. ESCs, on the other hand, are capable of larger power densities and have a lower ESR (thus a higher operating frequency) [64]. To improve capacity, ESCs rely on choosing materials with high dielectric constant, reducing the dielectric thickness and extending the surface of the capacitor. Two types of dielectrics can be distinguished: class 1 or direct/paraelectric, and class 2 or ferroelectric [65]. Ferroelectric materials have very large dielectric constants but are sensitive to electric field and to ambient conditions (humidity and temperature) [66]. On the other hand, paraelectric materials have lower dielectric constant but are more stable. Some class 1 dielectrics are silicon dioxide and other materials frequently used in semiconductor industry for gate insulator such as alumina (Al_2O_3), silicon nitride (Si_3N_4) or high-k materials [67]. Though using high-k materials seems the most obvious solution, it has to be considered that they also have lower breakdown electric field (E_{bd}) and larger leakage; thus, complex insulator layer stacks [68] may be needed to achieve the desired performance.

Having a thinner insulating layer between the capacitor plates will help increase the capacitance of the device. However, care must be exercised as leakage currents will increase substantially due to direct tunnelling for thin insulators [68]. Furthermore, a thinner insulating layer will also affect the maximum operating voltage. Finally, to maximize capacitance, the effective

[11]Modern implementations of such capacitors having *activated carbon* as electrodes offer large gains in capacity density and are referred as *electric double-layer capacitors* (EDLCs), *ultracapacitors* and commercially as *supercapacitors*.

plate area can be extended without increasing the footprint of the device. To achieve such goal, advanced devices have intricate surface shapes that extend into the volume of the capacitor plates [18, 61, 62, 69].

The use of PS in its various forms for high-value capacitor devices can be traced back to the early development of PS. MPS was first suggested for its use as capacitors with the work from Lehmann [61], and since then, the industry has shown interest in the research and development of high-value capacitors based on this material [61, 70]. The techniques used to build the MPS structures of the capacitors have been both PECE [61, 71] and RIE [70, 71]. The pores of MPS give an enlarged surface to create the plates of the capacitor device. Pores may be arranged randomly or in a pattern. Random MPS is simpler to fabricate, and capacitor devices with $C_{sp} = 4\ nF/mm^2$ have been reported in Ref. [69]. However, using random MPS will result in uneven current distribution, and sharp edges where the electric field will concentrate and possibly break the dielectric. Furthermore, if the devices are to be defined in specific zones, a mask is needed nonetheless, so not using a patterned pore array is less justified. The use of ordered MPS requires a more complex fabrication flow; however, it has the advantage of being well defined, so final capacitance can be accurately calculated and obtained, plus giving control over the actual shape of the pores.

MPS-based capacitors use the porous substrate as one of the plates, so the other electrode must be deposited over the substrate. Plates can be arranged in several ways. For instance, it is possible to deposit a single layer to create a two-plate capacitor [18, 69], or to use a multiple layer approach, stacking several dielectric-conductor coatings to create a MIM-like capacitor. In the work by Klootwijk [70], the MIMIMIM stack depicted in **Figure 13** results in a very large capacitance density device, with a specific capacitance $C_{sp} = 440\ nF/mm^2$ [70]. However, its performance is lower than expected, see **Table 2**. Using multilayer stacking greatly complicates the fabrication process, so the simpler approach of a two-plate capacitor may be desirable. In **Table 2**, a summary of the state-of-the-art supercapacitors based on MPS is given.

4.2.1. Fabrication

Once the MPS is etched, it may be necessary to perform a doping of the porous substrate. This is especially necessary in case of EE MPS as the Si substrate used has 'low' doping concentrations as seen in section *Electrochemical Etching*. In particular, for a 4 μm pitch MPS structure, an n-type phosphorous doping of $1.5 \times 10^{16}\ cm^{-3}$ is used in Ref. [18]. To improve ESR, a degenerate n^{++} layer deep enough is diffused surrounding all of the porous area. Additionally, a porous membrane can be formed as described in section *Porous Membranes*. This step can be done after the doping step, though it is preferably done before it to better control the doping of the MPS structure.

A common issue found in such large surface capacitors is the elevated leakage currents,[12] as seen in **Table 2**. To keep them at reasonable levels is mandatory to have a very high quality insulating layer. The predominantly used insulating materials have been those common in CMOS: SiO_2, Si_3N_4, and Al_2O_3. Thicknesses used have been around 10 nm to 50 nm. However, most works noted a larger than expected leakage, and this has been primarily attributed to the

[12]Leakage current density is calculated considering the footprint area of the device.

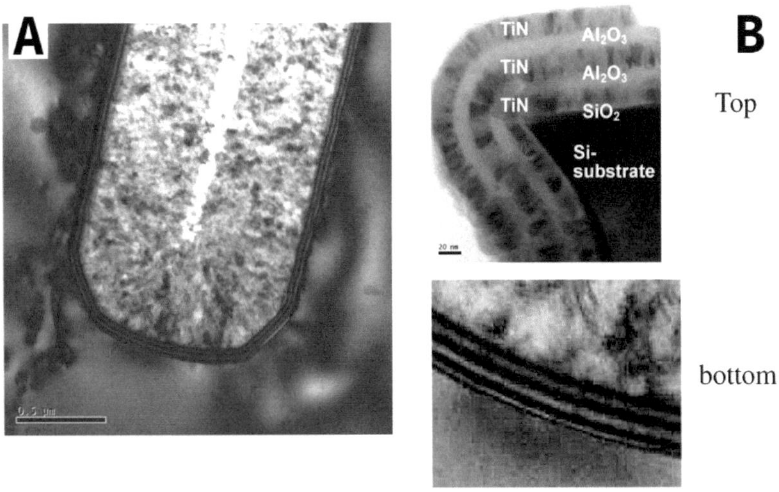

Figure 13. Cross-sectional TEM view for MIM-type MPS supercapacitors. (a) shows an overview of the capacitor MIM stack at the bottom of the pore. In (b), a detail view of the top and bottom of the pore is shown. © 2008 IEEE. Reprinted, with permission, from [70].

Refs.	**Method**	**Configuration**	**Depth (μm)**	**Density (pore/μm²)**	**Specific capacitance C_{sp} (nF/mm²)**	**Breakdown voltage V_{br} (V)**	**ESR (Ω)**	**Leakage I_{leak} (mA/cm²)**
[71]	EE	ONO/polySi	100	0.094	100	–	0.1	–
[72]	RIE	SiO_2/TaN	0.1	721.69	31.3	–†	–	1 @ 1V
[44]	EE	Al_2O_3/ZnO : Al	50	0.16	25	~8*	–	1.5 @ 2.5V
[70]	DRIE	$SiO_2/[TiN/Al_2O_3]_3$	30	–	440	6	10	1 @ 3V
[73]	EE	ON/polySi	72	1	700	8	20	1 @ 3V
[18]	EE	SiO_2/Ni	130	0.0625	110	20	1.1	≪0.001@ 1V

*Not reported but inferred from **Figure 4**.
†From **Figure 3**, the reported device should be able to operate in the ±3 V range.

Table 2. MPS supercapacitors state of the art. Best figures reported.

increased E field strength in the curved regions (tip and edges) of the pores. To achieve the remarkably low leakage current in [18], the authors note that especial care of surface cleanliness was needed; also after the membrane opening and substrate doping, it was found best no to completely fill the pores.

After the dielectric layer is placed, the pores are filled to define the other plate. This can be done with nickel following the procedures described in section *pore filling*. The use of a metal electrode helps in achieving a very low ESR for the final device. The fabrication process flow is schematically depicted in **Figure 14**.

Electrical characterization of the devices can be approximated by a simple RC model. As the main contributor to capacitance is due to the cylindrical pores, the capacitance value can be predicted by the formula of the cylindrical capacitor $C_{\text{unit}} = 2\pi\varepsilon_0\varepsilon_r l_{fill}/\ln(1 + t_{ox}/r_{pore})$. As seen from **Figure 15**, the simple RC model is reasonable at low frequencies; however, the capacity is shown to be not ideal and is presumed to be due to relaxation in the dielectric interfaces [74].

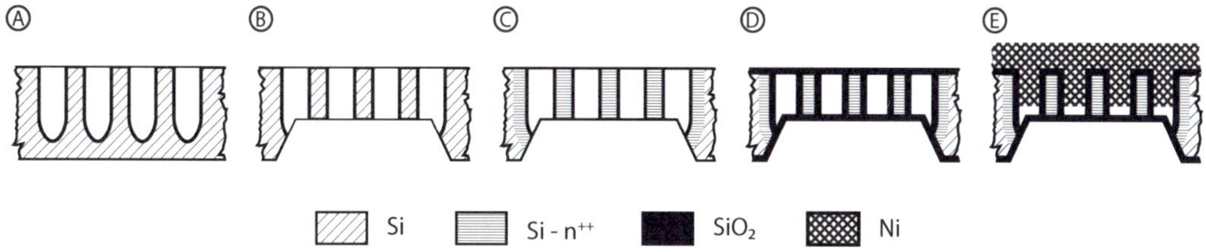

Figure 14. Fabrication flow for the capacitor described in Ref. [18]. (a) Etching of the pores, (b) membrane opening, (c) substrate doping, (d) thermal SiO_2 layer growth and (e) pore filling and electrode connection.

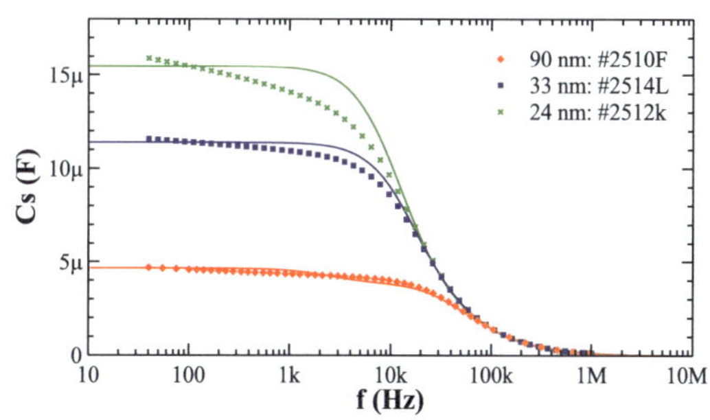

Figure 15. Series capacitance (symbols) and the fitted RC model (solid line) for the MPS capacitors of [18]. © 2014 IEEE. Reprinted, with permission, from [18].

4.3. Through silicon vias

Nowadays, the microelectronics industry is highly interested in 3-d (monolithic) integration looking for improvements in package integration, the reduction of interconnect delays, cost savings and system scaling [75]. Among the different technological options, through silicon vias (TSVs) and silicon interposers have been shown to be an enabling technology that has recently reached the consumer market in the form of ultra-high-density memory modules [76]. The most common method for TSV fabrication found in the literature is RIE etching of the silicon to create the vias, then pore filling with copper by a dual damascene process, surface planarization, and pad definition (and repeat the process for the other side) [75]. Pore surface roughness may impose certain complications when filling the pores [77], as well as pore shape has an impact on reliability by thermal cycling [77].

EE MPS may offer a simplified approach to TSV fabrication. The pore shape is easily controlled by the current waveform, and deep pores are effortlessly etched. The obtained surfaces can be made smooth by a short oxidation and oxide stripping (see *Surface treatment and functionalization*), and pores can be easily filled. An example of MPS for TSVs can be found in the work by Defforge [78].

4.4. Microneedles

For biomedical applications, microfabrication of miniaturized complete analysis platforms (lab-on-a-chip) is being vigorously researched [79]. In particular, the extraction and injection of fluids to tissues are of great interest. Microneedles provide a way to simplify this procedure while minimizing some nuisances like pain [80] or risk of infection, being minimally invasive[13] [80]. Microneedles have been reported of several materials and features, as well as fabrication methods [80]. MPS has also been shown to be suitable to be used in such applications. Furthermore, MPS-based microneedles have shown distinct advantages over 'conventional' ones, like smaller diameter and higher density, and foremost a simpler fabrication process.

The first reference to MPS fabricated by EE applied for microneedles is found in the work by Rodríguez et al. [17] and Rajaraman [81]. In Rodriguez et al. work, MPS is used to create arrays of SiO_2 microneedles with diameters ranging from 2 to 5 μm, up to 140 μm in length, and wall thickness about 100 nm. The process and results obtained are shown in **Figure 16**. The fabrication process consists of creating a MPS membrane, and once the pore tips are exposed, remove them by a rapid dip in HF (see **Figure 16c**). To create the needles, the membrane is further processed in TMAH to remove silicon from the backside (see **Figure 16d**). At this stage, the microneedles are prepared and just remain attaching the fluid handling fixtures. Rajaraman's work is essentially identical, though a slightly more complex process is described, as needles are formed in the front face.

A similar approach is followed by Barillaro and his group [82]. In this case, instead of forming the needles by removing silicon from the backside, they use the back-face

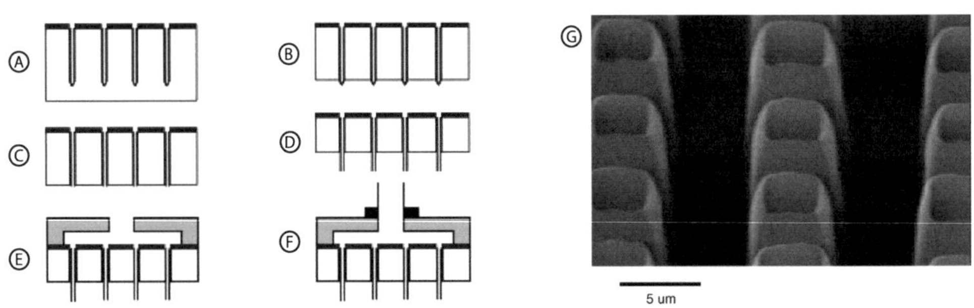

Figure 16. Silicon dioxide microneedles fabricated from MPS. From the starting MPS in (A), a membrane is done (B) and the oxidized pore tips are etched in HF to open the membrane (C). The membrane is further processed by removing silicon from the backside with TMAH (D) and revealing the SiO_2 microneedles. A reservoir is glued to the prepared device (E) and a tube for fluid delivery can be attached (F). In (G), a close-up of the needles tip is shown. Adapted from [17], Copyright 2005, with permission from Elsevier.

---

[13]Microneedles are short and only penetrate the epidermis [91], avoiding the dermis where nerve terminations are located.

membrane cavity to create a reservoir and subsequently etch the frontside to create the needles. Furthermore, Barillaro et al. conducted skin penetration tests showing that 1 μm wall thick can withstand repeated skin insertion without breaking. Other studies confirm these findings and show that SiO_2 microneedles require less force for effective penetration [83].

Besides hollow microneedles as the ones previously described, MPS has also been used to create high AR SiO_2 pillars [84]. Basically, the fabrication process is the same as for the microneedles, described earlier (see **Figure 16**), but stops after the membrane creation. The released pillars can be further used for sensing applications, drug delivery by coating the pillars [80], skin pre-perforation [80] or microreactors [45].

Microneedles can also be applied to other uses in addition to biomedicine. For example, for inkjet printing [85], micronozzles have been shown to be able to print extremely fine detail lines.

4.5. Other MEMS devices

MPS technology can be adapted for the fabrication of complex shapes and structures with uses in other MEMS applications. For example, EE micromachining has been proposed as way for fabrication of embedded heat sinks [86]. The idea is to create a free-standing MPS structure by in situ membrane etching and later filling the porosified volume with copper by electrodeposition. Thanks to the higher thermal conductivity of Cu, higher efficiency in thermal dissipation can be achieved for dense VLSI circuits.

However, it is in microfluidics and lab-on-chip applications that MPS is most interesting. Pore diameter can be modulated and adjusted for filtering applications [46, 54], such that using the appropriate pore size one can selectively allow the passage of particles or biological species. Furthermore, the device can easily be protected against harsh environments by oxidation or other surface treatment. Particular mention must be made to the work of Barillaro's group [22]. Some outstanding examples are a capillarity-driven (self-powered) microfluidic system based on MPS [87] and a MPS microstructures for label-free optical detection of cells [88].

6. Conclusion

This review has shown several applications of macroporous silicon to a wide range of fields, besides the optical and photonic, that are also of utmost relevance. In particular, it has been shown that MPS has many uses the fabrication of electronic devices, such as in energy storage (supercapacitors). Other important areas of application of MPS are microfluidics and lab-on-a-chip, where the freedom in structure micromachining allows creating complex structures capable of being integrated in a complete sensing platform. The use of MPS has helped to improve existing devices and has allowed to develop new ones. Also, MPS has been applied in MEMS device fabrication. The applications of MPS represent a vast field that is still in active development.

Author details

Didac Vega Bru* and Ángel Rodríguez Martínez

*Address all correspondence to: didac.vega@upc.edu

Department of Electronic Engineering (EEL), Universitat Politècnica de Catalunya (UPC), Barcelona, Spain

References

[1] Lehmann V, Föll H. Formation mechanism and properties of electrochemically etched trenches in n-type silicon. J Electrochem Soc 1990;137:653–9. doi:10.1149/1.2086525.

[2] Lehmann V. The physics of macropore formation in low doped n-type silicon. J Electrochem Soc 1993;140:2836. doi:10.1149/1.2220919.

[3] Lehmann V. The physics of macroporous silicon formation. Thin Solid Films 1995;255:1–4. doi:10.1016/0040-6090(94)05620-S.

[4] Zhang XG. Mechanism of pore formation on n-type silicon. J Electrochem Soc 1991;138:3750. doi:10.1149/1.2085494.

[5] Levy-Clement C, Lagoubi A, Tenne R, Neumann-Spallart M. Photoelectrochemical etching of silicon. Electrochim Acta 1992;37:877–88. doi:10.1016/0013-4686(92)85039-N.

[6] Rodriguez A, Vega D, Najar R, Pina M. Novel electronic devices in macroporous silicon: design of FET transistors for power applications. IEEE Trans Electron Devices 2011;58:3065–71. doi:10.1109/TED.2011.2159508.

[7] Vega D, Marti F, Rodriguez A, Trifonov T. Macroporous silicon for spectroscopic CO_2 detection. IEEE SENSORS 2014 Proc., Valencia: IEEE; 2014, pp. 1061–4. doi:10.1109/ICSENS.2014.6985187.

[8] Schilling J, Scherer A. 3D photonic crystals based on macroporous silicon: Towards a large complete photonic bandgap. Photonics Nanostruct Fundam Appl 2005;3:90–5. doi:10.1016/j.photonics.2005.09.015.

[9] Salman KA, Omar K, Hassan Z. The effect of etching time of porous silicon on solar cell performance. Superlattices Microstruct 2011;50:647–58. doi:10.1016/j.spmi.2011.09.006.

[10] Long JW, Dunn B, Rolison DR, White HS. Three-dimensional battery architectures. Chem Rev 2004;104:4463–92. doi:10.1021/cr020740l.

[11] Gautier G, Kouassi S, Desplobain S, Ventura L. Macroporous silicon hydrogen diffusion layers for micro-fuel cells: from planar to 3D structures. Microelectron Eng 2012;90:79–82. doi:10.1016/j.mee.2011.04.003.

[12] López E, Irigoyen A, Trifonov T, Rodríguez A, Llorca J. A million-channel reformer on a fingertip: moving down the scale in hydrogen production. Int J Hydrogen Energy 2010;35:3472–9. doi:10.1016/j.ijhydene.2010.01.146.

[13] Uhlir A. Electrolytic shaping of germanium and silicon. Bell Syst Tech J 1956;35:333–47. doi:10.1002/j.1538-7305.1956.tb02385.x.

[14] Turner DR. Electropolishing silicon in hydrofluoric acid solutions. J Electrochem Soc 1958;105:402–8. doi:10.1149/1.2428873.

[15] Gerischer H, Mindt W. The mechanisms of the decomposition of semiconductors by electrochemical oxidation and reduction. Electrochim Acta 1968;13:1329–41. doi:10.1016/0013-4686(68)80060-X.

[16] Llorca J, Casanovas A, Trifonov T, Rodriguez A, Alcubilla R. First use of macroporous silicon loaded with catalyst film for a chemical reaction: a microreformer for producing hydrogen from ethanol steam reforming. J Catal 2008;255:228–33. doi:10.1016/j.jcat.2008.02.006.

[17] Rodríguez A, Molinero D, Valera E, Trifonov T, Marsal LF, Pallarès J, et al. Fabrication of silicon oxide microneedles from macroporous silicon. Sensors Actuators, B Chem 2005;109:135–40. doi:10.1016/j.snb.2005.03.015.

[18] Vega D, Reina J, Pavon R, Rodriguez A. High-density capacitor devices based on macroporous silicon and metal electroplating. IEEE Trans Electron Devices 2014;61:116–22. doi:10.1109/TED.2013.2290065.

[19] Lehmann V. Porous silicon-a new material for MEMS. Proc Ninth Int Work Micro Electromechanical Syst 1996. pp. 1–6. doi:10.1109/MEMSYS.1996.493820.

[20] Rouquerol J, Avnir D, Fairbridge CW, Everett DH, Haynes JM, Pernicone N, et al. Recommendations for the characterization of porous solids (Technical report). Pure Appl Chem 1994;66:1739–58. doi:10.1351/pac199466081739.

[21] Artillan P, Brunet M, Bourrier D, Laur JP, Mauran N, Bary L, et al. Integrated LC filter on silicon for DC–DC converter applications. IEEE Trans Power Electron 2011;26:2319–25. doi:10.1109/TPEL.2010.2102776.

[22] Barillaro G, Nannini A, Piotto M. Electrochemical etching in HF solution for silicon micromachining. Sensors Actuators A Phys 2002;102:195–201. doi:10.1016/S0924-4247(02)00385-0.

[23] Trifonov T, Rodríguez A, Marsal LF, Pallarès J, Alcubilla R. Macroporous silicon: a versatile material for 3D structure fabrication. Sensors Actuators A Phys 2008;141:662–9. doi:10.1016/j.sna.2007.09.001.

[24] Vázsonyi É, Szilágyi E, Petrik P, Horváth Z, Lohner T, Fried M, et al. Porous silicon formation by stain etching. Thin Solid Films 2001;388:295–302. doi:10.1016/S0040-6090(00)01816-2.

[25] Huang Z, Geyer N, Werner P, de Boor J, Gösele U. Metal-assisted chemical etching of silicon: a review. Adv Mater 2011;23:285–308. doi:10.1002/adma.201001784.

[26] Lewis JA, Gratson GM. Direct writing in three dimensions. Mater Today 2004;7:32–9. doi:10.1016/S1369-7021(04)00344-X.

[27] Qi M, Lidorikis E, Rakich PT, Johnson SG, Joannopoulos JD, Ippen EP, et al. A three-dimensional optical photonic crystal with designed point defects. Nature 2004;429:538–42. doi:10.1038/nature02575.

[28] George D, Lutkenhaus J, Lowell D, Moazzezi M, Adewole M, Philipose U, et al. Holographic fabrication of 3D photonic crystals through interference of multi-beams with 4 + 1, 5 + 1 and 6 + 1 configurations. Opt Express 2014;22:22421–31. doi:10.1364/OE.22.022421.

[29] Gu X, Gunkel I, Russell TP. Pattern transfer using block copolymers. Philos Trans A Math Phys Eng Sci 2013;371:20120306. doi:10.1098/rsta.2012.0306.

[30] Summers MA, Tabunshchyk K, Kovalenko A, Brett MJ. Fabrication of 2D–3D photonic crystal heterostructures by glancing angle deposition. Photonics Nanostructures—Fundam Appl 2009;7:76–84. doi:10.1016/j.photonics.2008.12.001.

[31] Laermer F, Franssila S, Sainiemi L, Kolari K. Deep reactive ion etching. In: Lindroos V, Tilli M, Lehto A, Motooka T, editors. Handbook of Silicon Based MEMS Materials and Technologies., William Andrew Publishing, Boston; ISBN 9780815515944, 2010, pp. 349–74. doi:10.1016/B978-0-8155-1594-4.00023-1.

[32] Zhang XG. Morphology and formation mechanisms of porous silicon. J Electrochem Soc 2004;151:C69. doi:10.1149/1.1632477.

[33] Kolasinski KW. Etching of silicon in fluoride solutions. Surf Sci 2009;603:1904–11. doi:10.1016/j.susc.2008.08.031.

[34] Garín M, Trifonov T, Hernández D, Rodriguez A, Alcubilla R. Thermal emission of macroporous silicon chirped photonic crystals. Opt Lett 2010;35:3348–50. doi:10.1364/OL.35.003348.

[35] Vega Bru D, Cardador Maza D, Trifonov T, Garin Escriva M, Rodriguez Martinez A. The effect of absorption losses on the optical behaviour of macroporous silicon photonic crystal selective filters. J Light Technol 2016;34:1281–7. doi:10.1109/JLT.2015.2503354.

[36] Hippo D, Nakamine Y, Urakawa K, Tsuchiya Y, Mizuta H, Koshida N, et al. Formation mechanism of 100-nm-scale periodic structures in silicon using magnetic-field-assisted anodization. Jpn J Appl Phys 2008;47:7398–402. doi:10.1143/JJAP.47.7398.

[37] Laffite G, Roumanie M, Gourgon C, Perret C, Boussey J, Kleimann P. Formation of submicrometer pore arrays by electrochemical etching of silicon and nanoimprint lithography. J Electrochem Soc 2011;158:D10–4. doi:10.1149/1.3514584.

[38] Geppert T, Schweizer SL, Gösele U, Wehrspohn RB. Deep trench etching in macroporous silicon. Appl Phys A Mater Sci Process 2006;84:237–42. doi:10.1007/s00339-006-3628-7.

[39] Debarge L, Stoquert J, Slaoui A, Stalmans L, Poortmans J. Rapid thermal oxidation of porous silicon for surface passivation. Mater Sci Semicond Process 1998;1:281–5. doi:10.1016/S1369-8001(98)00039-0.

[40] Trifonov T, Garin M, Rodriguez A, Marsal LF, Pallares J, Alcubilla R. Towards more complex shapes of macroporous silicon. 2007 Spanish Conf. Electron Devices, IEEE; 2007, pp. 9–12. doi:10.1109/SCED.2007.383984.

[41] Fan Y, Ju J, Zhang W, Xia Y, Wang Z, Fang Z, et al. A new passivation method for porous silicon. Solid State Commun 2001;120:435–7. doi:10.1016/S0038-1098(01)00423-9.

[42] Björkqvist M, Salonen J, Paski J, Laine E. Characterization of thermally carbonized porous silicon humidity sensor. Sensors Actuators A Phys 2004;112:244–7. doi:10.1016/j.sna.2004.01.002.

[43] Wang Y, Park S, Yeow JTW, Langner A, Müller F. A capacitive humidity sensor based on ordered macroporous silicon with thin film surface coating. Sensors Actuators, B Chem 2010;149:136–42. doi:10.1016/j.snb.2010.06.010.

[44] Kemell M, Ritala M, Leskelä M, Ossei-Wusu E, Carstensen J, Föll H. Si/Al2O3/ZnO:Al capacitor arrays formed in electrochemically etched porous Si by atomic layer deposition. Microelectron Eng 2007;84:313–8. doi:10.1016/j.mee.2006.10.085.

[45] Urbiztondo M, Valera E, Trifonov T, Alcubilla R, Irusta S, Pina M, et al. Development of microstructured zeolite films as highly accessible catalytic coatings for microreactors. J Catal 2007;250:190–4. doi:10.1016/j.jcat.2007.05.022.

[46] Matthias S, Müller F. Asymmetric pores in a silicon membrane acting as massively parallel brownian ratchets. Nature 2003;424:53–7. doi:10.1038/nature01736.

[47] Lindroos V, Tilli M, Lehto A, Motooka T, editors. Handbook of silicon based MEMS materials and technologies. 1st ed. William Andrew Publishing, Boston; ISBN 978-0-81551-594-4, 2010. doi:10.1016/B978-0-8155-1594-4.00055-3.

[48] Zubel I. Silicon anisotropic etching in alkaline solutions II on the influence of anisotropy on the smoothness of etched surfaces. Sensors Actuators A Phys 1998;70:260–8. doi:10.1016/S0924-4247(98)00142-3.

[49] Jilei Lin, Xiaoming Chen, Shaohui Xu, Peisheng Xin, Lianwei Wang. Investigation of the formation of undercut during the fabrication of silicon microchannels by electrochemical etching. 2008 3rd IEEE Int. Conf. Nano/Micro Eng. Mol. Syst., IEEE; 2008, pp. 74–7. doi:10.1109/NEMS.2008.4484289.

[50] Steiner P, Lang W. Micromachining applications of porous silicon. Thin Solid Films 1995;255:52–8. doi:10.1016/0040-6090(95)91137-B.

[51] Valera E, Duch M, Rodriguez A, Esteve I. Microporous silicon for CMOS compatible MST. Conf. Electron Devices, 2005 Spanish, IEEE; 2005, pp. 481–3. doi:10.1109/SCED.2005.1504489.

[52] Lammel G, Renaud P. 3D flip-up structure of porous silicon with actuator and optical filter for microspectrometer applications. Proc IEEE Thirteen Annu Int Conf Micro Electro Mech Syst. (Cat. No.00CH36308), IEEE; 2000, pp. 132–7. doi:10.1109/MEMSYS.2000.838503.

[53] Baratto C, Faglia G, Sberveglieri G, Boarino L, Rossi A, Amato G. Front-side micromachined porous silicon nitrogen dioxide gas sensor. Thin Solid Films 2001;391:261–4. doi:10.1016/S0040-6090(01)00992-0.

[54] Pagonis DN, Nassiopoulou AG. Free-standing macroporous silicon membranes over a large cavity for filtering and lab-on-chip applications. Microelectron Eng 2006;83:1421–5. doi:10.1016/j.mee.2006.01.065.

[55] Lammel G, Renaud P. Free-standing, mobile 3D porous silicon microstructures. Sensors Actuators A Phys 2000;85:356–60. doi:10.1016/S0924-4247(00)00382-4.

[56] Gatzen HH, Saile V, Leuthold J. Micro and nano fabrication. Berlin, Heidelberg: Springer Berlin Heidelberg; 2015. doi:10.1007/978-3-662-44395-8.

[57] Robert D, Yoshio N, editors. Handbook of semiconductor manufacturing technology. 2nd ed. CRC Press, Boca Raton, FL; ISBN: 978-1-57444-675-3, 2007. doi:10.1201/9781420017663.fmatt.

[58] Bracht H. Copper related diffusion phenomena in germanium and silicon. Mater Sci Semicond Process 2004;7:113–24. doi:10.1016/j.mssp.2004.06.001.

[59] Abbott AP, McKenzie KJ. Application of ionic liquids to the electrodeposition of metals. Phys Chem Chem Phys 2006;8:4265. doi:10.1039/b607329h.

[60] Fujihira T. Theory of semiconductor superjunction devices. Jpn J Appl Phys 1997;36:6254–62. doi:10.1143/JJAP.36.6254.

[61] Lehmann V, Hönlein W, Reisinger H, Spitzer A, Wendt H, Willer J. A novel capacitor technology based on porous silicon. Thin Solid Films 1996;276:138–42. doi:10.1016/0040-6090(95)08038-4.

[62] Geiss PJ, Kenney DM. Porous silicon trench and capacitor structures, 1997.

[63] Scheffler M, Troster G. Assessing the cost effectiveness of integrated passives. Proc Des Autom Test Eur Conf Exhib. 2000, IEEE Comput. Soc; 2000, pp. 539–43. doi:10.1109/DATE.2000.840838.

[64] Sharma P, Bhatti TS. A review on electrochemical double-layer capacitors. Energy Convers Manag 2010;51:2901–12. doi:10.1016/j.enconman.2010.06.031.

[65] Ulrich R, Schaper L. Materials options for dielectrics in integrated capacitors. Proc Int Symp Adv Packag Mater Process Prop Interfaces IMAPS Int Microelectron Packaging Soc; 2000, pp. 38–43. doi:10.1109/ISAPM.2000.869240.

[66] Pan MJ, Randall CA. A brief introduction to ceramic capacitors. IEEE Electr Insul Mag 2010;26:44–50. doi:10.1109/MEI.2010.5482787.

[67] Choi JH, Mao Y, Chang JP. Development of hafnium based high-k materials—A review. Mater Sci Eng R Reports 2011;72:97–136. doi:10.1016/j.mser.2010.12.001.

[68] Zhang J, Yuan J, Ma Y, Oates A. Design optimization of stacked layer dielectrics for minimum gate leakage currents. Solid State Electron 2000;44:2165–70. doi:10.1016/S0038-1101(00)00185-4.

[69] Sancho A, Arizti F, Gracia FJJ. Porous silicon for the development of capacitive microstructures. Microelectron Eng 2009;86:2144–8. doi:10.1016/j.mee.2009.02.031.

[70] Klootwijk JH, Jinesh KB, Dekkers W, Verhoeven JF, Heuvel FC van den, Kim HD, et al. Ultrahigh capacitance density for multiple ALD-grown MIM capacitor stacks in 3-D silicon. IEEE Electron Device Lett 2008;29. doi:10.1109/LED.2008.923205.

[71] Roozeboom F, Elfrink RJG, Rijks T, Verhoeven J, Kemmeren A, Van den Meerakker J. High-density, low-loss MOS capacitors for integrated RF decoupling. Int J Microcircuits Electron Packag 2001;24:182–96.

[72] Black CT, Guarini KW, Zhang YZY, Kim HKH, Benedict J, Sikorski E, et al. High-capacity, self-assembled metal-oxide-semiconductor decoupling capacitors. IEEE Electron Device Lett 2004;25. doi:10.1109/LED.2004.834637.

[73] Brunet M, Kleimann P. High-density 3-D capacitors for power systems on-chip: evaluation of a technology based on silicon submicrometer pore arrays formed by electrochemical etching. IEEE Trans Power Electron 2013;28:4440–8. doi:10.1109/TPEL.2012.2233219.

[74] Shoar Abouzari MR, Berkemeier F, Schmitz G, Wilmer D. On the physical interpretation of constant phase elements. Solid State Ionics 2009;180:922–7. doi:10.1016/j.ssi.2009.04.002.

[75] Khan N, Rao VS, Lim S, Ho Soon We, Lee V, Xiaowu Zhang, et al. Development of 3-D silicon module with TSV for system in packaging. IEEE Trans Components Packag Technol 2010;33:3–9. doi:10.1109/TCAPT.2009.2037608.

[76] Beica R. 3D integration: Applications and market trends. 2015 Int 3D Syst Integr Conf IEEE; 2015, pp. TS5.1.1–1.7. doi:10.1109/3DIC.2015.7334567.

[77] Murayama T, Morikawa Y. TSV etching and VDP process integration for high reliability. 2015 Int 3D Syst Integr Conf IEEE; 2015, pp. TS8.11.1–11.4. doi:10.1109/3DIC.2015.7334580.

[78] Defforge T, Coudron L, Ménard O, Grimal V, Gautier G, Tran-Van F. Copper electrodeposition into macroporous silicon arrays for through silicon via applications. Microelectron Eng 2013;106:160–3. doi:10.1016/j.mee.2013.01.014.

[79] Ríos Á, Zougagh M, Avila M. Miniaturization through lab-on-a-chip: Utopia or reality for routine laboratories? A review. Anal Chim Acta 2012;740:1–11. doi:10.1016/j.aca.2012.06.024.

[80] Tuan-Mahmood TM, McCrudden MTC, Torrisi BM, McAlister E, Garland MJ, Singh TRR, et al. Microneedles for intradermal and transdermal drug delivery. Eur J Pharm Sci 2013;50:623–37. doi:10.1016/j.ejps.2013.05.005.

[81] Rajaraman S, Henderson HT. A unique fabrication approach for microneedles using coherent porous silicon technology. Sensors Actuators B Chem 2005;105:443–8. doi:10.1016/j.snb.2004.06.035.

[82] Strambini LM, Longo A, Diligenti A, Barillaro G. A minimally invasive microchip for transdermal injection/sampling applications. Lab Chip 2012;12:3370–9. doi:10.1039/c2lc40348j.

[83] Kim S, Shetty S, Price D, Bhansali S. Skin penetration of silicon dioxide microneedle arrays. 2006 Int Conf IEEE Eng Med Biol Soc IEEE; 2006, pp. 4088–91. doi:10.1109/IEMBS.2006.260142.

[84] Trifonov T, Rodríguez A, Servera F, Marsal LF, Pallarès J, Alcubilla R. High-aspect-ratio silicon dioxide pillars. Phys Status Solidi Appl Mater Sci. 2005:202;1634–8. doi:10.1002/pssa.200461205.

[85] Ohigashi R, Tsuchiya K, Mita Y, Fujita H. Electric ejection of viscous inks from MEMS capillary array head for direct drawing of fine patterns. J Microelectromechanical Syst 2008;17:272–7. doi:10.1109/JMEMS.2007.911374.

[86] Zacharatos F, Nassiopoulou AG. Copper-filled macroporous Si and cavity underneath for microchannel heat sink technology. Phys Status Solidi 2008;205:2513–7. doi:10.1002/pssa.200780161.

[87] Surdo S, Carpignano F, Strambini LM, Merlo S, Barillaro G. Capillarity-driven (self-powered) one-dimensional photonic crystals for refractometry and (bio)sensing applications. RSC Adv 2014;4:51935–41. doi:10.1039/C4RA09056J.

[88] Merlo S, Carpignano F, Silva G, Aredia F, Scovassi AI, Mazzini G, et al. Label-free optical detection of cells grown in 3D silicon microstructures. Lab Chip 2013;13:3284–92. doi:10.1039/c3lc50317h.

[89] Braun PV, Rinne SA, García-Santamaría F. Introducing defects in 3D photonic crystals: state of the art. Adv Mater 2006;18:2665–78. doi:10.1002/adma.200600769.

[90] Grishina DA, Harteveld CAM, Woldering LA, Vos WL. Method for making a single-step etch mask for 3D monolithic nanostructures. Nanotechnology 2015;26:505302. doi:10.1088/0957-4484/26/50/505302.

[91] Sivamani RK, Stoeber B, Liepmann D, Maibach HI. Microneedle penetration and injection past the stratum corneum in humans. J Dermatolog Treat 2009;20:156–9. doi:10.1080/09546630802512679.

Chapter 8

Silicon-Germanium (SiGe) Nanostructures for Thermoelectric Devices: Recent Advances and New Approaches to High Thermoelectric Efficiency

Jaime Andrés Pérez-Taborda,
Olga Caballero-Calero and Marisol Martín-González

Additional information is available at the end of the chapter

http://dx.doi.org/10.5772/67730

Abstract

Silicon and germanium present distinct and interesting transport properties. However, composites made of silicon-germanium (SiGe) have resulted in a breakthrough in terms of their transport properties. Currently, these alloys are used in different applications, such as microelectronic devices and integrated circuits, photovoltaic cells, and thermoelectric applications. With respect to thermoelectricity, in the last decades, $Si_{0.8}Ge_{0.2}$ has attracted significant attention as an energy harvesting material, for powering space applications and other industrial applications. This chapter focuses on the recent advances and new approaches in silicon-germanium ($Si_{1-x}Ge_x$) nanostructures for thermoelectric devices with high thermoelectric efficiency obtained through magnetron sputtering.

Keywords: silicon-germanium nanostructures, magnetron sputtering deposition, thin films, nanomesh, thermoelectric, Raman spectroscopy

1. Introduction

Several chapters of this book describe different approaches, properties, and applications of silicon and its undeniable impact on our culture, technology, and commerce. Its usefulness has made us talk about the silicon era [1]. In this chapter, we are going to focus on the use of silicon-based materials in one of the main pillars of our life nowadays: the obtainment of energy to power up all the resources in which our society is based (transport, communications, and human infrastructures in general).

Although silicon is mainly associated with microchip devices and advances in computing, the alloy that silicon forms with germanium can be used as a thermoelectric material, which is, in the presence of a gradient of temperature, able to generate an electrical voltage and *vice versa*. This thermoelectric effect has been long known. Nevertheless, it has not been widely used because of its modest efficiency. In recent years, the interest on thermoelectricity has revamped due to the use of thermoelectric devices for micro-energy harvesting or as a large-scale conversion of residual heat into electricity. This increase in the research on thermoelectrics is mostly due to the impact nanostructuration has on improving the efficiency of these materials, which has been increased almost a factor of three over the last 20 years. The purpose of this chapter is to highlight the ways in which silicon-germanium has improved its efficiency by nanostructuration.

Considering the decreasing fossil fuels and increasing energy demand worldwide, a pressing need for improved direct thermal (wasted heat) into electrical energy conversion is imposed. The wasted heat comes from the energy transportation, vehicles, electricity generating sources, industry, etc., which tampers the actual efficiency of the initial resources. For instance, around 30% of the energy obtained from the fuel of a car is actually used in its movement. The other 70% is lost in the form of heat, friction, and cooling the car. Furthermore, it is completely reasonable to look for alternative energy technologies to reduce our dependence on fossil fuels and greenhouse gas effects. This necessity has fostered multiple lines of research, including the conversion of thermal energy through thermoelectricity [2]. As an example, the most recent International Energy Outlook 2016 (IEO2016) [3] prepared by the USA Energy Information Administration shows the energy production predictions for the year 2040, based on the data recorded previously (**Figure 1a**). It is shown that the total world consumption of energy will increase a 48% from 2012 to 2040. Renewable energies are the fastest-growing energy sources over the predicted period, with a foreseen increase in their consumption of around 2.6%/year between 2012 and 2040. In **Figure 1a**, CPP refers to a Clean Power Plan (CPP), which is a USA regulation that aims to reduce carbon dioxide emissions from electric power generation by 32% within 25 years, relative to the levels of 2005 in the USA.

Focusing on the future of the different sources of energies, that is **Figure 1b**, world net electricity generation is envisioned to increase by 69%, in 2040, going from 21.6 trillion kilowatt hours (10^{12} kWh) registered in 2012 to 25.8 trillion kWh predicted for 2020 and to 36.5 trillion kWh in 2040. It is worth noting that, even with initiatives as the CPP, or the development predicted for renewable energies, fossil fuels will still account for a 78% of the energy used in 2040 [3].

For these reasons, in late 2015, representatives from 185 countries and the European Union (EU) met in Paris to reach a commitment to addressing climate change, called Paris-COP21. This worldwide engagement is expected to drive innovation in renewable energies, battery storage, energy efficiency, and energy recovery. One of the main conclusions obtained in the conference is that climate change is often discussed as a single problem, but solving it will require a wide variety of solutions [4]. The EU budget for low carbon-related research under Horizon 2020 has been effectively doubled for the period 2014–2020, and the EU has promised to invest at least 35% of Horizon 2020 resources into climate-related activities [5]. In the United States, hundreds of major companies, including energy-related companies such as ExxonMobil, Shell, DuPont, Rio Tinto, Berkshire Hathaway Energy, Cal-pine, and Pacific Gas and Electric Company, have supported the Paris-COP21 [6]. In the coming decades, there will be a need for more energy-efficient technologies, easily compatible with the non-renewable

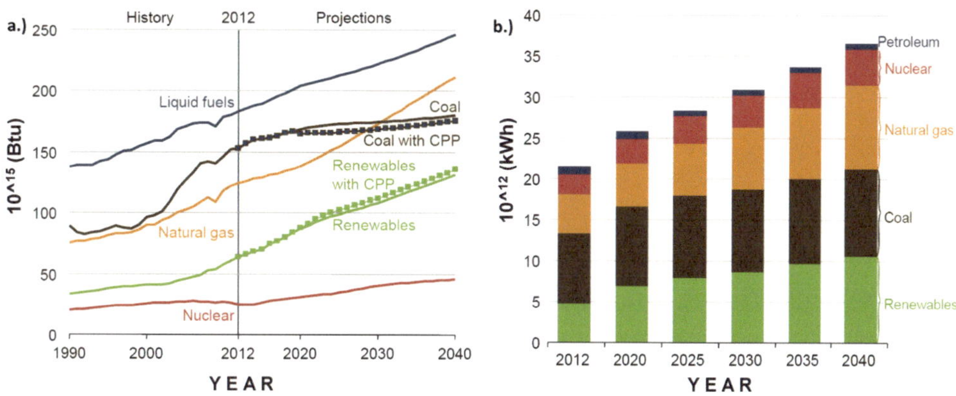

Figure 1. (a) Total world energy consumption sorted by energy source between the period 1990 and 2040. Dotted lines for coal (black) and renewables (green) show the predicted effects of the USA Clean Power Plan (CPP) regulation. (b) World net electricity generation predictions sorted by energy source, for the period of 2012–2040. Both figures are reprinted with permission from Ref. [3]. Copyright 2016.

energies (that will not disappear in the near future as it can be seen in **Figure 1b**). Certainly, thermoelectric materials and especially thin films are interesting players in this scenario. Its ability to convert waste heat into electricity regardless of the source of heat generation, stability over time, and the ability to generate electricity locally without the need for transportation are some of its many advantages.

Likewise, **Figure 2** and **Table 1** show some of the most outstanding historical facts and current state of the art of Si and SiGe in thermoelectric, microelectronic, and photovoltaic applications.

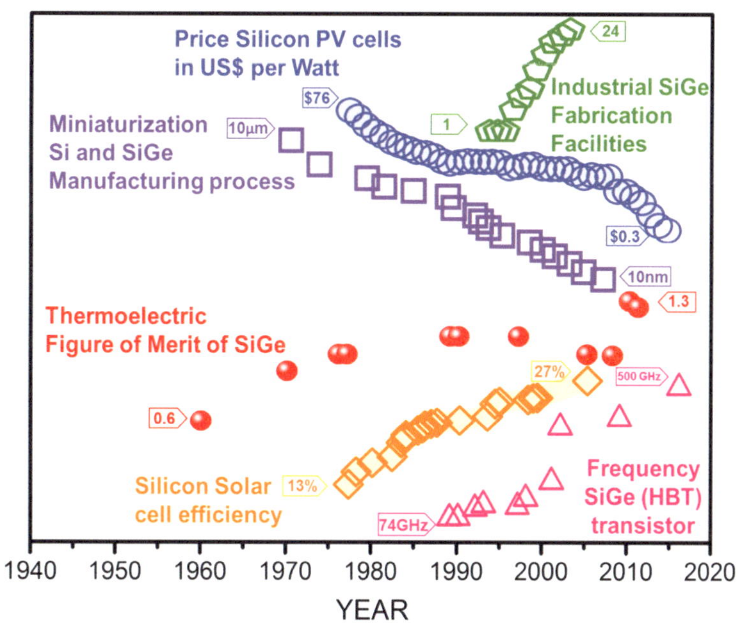

Figure 2. Timeline of some breakthrough or historical event in Si-Ge in thermoelectric, photovoltaic cells and microelectronics. References in **Table 1**.

	Some breakthrough or historical event in SiGe	Year	Refs
Microelectronics and Manufacturing	First epitaxial silicon transistors	1960	[7]
	First oxidation study of SiGe	1971	[8]
	First SiGe n-type MODFET	1986	[9]
	First SiGe p-type MODFET	1986	[10]
	First SiGe photodetector	1986	[11]
	First SiGe HBT (heterojunction bipolar transistor)	1987	[12]
	First SiGe hole RTD (resonant-tunneling diode)	1988	[13]
	First SiGe (BiCFET) (bipolar inversion channel FET)	1989	[14]
	First SiGe HBT grown by CVD (chemical vapor deposition)	1989	[15]
	First SiGe gate (CMOS) technology	1990	[16]
	First SiGe waveguide	1990	[17]
	First SiGe LED	1991	[18]
	First SiGe solar cell	1992	[19]
	First SiGe phototransistor	1993	[20, 21]
	First SiGe HBT with peak cutoff frequency above 100 GHz	1993	[22]
	First SiGe HBT with peak cutoff frequency above 200 GHz	2001	[23]
	First SiGe HBT with peak cutoff frequency above 300 GHz	2002	[7]
	Current Record SiGe HBT with peak cutoff frequency 500 GHz	2016	[24]
Thermoelectric figure of merit	SiGe radioisotope thermoelectric generators (RTGs) Mission LES 8, 9	1976	*[25–28]*
	SiGe (RTGs) in mission Voyager 1 and 2 spacecraft	1977	[25–27]
	SiGe (RTGs) in mission Galileo spacecraft	1989	[25–27, 29]
	SiGe (RTGs) in mission Ulysses spacecraft	1990	[25–27, 29]
	SiGe (RTGs) in mission Cassini spacecraft	1997	[25, 26]
	SiGe (RTGs) in mission New Horizons spacecraft	2005	[25–27]
	Bulk material (zT)~ 1.3 at 1073 K	2014	[30]
	Historical evolution zT SiGe	2016	[31, 32]

http://dx.doi.org/10.5772/67730

	Some breakthrough or historical event in SiGe	Year	Refs
✷	% Solar cell efficiency	1998	[33]
		2003	[34]
		2014	[35]
		2015	[36, 37]
		2016	[38]
$	Price history photovoltaic cells in US$ per watt	2012	[39–41]
		2015	[42, 43]
↓	Recent progress of the miniaturization of semiconductors Si and SiGe	1988	[44]
		2000	[45]
		2004	[46]
		2010	[47]
↑	Number of industrial SiGe and strained Si fabrication facilities	2000	[48]
		2007	[49]

Table 1. This table highlights historical events and the latest advances in silicon and silicon-germanium in thermoelectric, microelectronic, and photovoltaic applications.

2. Thermoelectric concepts: current overview and strategies for improving the thermoelectric efficiency

The efficiency of a thermoelectric material is controlled by its figure of merit, denoted as zT. This parameter is defined as follows:

$$zT = \alpha^2 \cdot \sigma \cdot T \cdot \kappa^{-1} \tag{1}$$

where the parameters are the following: the square of the Seebeck coefficient, α, times the electrical conductivity, σ, times the operating temperature, T (in Kelvin), divided by the thermal conductivity, κ. The thermal conductivity itself is a sum of its lattice and electronic contributions, κ_L and κ_e, respectively.

Most of these parameters are heavily interdependent [50–53], as it is shown in **Figure 3**. If one takes into account the material properties in classical physics, large α usually results in a low σ, and a large σ increases κ_e, given that these parameters depend on the carrier concentration. Therefore, the fabrication of materials with high power factor ($\alpha^2 \cdot \sigma$) and low thermal conductivity (κ) necessary for obtaining a high zT is quite challenging. The energy conversion efficiency (η_{max}) of thermoelectric devices is determined by Eq. (2), with T_H and T_C being the hot and cold temperatures, respectively.

$$n_{max} = \frac{T_H - T_C}{T_H} \frac{\sqrt{1+ZT} - 1}{\sqrt{1+ZT} + \frac{T_C}{T_H}} \tag{2}$$

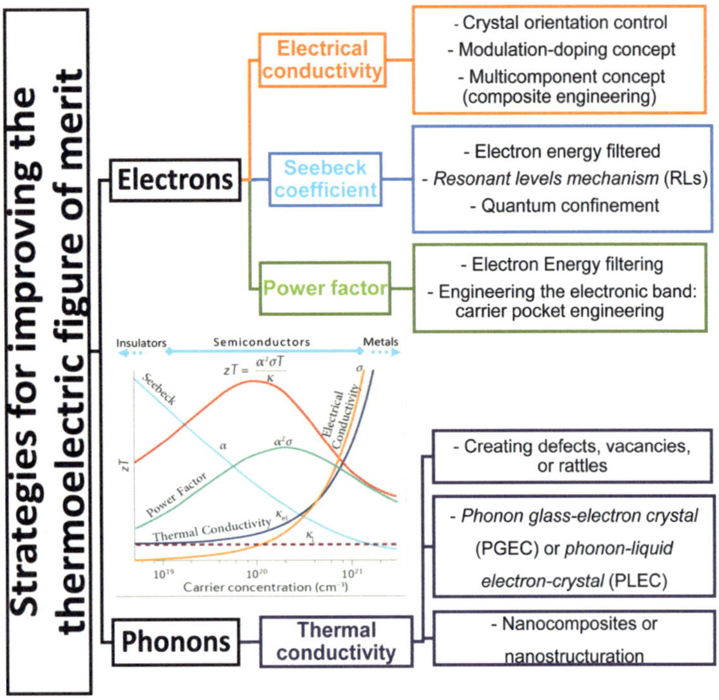

Figure 3. Schematic diagram that briefly summarizes some of the main strategies for the improvement of the figure of merit through the increase in the power factor and the decrease in the thermal conductivity. The graph shows the behavior of the Seebeck coefficient, electrical conductivity, and thermal conductivity versus carrier concentration. This figure is adapted from Ref. [50].

As it can be seen from the definition of the figure of merit, a large value of zT can be obtained by having a high power factor ($\alpha^2 \cdot \sigma$) and a low thermal conductivity (κ). In **Figure 3**, there is also a scheme with some of the strategies that are being used nowadays to improve the figure of merit.

As it can be seen in **Figure 3**, there are two main routes to improve the thermoelectric figure of merit, which are tailoring to improve the power factor and lowering the thermal conductivity. In the first case, there have been different ideas proposed recently. Some of them are aimed to increase the Seebeck coefficient, such as quantum confinement in low-dimensional structures, which was first proposed by Hicks and Dresselhaus in 1993 [54]. It is based on the dependence of the Seebeck coefficient on the gradient of the density of states (DOS) with energy. Then, given that very sharp DOS would be found in quantum confined structures, the Seebeck coefficient would be greatly enhanced. Other approaches to obtain higher Seebeck coefficients are electron energy filtering [55], which proposes the filtering of the electrons with the lowest mean energy, and resonant scattering [56] by introducing distortions into the DOS. In the case of the electrical conductivity, modulation doping has been used to improve carrier mobility [57]. Also, controlling the crystal orientation or composite engineering has shown results in this sense [58]. The main problem is that an increase in the Seebeck coefficient comes along with a decrease in electrical conductivity, as it is the case in energy filtering.

Therefore, other routes to obtain both an increase in the Seebeck coefficient and electrical conductivity have been proposed, such as band engineering [59], and electron energy filtering are nowadays under study. A recent review on all these strategies can be found in Ref. [58].

The other mentioned route to improving the thermoelectric performance is to engineering the structure of the material to reduce lattice thermal conductivity, what is called phononics engineering [60–62]. This last approach can be understood if one takes into account that classical thermoelectric materials are usually semiconductors. Indeed, for metals, κ is dominated by free electrons, whereas in semimetals and heavily doped semiconductors, both κ_L (lattice thermal conductivity) and κ_e (electron thermal conductivity) play an important role in the total thermal conductivity. In particular, in the case of semiconductors, heat is conducted primarily by the acoustic phonons [51, 52, 63]. Undoubtedly, in recent years, there has been an explosion in the research and understanding of the tailoring of thermal conductivity through nanostructure fabrication [51, 52, 64]. In these cases, low thermal conductivity can be achieved by inhibiting the transport of heat through the lattice vibrations, which are called phonons. Phonons can be divided into those having low, medium, or long wavelengths. **Figure 4** depicts how the nano-inclusions, defects, or vacancies significantly reduce the mean free path of the different phonons, thereby reducing the lattice thermal conductivity [64, 65]. In pure materials (non-alloys or doped), the dominant phonon scattering mechanisms go from boundary scattering to phonon-phonon Umklapp scattering with increasing temperature. Then, in order to reduce the thermal conductivity, point inhomogeneities are usually introduced, such as alloy atoms, dopants, isotope variations, rattlers, and point defects. Through these mechanisms, not only phonons, but also electrons are scattered, and thus, the κ is reduced [51, 52, 62, 66]. In the case of nanostructure fabrication, the idea is to form structures with smaller sizes than the phonon mean free paths, but greater than the electron or hole mean free paths, given that phonons are more strongly scattered by the interfaces than are electrons or holes [67], giving

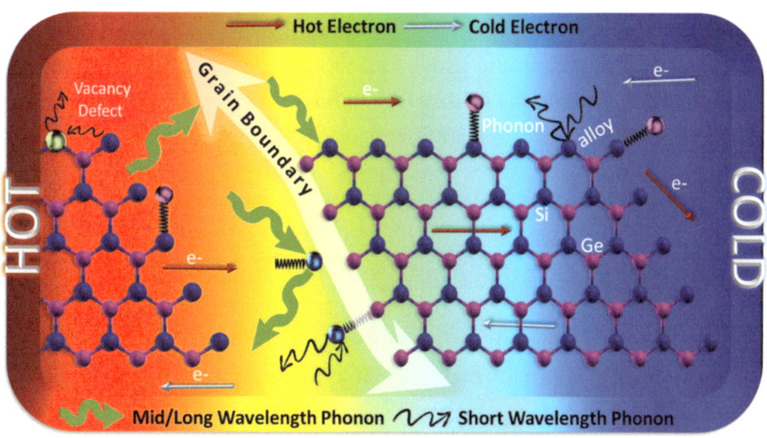

Figure 4. Scheme of the most used strategies for reducing thermal conductivity and their effect on phonon scattering. Grain boundaries scatter mid-long wavelength phonons at their interfaces, while alloy atoms, dopants, defects, lattice vibrations, and nano-inclusions scatter short-wavelength phonons. The electrons, which are depicted as arrows in the figure, are supposedly not scattered and thus electrical conductivity is not altered.

rise to phonon glass-electron crystals (PGEC). As it was said before, the lower the thermal conductivity is, the higher and longer the temperature gradient is maintained and thus more efficient the material results.

3. Thermoelectric properties of silicon-germanium alloys

Silicon-germanium alloys (SiGe) have played a primary role in thermoelectricity in the last decades, although its potential for thermoelectric application was first shown in the 1950s [68–70]. Less than 10 years later, in 1964, a study on how to improve silicon-rich SiGe alloys was published [71, 72], and, the next year, they were used for the first time in a spatial mission from NASA (the SNAP-10 [71]). From that moment onward, they have been successfully used in radioisotope thermoelectric generators (RTGs) for deep-space NASA missions. Bulk silicon-germanium nanostructures (that is, compacted nanograins) are used in the RTGs that power different spacecraft's such as Voyager 1, Voyager 2, Galileo, Ulysses, Cassini, and New Horizons missions [25, 26, 73]. For instance, missions Voyager and Cassini spacecrafts are equipped with RTGs that use a pellet of $^{238}PuO_2$ as the thermal energy source and SiGe as the thermoelectric conversion material. In addition to having very attractive thermoelectric and physical properties, SiGe devices can operate at temperatures up to about 1050°C without significant degradation [25, 73]. For high-temperature applications (above 600°C), SiGe alloys have a high thermoelectric efficiency and have been the type of conduction and the carrier concentration in SiGe can be controlled by doping with phosphorous (*n*-type) or boron (*p*-type). As a consequence, a total of 28 USA space missions have safely flown powered by RTGs [26, 73]. In this field, SiGe used as thermoelectric conversion material has accumulated over 250 million devices working hours in space applications (running for over 40 years in Voyager missions) without failure [25, 26, 74].

In all these years, different studies on how to increase the efficiency of these materials, such as the use of grain-refined alloys [75, 76], nano-inclusions [77], SiGe superlattices fabrication [78], and understanding how the grain size affects thermal conductivity [79], were performed. Also, novel methods for the fabrication of SiGe, such as the chill casting method [80], milling and sintering techniques [81], high-energy ball milling [82], spark plasma sintering [83, 84], and mechanical alloying, were developed. The improvements achieved in SiGe were all related to nanostructuring the material and reducing the lattice thermal conductivity. In 2008, a theoretical work proposed that the introduction of silicide nanoparticles into the SiGe matrix would reduce drastically the thermal conductivity [85]. That is, if the grain size is smaller than the mean free path of the phonons, the total effect is a reduction in the effective mean free path and thus a reduction in the thermal conductivity. Another route studied has been the enhancement of the power factor in SiGe through the concept modulation doping [86]. In this case, a 40% power factor enhancement in $Si_{80}Ge_{20}$ bulk nanocomposites has been reported, and it was a direct result of the enhanced mobility due to this modulated doping [86]. With all these advances, zT values for nanostructured bulk SiGe as high as 1.3 for *n*-type and 0.95 for *p*-type have been measured [87–89].

Apart from these successes in the increase of thermoelectric efficiency and the space applications of SiGe, there is another outstanding property that makes SiGe appealing for many other applications, which is the possibility of integration (compatibility) in the technology of semiconductors based on silicon. This can be made through thin films fabrication of SiGe on silicon like it is done in the complementary metal-oxide semiconductors (CMOS) industry [90]. In general, thermoelectricity struggles with the lack of cheap, abundant, and environmentally friendly materials. Recent works have emphasized the importance of considering the relationship between material's price, manufacturing costs, and efficiency to consider different thermoelectric materials [91]. Silicon-germanium could overcome this deficiency by proposing high harvested power density, abundant on earth, low toxicity, and cost-efficiency. These characteristics increase the interest of SiGe among other thermoelectric materials [90, 91].

3.1. Some strategies for reducing the thermal conductivity of silicon-germanium

The challenge of obtaining ultra-low thermal conductivities in silicon-germanium, in particular, for thermoelectric applications, is not recent. **Figure 5** represents the different strategies that have been followed to fabricate nanostructures with reduced thermal conductivity in SiGe.

In the case of pure silicon and germanium, measurements in bulk, the room temperature thermal conductivities are ~140 W K^{-1} m^{-1} [67] and ~60 W K^{-1} m^{-1}, respectively [30]. However, SiGe alloys provide a significant reduction in thermal conductivity versus the above-mentioned values. Depending on the germanium content, values ranging from ~20 to ~9 W K^{-1} m^{-1} have been measured in bulk [30]. The lowest value of room temperature thermal conductivity has been achieved for a stoichiometry of $Si_{0.8}Ge_{0.2}$ (~9 W K^{-1} m^{-1}), which is still large for thermoelectric applications. Nevertheless, an even lower value (<1 W K^{-1} m^{-1}) has been measured for films grown by sputtering with a $Si_{0.8}Ge_{0.2}$ stoichiometry [92]. The difference with the previous case is that these films were grown through metal-induced crystallization (MIC),

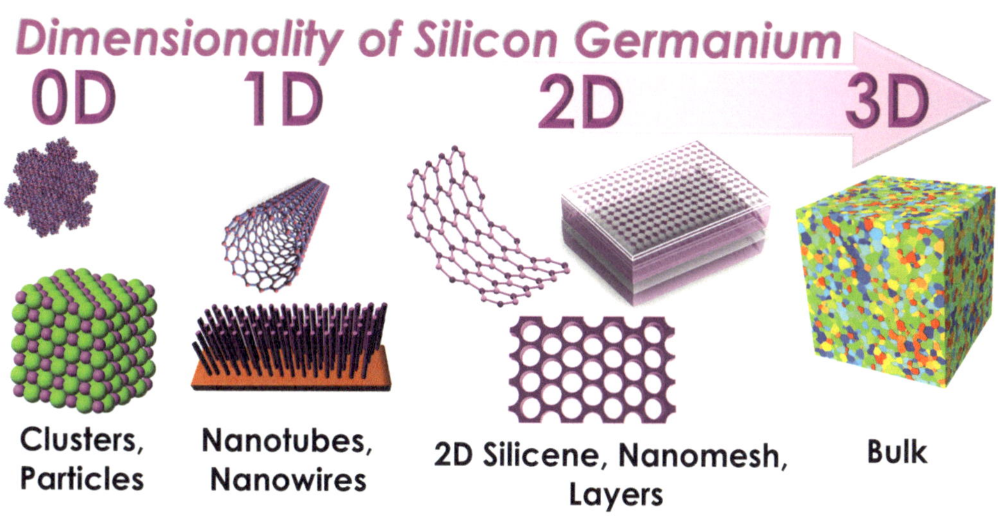

Figure 5. One of the strategies that has been proven to be useful in improving thermoelectric performance is to reduce dimensionality. Here, different configurations that the silicon-germanium has been fabricated at the nanometric scale to improve its thermoelectric properties are shown.

which allows the reduction of the crystallization temperature of the SiGe. With this technique, films with thermal conductivity values down to ~1.2 W K^{-1} m^{-1} at room temperature [92, 93] have been obtained. (See Section 4 for more details.)

At the same time, the growing interest in 2D materials has also triggered the study of 2D silicene [94], which has low thermal conductivity and high power factor. Although the thermal conductivity of silicene has not been measured experimentally due to the difficulty of synthesizing freestanding silicene (and also the complication to carry out thermal measurements in the in-plane direction), several numerical simulations have predicted a thermal conductivity of silicene at room temperature from 5 to 70 W K^{-1} m^{-1} [94–97]. Also graded $Si_{1-x}Ge_x$/Si superlattice structures have been theoretically proposed and fabricated. The idea behind these structures was to demonstrate a thermal rectification effect derived from a theoretical model (the kinetic collective model), which showed that the thermal boundary resistance of a $Si_{1-x}Ge_x$/Si depends on the direction of the heat flow if the structure is symmetric. The predicted effect would cause around 40% difference depending on the heat flow direction. Experimentally, these graded superlattices were fabricated via molecular beam epitaxy (MBE) on silicon substrates, and further studies on the impact of the composition, strain, or alloy inhomogeneities [98] showed that the transport properties could be engineered, obtaining values for the thermal conductivity as low as 2.2 W m^{-1} K^{-1} [99].

Another 2D structure that has recently been developed is the fabrication of nanomeshes (nanoporous or holey silicon or SiGe membranes). These structures can be fabricated by sputtering deposition in large areas [100], which offers the advantages of scalability and flexibility required for real applications [100–102]. Moreover, the variation of the geometry of the mesh influences its thermal conductivity [100, 102], allowing a further control on this parameter. In particular, the thermal conductivity of the nanomeshes was reduced as the diameter of the pores became smaller, achieving values that varied from $\kappa = 1.54 \pm 0.27$ W K^{-1} m^{-1}, down to the ultra-low $\kappa = 0.55 \pm 0.10$ W K^{-1} m^{-1} value [100]. The latter is well below the amorphous limit, while the Seebeck coefficient and electrical conductivity of the material were retained [100]. (More details of these nanomeshed structures will be given in Section 5.)

In addition to phonon transport engineering [59, 60], different technological strategies such as the fabrication multilayers [103] and channels with reduced dimensionality such as 1D nanotubes [104] and 1D nanowires [105] have achieved a significant reduction in the thermal conductivity. Furthermore, several authors have demonstrated that the obvious reduction in cross-plane thermal conductivity in SiGe 0D cluster—particle (quantum dots) [106] superlattices is primarily due to the increased physical roughness at the superlattice interfaces and not due to quantum confinement effects [107, 108].

Figure 6 summarizes the current state of the art for silicon-germanium in terms of thermoelectric properties. Here, the most promising materials in the form of bulk, thin films, nanomeshes, nanowires, and nanotubes are shown. It is worth noting in these figures that the results of our works, which will be explained later, namely the MIC films and the nanomeshes, are among the best-performing materials. In terms of Seebeck coefficient

http://dx.doi.org/10.5772/67730

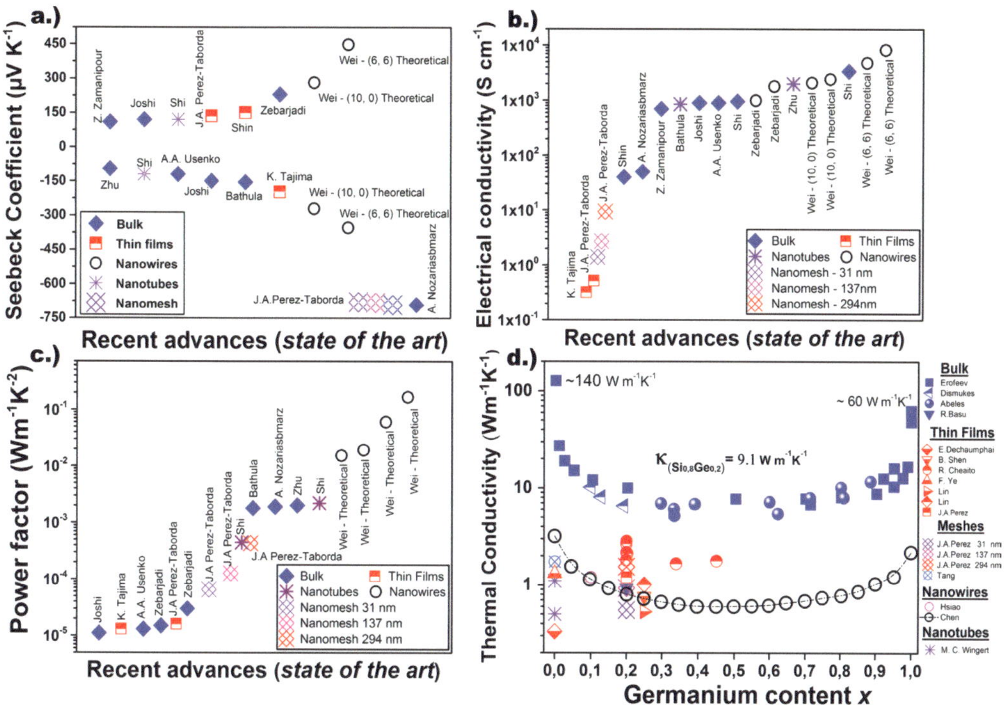

Figure 6. A summary of the latest reported measurements of different structures of Si_xGe_{1-x} is presented, shows (a) Seebeck coefficient, (b) electrical conductivity, and (c) power factor reported for bulk, thin films, nanomeshes, nanowires, and nanotubes. (d) The thermal conductivity for different Si_xGe_{1-x} nanostructures and bulk samples as a function of the alloy composition. This figure is adapted from Ref. [100].

(see **Figure 6a**), the values measured for MIC SiGe thin films are comparable to other values measured in different thin films fabricated with other techniques, while the nanomeshes present the highest Seebeck coefficients, only comparable to values measured in bulk. Nevertheless, the values of electrical conductivity (**Figure 6b**) are quite low, when compared to the values of bulk or nanotubes, but in the order or even better than the values given for thin films. The enhancement of the electrical conductivity within those structures is one of the improvements that are being studied nowadays.

On a whole, the power factor (presented in **Figure 6c**), which takes into account the square of the Seebeck coefficient times the electrical conductivity, shows that the MIC films have power factors compared to other thin films (even higher) and the values achieved in nanomeshes are only overridden by bulk materials and nanotubes [109] (note that the black circles are theoretical calculations, not actual measurements). The last data show the thermal conductivity of different alloy compositions for different kinds of structures (**Figure 6d**). Here, it is worth noting that the values measured for both MIC films and nanomeshes are well below the values measured for crystalline bulk SiGe and among the lowest ever recorded, comparable with the value of the amorphous material (which is 1 W m^{-1} K^{-1}).

4. Thin films: improvement of the thermoelectric performance through the reduction of thermal conductivity of nanocrystalline $Si_{0.8}Ge_{0.2}$ films by sputtering deposition

Silicon-germanium thin films can be easily *p*- or *n*-type doped at room temperature when the material is amorphous. Nevertheless, doping is particularly difficult when the material is crystalline, given that it is usually crystallized at high temperatures. Si_xGe_{1-x} films grown by different techniques, such as low-pressure chemical vapor deposition (LPCVD) or sputtering, turn out to be amorphous unless the deposition itself is performed at very high temperatures [110–112]. Certainly, amorphous SiGe layers are not an option to be used as thermoelectric materials, given their low Seebeck and low electrical conductivity. Therefore, the main challenge with these Si_xGe_{1-x} alloys, which are to be applied in large-scale practical applications, has not yet been overcome due to the difficulties in the growth of high-quality, highly crystalline, low-cost, and appropriately doped films. On that sense, some examples that can be found in the literature obtained in our lab are compiled in this section.

Recently, metal-induced crystallization (MIC) [92, 113, 114] has proved to be an interesting alternative to reduce the crystallization temperature required for SiGe. This process is based on the growth of the films on substrates with Au [115], Ag [116], Al [117], Ni [118], Cr [119], or Sn [120] layer. Then, an appropriate heat treatment is performed, allowing the gold from the film to migrate through the semiconductor film all the ways to the surface. This gives rise to a eutectic mixture. The Au-Si eutectic temperature occurs around 350°C, Au-Ge being at around 361°C [121]. Using this MIC technique, quite promising results have been reported recently for thin films of boron-doped $Si_{0.8}Ge_{0.2}$ (*n*-type) grown by sputtering, resulting in films with a good power factor and a very low thermal conductivity [92]. In that work, two different approaches were followed: (i) *in situ* MIC (depositing the films at different controlled temperatures during the sputtering process) and (ii) *ex situ* MIC (deposition of the films at room temperature in the sputtering chamber and subsequently post-annealing in an external furnace under a controlled atmosphere) [92].

The structural evolution from amorphous to crystalline as a function of the different treatment temperatures can be observed in **Figure 7** through the Raman spectra, both for *in situ* (left hand, red color) and *ex situ* (right hand, blue color) MIC films for different temperatures. It can be seen that the vibration modes appear as broad bands for room temperature treatments (see **Figure 7a** and **e**), which means that the material is amorphous. Then, the peaks become narrower as the treatment temperatures increase. Moreover, the Si-Si vibrational peak shows a clear red shift for the highest temperature (500°C) of the *ex situ* samples (**Figure 7d**). This shift may be related to the formation of silicon-rich clusters. Moreover, the relative intensities and frequencies corresponding to the main peaks present in the Raman spectra are strongly dependent on the alloy composition.

A closer look at the Si-Si peak reveals that it is, in fact, a convolution of two peaks; a very narrow peak corresponding to the Raman spectrum of crystalline silicon-rich SiGe and a broader, smaller peak corresponding to the Si-Si vibrations are typically found in $Si_{0.8}Ge_{0.2}$. This could be an indication that silicon is partially segregated in the *ex situ* samples. The peaks observed in the *in situ* samples are narrower than those of the *ex situ* annealed samples, which confirms the high

http://dx.doi.org/10.5772/67730

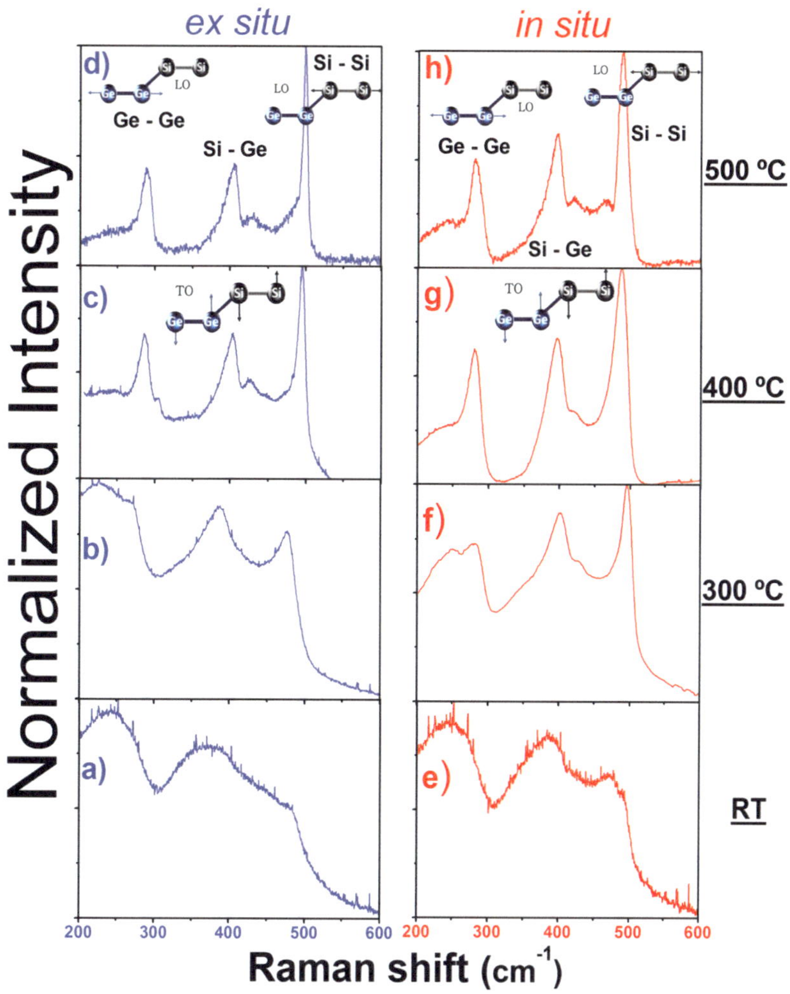

Figure 7. Raman spectra of thin films deposited on gold/glass substrates: *ex situ* thermal treatment (in blue: a, b, c, and d) and *in situ* thermal treatment (in red: e, f, g, and h) for samples treated at RT, 300, 400, and 500°C, respectively. The expected vibrational bands (schematically represented) corresponding to Ge-Ge, Si-Si, and Si-Ge bonds are marked on the figure. With permission from Ref. [92].

degree of crystalline order. Furthermore, these results clearly indicate that whereas, in the *in situ* treatment, the crystallization starts at 300°C, the crystallization onset is lower for *ex situ* treatments. It is interesting to note that in the 400–500 cm^{-1} region, secondary modes start to appear at high temperatures. These modes might be associated with the formation of a compositional gradient due to segregation of Si and Ge, which promotes predominantly Si cluster formation. These clusters would remain embedded in the SiGe matrix when the post-annealing is performed.

Then, the structural analysis by synchrotron radiation-grazing incidence X-ray diffraction (SR-GIXRD) for samples treated at 500°C is shown in **Figure 8**. In order to perform the XRD study, the gold layer was also selectively removed by potassium iodide etching. Nevertheless, gold diffraction maxima are dominant in the *ex situ* treated sample, which means that not all the gold was removed. This indicates that the gold, instead of migrating completely to the

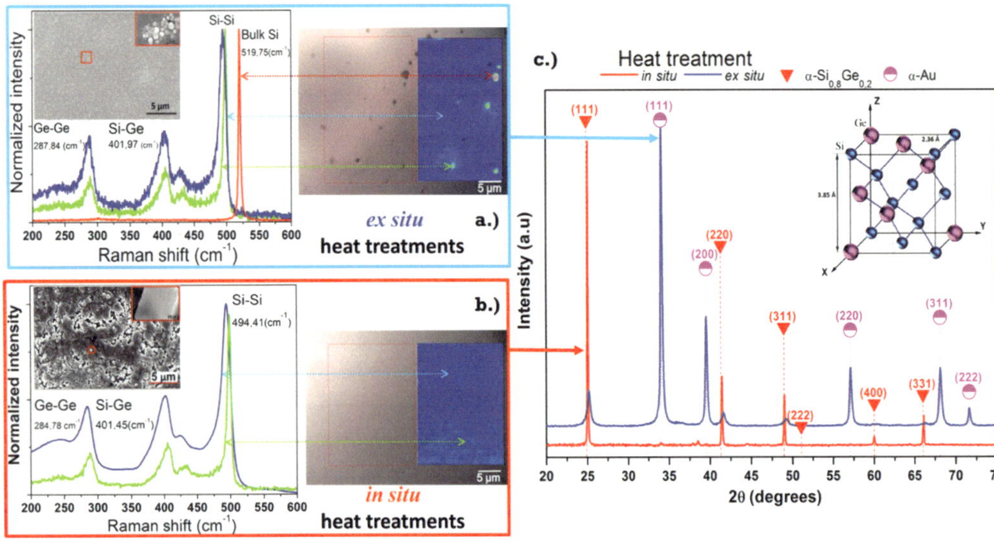

Figure 8. (a) and (b) present different measurements for the *ex situ* (a) and *in situ* (b) MIC fabricated films with 500°C treatment temperatures. On the right side, the optical image of the surface along with a Raman mapping of the surface is presented. In the left side, the different Raman spectra collected corresponding to $Si_{0.8}Ge_{0.2}$ (blue color), nano-$Si_{1-x}Ge_x$ (green color), and pure silicon (red color) are shown (note that for the *in situ* film, b), there is no evidence of silicon segregation). The inset at the left side presents SEM image of the film surface. (c) Synchrotron radiation SR-GIXRD diffractograms measured at 1.3775 Å wavelength for *ex situ* (blue color) and *in situ* (red color) MIC fabricated films, with heat treatments at 500°C. The heights of the intensities in dotted lines correspond to the Si-Ge phase intensity values given in the JCPDS 04-016-6750 data sheet. The inset shows the calculated lattice parameters for the Si-Ge films. This figure is adapted from Ref. [92].

surface during the *ex situ* MIC treatment, part of it stayed, trapped inside the film. In the case of samples deposited *in situ* at temperatures of 500°C, the diffraction peak intensities at (111), located at 2θ=25.33° for the synchrotron radiation source, are higher than the intensities of samples treated *ex situ*.

The low values of thermal conductivities (1.13 and 1.23 W m^{-1} K^{-1} for *in situ* and *ex situ* thermal treated at 500°C, respectively) obtained in Ref. [92] have been associated with the formation of Si-rich SiGe and Si clusters during the gold-induced crystallization, which creates plenty of phonon scattering sites at the grain boundaries. The best power factors were achieved for samples grown at 500°C, that is, *in situ* MIC. The results indicate a maximum of 16 μW m^{-1} K^{-2} at 315°C, which is the best-reported value, to date, for SiGe films grown by DC sputtering with Au-MIC—similar to the state-of-the-art values available in the literature for Si-Ge bulk samples. This is due to the fact that this sample is not contaminated with gold and also that the doping has not been lost by this thermal treatment.

In the same way, these results also suggest two different mechanisms of induced crystallization dependent on the type of heat treatment (*ex situ* and *in situ* MIC). For the *ex situ* samples, the gold layer travels through the Si-Ge film grown at RT when heated afterward at 500°C, while in the case of *in situ* treatment, a eutectic is formed and the nanocrystalline Si-Ge film seems to be formed underneath.

5. Nanomeshes: record low thermal conductivities in large-area nanoporous $Si_{0.8}Ge_{0.2}$ for enhanced thermoelectric applications

Another recent example of a reduction in thermal conductivity by using the low-dimensional concept has been recently reported [100]. These large-area nanomeshed films were fabricated by DC sputtering of $Si_{0.8}Ge_{0.2}$ on highly ordered porous alumina matrices (see **Figure 9a**), in such a way that the formed $Si_{0.8}Ge_{0.2}$ film replicated the porous alumina structure, resulting in the nanomeshed films shown in **Figure 9b**. A very good control of the nanomesh geometrical features (pore diameter, pitch, and neck) was achieved thanks to the alumina templates used, with pore diameters ranging from 294 ± 5 nm down to 31 ± 4 nm. The method developed is able to provide large areas of nanomeshes in a straightforward and reproducible way, being easily scalable for industrial applications.

Most importantly, as shown in **Figure 10a**, the thermal conductivity of the films was reduced as the diameter of the porous became smaller, achieving values that varied from $\kappa = 1.54 \pm 0.27\ W\ K^{-1}\ m^{-1}$, down to the record low $\kappa = 0.55 \pm 0.10\ W\ K^{-1}\ m^{-1}$ value. The latter is well below the amorphous limit, while both the Seebeck coefficient and electrical conductivity of the material were maintained (see **Figure 10b**).

Likewise, as in the previous case for the nanocrystalline $Si_{0.8}Ge_{0.2}$ films grown by sputtering deposition, the nanomeshed SiGe films were oriented along the [111] direction, as revealed by XRD measurements (see **Figure 11a**). Raman spectra showed the three characteristic

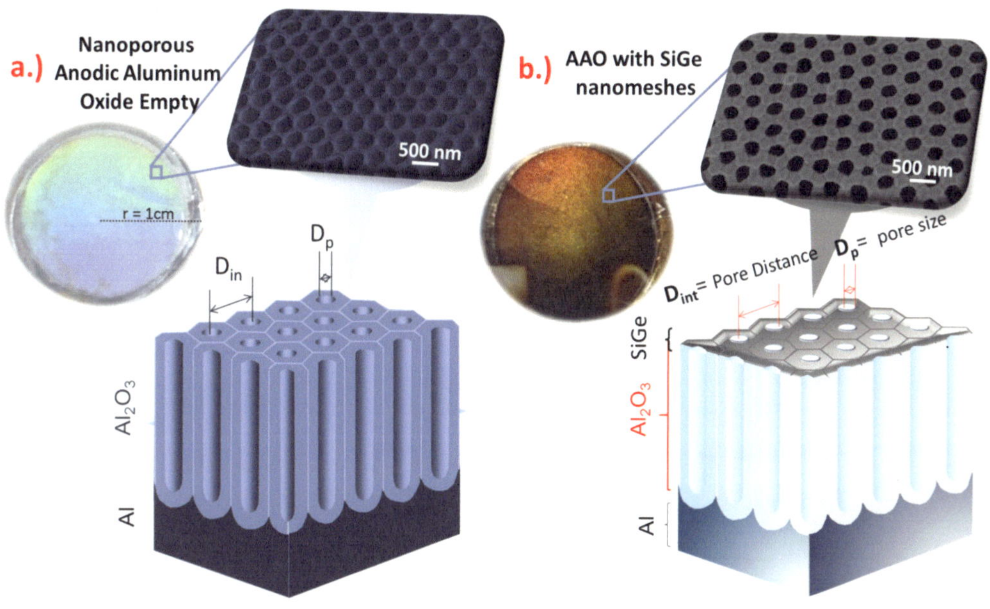

Figure 9. (a) Sketch and optical image of a porous alumina template and (b) the SiGe film nanomesh fabricated on top of it.

vibrational modes obtained for polycrystalline SiGe (see **Figure 11b**) and showed a homogenous phase in all the film.

Additionally, the chemical/surface potential of the $Si_{0.8}Ge_{0.2}$ nanomeshed films was studied by Kelvin probe microscopy (KPM). **Figure 12a** and **b** shows the SEM image and the AFM surface topography of a 294 ± 5 nm porous size nanomeshed film, which presents a homogeneous profile of the surface potential (see **Figure 12c**). This indicates that the work function of the films is homogeneous, confirming the homogeneity in the chemical composition obtained by Raman, and no potential drop is observed at the grain boundaries.

On the one hand, as far as the thermal conductivity is concerned, it is highly reduced when compared to bulk or thin films. This reduction is due to alloying, phonon boundary scattering on the upper/lower boundaries, and crystallite boundaries within the nanomeshes. Moreover, the measurements showed that the smaller the pore diameter is, the larger the thermal conductivity reduction in the $Si_{0.8}Ge_{0.2}$ nanomesh. This can be understood as a result of the enhanced scattering on the pore boundaries, along with the higher disorder or even coherent phonon effects that could be playing a role in the nanomeshed structures, when compared to plain films. Using this approach, it is possible to control thermal transport of these films

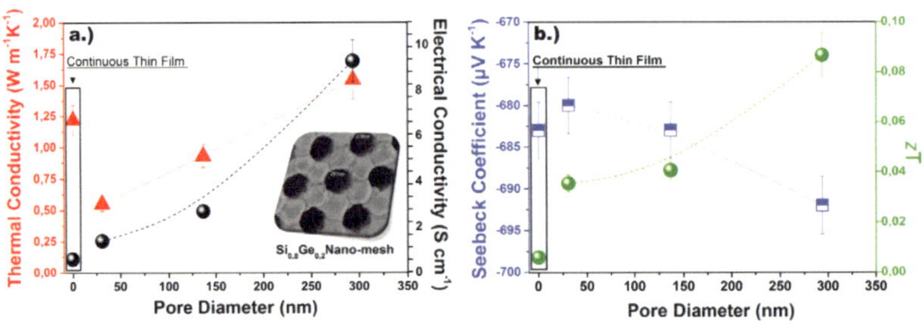

Figure 10. (a) Thermal conductivity (*κ*, red triangles) and electrical conductivity (*σ*, black spheres) and (b) Seebeck coefficient (*S*, blue squares) and figure of merit (*zT*, green spheres) plotted versus the pore diameter of the nanomesh. The transport properties obtained for a $Si_{0.8}Ge_{0.2}$ film grown under the same conditions are also plotted for comparison (inside the rectangle on the left of each graph, corresponding to continuous thin film). This figure is adapted from Ref. [100].

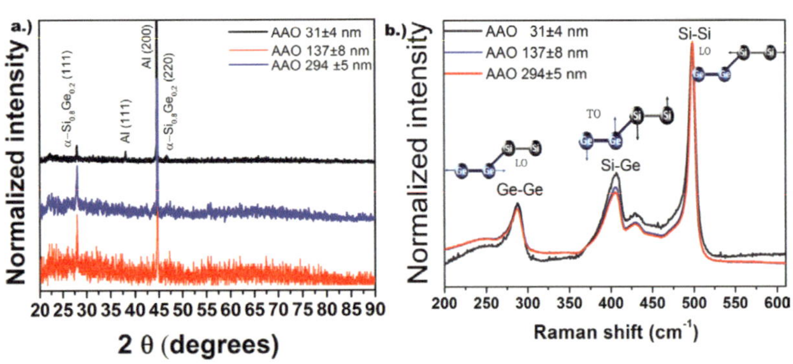

Figure 11. (a) X-ray diffraction and (b) Raman spectra of a $Si_{0.8}Ge_{0.2}$ grown on nanomeshes with a pore diameter of 31 nm (black line) 137 nm (blue line) and 294 nm (red line). This figure is adapted from Ref. [100].

http://dx.doi.org/10.5772/67730

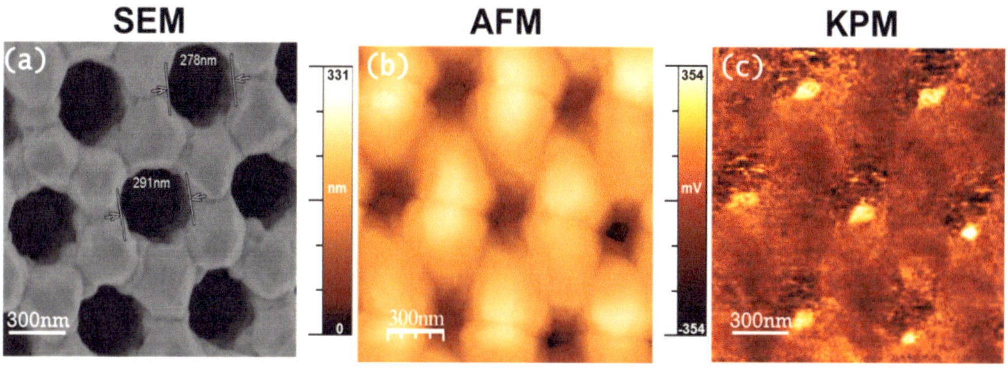

Figure 12. (a) SEM image of a $Si_{0.8}Ge_{0.2}$ nanomeshed film of 294 ± 5 nm pore size. (b) Topography image by AFM and (c) surface potential image by KPM. The uniformity in the contrast of the KPM image reveals homogeneity in the surface potential of the film. This figure is adapted from Ref. [100].

through nano-engineering. Moreover, the power factors of the nanomeshes are higher in the structures with larger pores (and larger distances between the pores), and consequently, they are found smaller in the more disordered structures, which comprise denser pore structure and smaller pore diameters.

The power factors are found to be between 65 and 455 μW K^{-2} m^{-1}, which seem to be as large as some of the last reported values for bulk $Si_{0.8}Ge_{0.2}$. This is attributed to the fact that the electrical conductivity in the nanomeshes with large interpore distance is much larger than the more dense nanomeshes, whereas the Seebeck coefficient remains almost the same [100]. while still retaining reasonable power factors, which opens the door for efficient thermoelectric applications for this alloy.

6. Concluding remarks and future directions

In summary, this chapter has shown how the nanostructuration of SiGe takes advantage of the reduction in the lattice thermal conductivity while maintaining the thermoelectric properties of the material, which makes the material quite competitive with others conventionally used. Moreover, the two examples of nanostructuration that have been described here in more detail, namely the sputtered MIC films and the nanomeshes present several advantages over other techniques, such as the possibility of coating large areas thanks to the sputtering process, which is also industrially scalable. Furthermore, the sputtering onto alumina templates, which gives rise to the nanomeshes, can also cover large areas, given that the aluminum oxide templates can be fabricated over large-area aluminum substrates. In the case of nanomeshes, the drastic reduction in the thermal conductivity achieved is due to alloying, phonon boundary scattering on the upper/lower boundaries, and crystallite boundaries. Therefore, this makes the method cost-effective to be scaled into the industry. Another key thing to remember is that both nanostructures (MIC films and nanomeshes) are compatible with silicon technology, opening the door to applications in electronic devices, which need to have thermal dissipation. This provides not only a novel approach to growing this kind of structures in a simple and reliable way, but also an important route toward achieving high conversion efficiency and highly scalable thermoelectric materials.

This chapter presents the most recent advancements in SiGe alloys for its use as efficient thermoelectric material. To this end, it is important to understand the mechanisms that govern the thermal conductivity, in order to engineer the material to reduce it as much as possible without affecting other thermoelectric properties. The thriving expansion of new capabilities of 1D and two-dimensional SiGe has progressed rapidly during the last few years. Although most of the two-dimensional materials have a simple honeycomb lattice structure, understanding the phonon transport mechanism in such atomic thin SiGe seems not an easy task. It is obviously important to recognize that no single technology can meet the world's energy needs in the twenty-first century; one needs a combination of many technologies in which the thermoelectric materials can undoutebtedly play a role. Further, these large-area films or nanomeshes provide a novel approach to growing nanostructured thermoelectric materials in a simple and reliable way.

Author details

Jaime Andrés Pérez-Taborda, Olga Caballero-Calero and Marisol Martín-González*

*Address all correspondence to: marisol@imm.cnm.csic.es

IMM-Instituto de Microelectrónica de Madrid (CNM-CSIC), Madrid, Spain

References

[1] D.C. Brock (Ed.), Moore's law at forty. *Understanding Moore's law: four decades of innovation*, Chemical Heritage Foundation (2006) last access: https://issuu.com/chemheritage/docs/understanding-moores-law?e=1220

[2] M. Martín-González, O. Caballero-Calero, P. Díaz-Chao, Renewable and Sustainable Energy Reviews **24**, 288 (2013).

[3] J. Conti, International energy outlook 2016 Report No. DOE/EIA-0484 US Energy Information Administration (eia), Washington DC (2016).

[4] P. Fragkos, N. Tasios, L. Paroussos, P. Capros, S. Tsani, Energy Policy **100**, 216 (2017).

[5] Communication from the Commission to the European Parliament, The Council. The Road from Paris: assessing the implications of the Paris Agreement and accompanying the proposal for a Council decision on the signing, on behalf of the European Union, of the Paris agreement adopted under the United Nations Framework Convention on Climate Change. European Commission, Brussels, 2.3.2016 COM(2016) 110 final last access: https://ec.europa.eu/transparency/regdoc/rep/1/2016/EN/1-2016-110-EN-F1-1.PDF

[6] B. Obama, The irreversible momentum of Clean energy, Science 355 (January (6321)) (2017) 126–129

[7] H. Theuerer, J. Kleimack, H. Loar, H. Christensen, Proceedings of the Institute of Radio Engineers **48**, 1642 (1960).

[8] P. Balk, Journal of the Electrochemical Society **118**, 494 (1971).

[9] H. Daembkes, H.-J. Herzog, H. Jorke, H. Kibbel, E. Kasper, IEEE Transactions on Electron Devices **33**, 633 (1986).

[10] T. Pearsall, J. C. Bean, IEEE Electron Device Letters **7**, 308 (1986).

[11] H. Temkin, T. Pearsall, J. Bean, R. Logan, S. Luryi, Applied Physics Letters **48**, 963 (1986).

[12] G. L. Patton, S. S. Iyer, S. Delage, S. Tiwari, J. Stork, IEEE Electron Device Letters **9**, 165 (1988).

[13] H. Liu, D. Landheer, M. Buchanan, D. Houghton, Applied Physics Letters **52**, 1809 (1988).

[14] R. Taft, J. D. Plummer, S. Iyer, IEEE Electron Device Letters **10**, 14 (1989).

[15] T.-J. King, J. R. Pfiester, J. D. Shott, J. P. McVittie, K. C. Saraswat, in *Electron Devices Meeting, 1990. IEDM'90. Technical Digest., International* (IEEE, 1990), p. 253.

[16] R. Soref, F. Namavar, J. Lorenzo, Optics Letters **15**, 270 (1990).

[17] D. Houghton, J.-P. Noël, N. Rowell, Materials Science and Engineering: B **9**, 237 (1991).

[18] D. Shen, J. Conde, V. Chu, S. Aljishi, J. Z. Liu, S. Wagner, IEEE Electron Device Letters **13**, 5 (1992).

[19] S.-B. Hwang, Y. Fang, K.-H. Chen, C.-R. Liu, J.-D. Hwang, M.-H. Chou, IEEE Transactions on Electron Devices **40**, 721 (1993).

[20] E. Kasper, A. Gruhle, H. Kibbel, *in Electron Devices Meeting, 1993. IEDM'93. Technical Digest., International* (IEEE, 1993), p. 79.

[21] E. Crabbe, B. Meyerson, J. Stork, D. Harame, in *Electron Devices Meeting, 1993. IEDM'93. Technical Digest., International* (IEEE, 1993), p. 83.

[22] S. Jeng et al., IEEE Electron Device Letters **22**, 542 (2001).

[23] J.-S. Rieh et al., in Electron Devices Meeting, 2002. IEDM'02. International (IEEE, 2002), p. 771.

[24] J. D. Cressler, C. Coen, S. Zeinolabedinzadeh, P. Song, R. Schmid, M. Oakley, P. Chakraborty, in *2016 IEEE Compound Semiconductor Integrated Circuit Symposium (CSICS)*2016), p. 1.

[25] R. R. Furlong, E. J. Wahlquist, Nuclear News **42**, 26 (1999).

[26] George R. Schmidt, Robert L. Wiley, Rebecca L. Richardson, and Richard R. Furlong, AIP Conference Proceedings 746, 429 (2005); doi: http://dx.doi.org/10.1063/1.1867159

[27] PATEL, M. R. *Spacecraft Power Systems* CRC-Press, Boca Raton, Florida/USA, 2004. ISBN 0-8493-2768-5.

[28] C. Kelly, in Space Nuclear Power Systems (1987).

[29] G. L. Bennett, C. W. Whitmore, W. R. Amos, "On the Development of the Power Sources for the Ulysses and Galileo Missions", Proceedings of the European Space Power Conference, 1989-October-2-6.

[30] R. Basu et al., Journal of Materials Chemistry A **2**, 6922 (2014).

[31] M. Haras, V. Lacatena, T. M. Bah, S. Didenko, J. F. Robillard, S. Monfray, T. Skotnicki, E. Dubois, IEEE Electron Device Letters **37**, 1358 (2016).

[32] X. Zhang, L.-D. Zhao, Journal of Materiomics **1**, 92 (2015).

[33] Goetberger, Adolf; Knobloch, Joachim; Voss, Bernhard. Crystalline silicon solar cells. editorial John Wiley & Sons Ltd, 1998, vol. 1.

[34] A. Martí, A. Luque, Next Generation Photovoltaics: High Efficiency through Full Spectrum Utilization, edited by A. Martí and A. Luque, Eds.; Institute of Physics, Bristol, 2003.

[35] L.M. Fraas Low-Cost Solar Electric Power Springer International Publishing, Cham (2014) http://dx.doi.org/10.1007/978-3-319-07530-3

[36] U. Rau and T. Kirchartz, in Photon Management in Solar Cell, edited by R. B. Wehrspohn, U. Rau , and A. Gombert (Wiley-VCH, Weinheim, 2015)

[37] M. A. Green, K. Emery, Y. Hishikawa, W. Warta, E. D. Dunlop, Progress in Photovoltaics: Research and Applications **23**, 1 (2015).

[38] Green, M. A., Emery, K., Hishikawa, Y., Warta, W., & Dunlop, E. D. (2015). Solar cell efficiency tables (Version 45). Progress in photovoltaics: research and applications, 23(1), 1–9.

[39] P. J. Reddy, *Solar Power Generation: Technology, New Concepts & Policy* (CRC Press, 2012).

[40] G. Carr, The Economist **11**, 21 (2012).

[41] D. M. Powell, M. T. Winkler, H. Choi, C. B. Simmons, D. B. Needleman, T. Buonassisi, Energy & Environmental Science **5**, 5874 (2012).

[42] D. Feldman et al., Photovoltaic System Pricing Trends: Historical Recent Near-Term Projections 2013 Edition, Oak Ridge, TN, USA:U.S. Dept. Energy, 2013.

[43] M. P. Paranthaman, W. Wong-Ng, R. N. Bhattacharya, *Semiconductor materials for solar photovoltaic cells* (Springer, 2015), Vol. 218.

[44] R. Keyes, IBM Journal of Research and Development **32**, 84 (1988).

[45] P. S. Peercy, Nature **406**, 1023 (2000).

[46] M. Ieong, B. Doris, J. Kedzierski, K. Rim, M. Yang, Science **306**, 2057 (2004).

[47] T. Ihn, *Semiconductor Nanostructures: Quantum States and Electronic Transport* (Oxford University Press, 2010).

[48] Y. Nishi, R. Doering, *Handbook of Semiconductor Manufacturing Technology* (CRC Press, 2000).

[49] J. Cressler, *in SiGe and Si Strained-Layer Epitaxy for Silicon Heterostructure Devices* (CRC Press, 2007).

[50] W. G. Zeier, J. Schmitt, G. Hautier, U. Aydemir, Z. M. Gibbs, C. Felser, G. J. Snyder, Nature Reviews Materials **1**, 16032 (2016).

[51] E. S. Toberer, L. L. Baranowski, C. Dames, Annual Review of Materials Research **42**, 179 (2012).

[52] A. Shakouri, Materials Research **41**, 399 (2011).

[53] A. Moure, M. Rull-Bravo, B. Abad, A. Del Campo, M. M. Rojo, M. H. Aguirre, A. Jacquot, J. F. Fernandez, M. Martin-Gonzalez, Nano Energy **31**, 393 (2017).

[54] L. Hicks, M. Dresselhaus, Physical Review B **47**, 12727 (1993).

[55] D. Vashaee, A. Shakouri, Physical Review Letters **92**, 106103 (2004).

[56] J. P. Heremans, V. Jovovic, E. S. Toberer, A. Saramat, K. Kurosaki, A. Charoenphakdee, S. Yamanaka, G. J. Snyder, Science **321**, 554 (2008).

[57] Q. Hou, B. Gu, Y. Chen, Y. He, J. Sun, Applied Physics A **114**, 943 (2014).

[58] A. M. Dehkordi, M. Zebarjadi, J. He, T. M. Tritt, Materials Science and Engineering: R: Reports **97**, 1 (2015).

[59] Y. Pei, H. Wang, G. Snyder, Advanced Materials **24**, 6125 (2012).

[60] W. Kim, Journal of Materials Chemistry C **3**, 10336 (2015).

[61] G. Zhang, Y.-W. Zhang, in *Nanoscale Thermoelectrics*, edited by X. Wang, and Z. M. Wang (Springer International Publishing, Cham, 2014), p. 185.

[62] Wan, C., Wang, Y., Wang, N., Norimatsu, W., Kusunoki, M., & Koumoto, K. (2010). Development of novel thermoelectric materials by reduction of lattice thermal conductivity. Science and Technology of Advanced Materials, 11(4), 044306.

[63] A. Cantarero, F. X. Alvarez, in *Nanoscale Thermoelectrics* (Springer, 2014), p. 1.

[64] M. G. Kanatzidis, Chemistry of Materials **22**, 648 (2009).

[65] N. Satyala, P. Norouzzadeh, D. Vashaee, in *Nanoscale Thermoelectrics* (Springer, 2014), p. 141.

[66] M. Zebarjadi, K. Esfarjani, M. Dresselhaus, Z. Ren, G. Chen, Energy & Environmental Science **5**, 5147 (2012).

[67] A. Minnich, M. Dresselhaus, Z. Ren, G. Chen, Energy & Environmental Science **2**, 466 (2009).

[68] M. Steele, F. Rosi, Journal of Applied Physics **29**, 1517 (1958).

[69] V. A. Johnson, K. Lark-Horovitz, Physical Review **92**, 226 (1953).

[70] A. Middleton, W. Scanlon, Physical Review **92**, 219 (1953).

[71] D. M. Rowe, *CRC Handbook of Thermoelectrics* (CRC Press, 1995).

[72] J. Dismukes, L. Ekstrom, E. Steigmeier, I. Kudman, D. Beers, Journal of Applied Physics **35**, 2899 (1964).

[73] G. R. Schmidt, R. D. Abelson, R. L. Wiley, in SPACE TECHNOLOGY AND APPLICATIONS INT. FORUM-STAIF 2005: Conf. Thermophys in Micrograv; Conf Comm/Civil Next Gen. Space Transp; 22nd Symp Space Nucl. Powr Propuls.; Conf. Human/Robotic Techn. Nat'l Vision Space Expl.; 3rd Symp Space Colon.; 2nd Symp. New Frontiers (AIP Publishing, 2005), p. 295.

[74] C. Vining, *CRC Handbook of Thermoelectrics*, ed. Rowe, DM, CRC Press, Boca Raton, 328 (1995).

[75] H. Goldsmid, A. Penn, Physics Letters A **27**, 523 (1968).

[76] J. Parrott, Journal of Physics C: Solid State Physics **2**, 147 (1969).

[77] Z. Tan, W. Jesser, F. Rosi, Materials Science and Engineering: B **33**, 195 (1995).

[78] Cook, J. Harringa, S. Loughin, Materials Science and Engineering: B **41**, 280 (1996).

[79] Bhandari, D. Rowe, Journal of Physics C: Solid State Physics **11**, 1787 (1978).

[80] G. S. Nolas, J. Sharp, J. Goldsmid, *Thermoelectrics: basic principles and new materials developments* (Springer Science & Business Media, 2013), Vol. 45.

[81] R. Bunce, D. Rowe, Journal of Physics D: Applied Physics **10**, 941 (1977).

[82] C. B. Vining, W. Laskow, J. O. Hanson, R. R. Van der Beck, P. D. Gorsuch, Journal of Applied Physics **69**, 4333 (1991).

[83] T. Noguchi, *in Thermoelectrics*, 1997. Proceedings ICT'97. XVI International Conference on (IEEE, 1997), p. 207.

[84] D. Thompson, D. Hitchcock, A. Lahwal, T. Tritt, Emerging Mater Res **1**, 299 (2012).

[85] N. Mingo, D. Hauser, N. Kobayashi, M. Plissonnier, A. Shakouri, Nano Letters **9**, 711 (2009).

[86] M. Zebarjadi et al., Nano Letters **11**, 2225 (2011).

[87] G. Joshi et al., Nano Letters **8**, 4670 (2008).

[88] X. Wang et al., Applied Physics Letters **93**, 193121 (2008).

[89] A. Minnich, H. Lee, X. Wang, G. Joshi, M. Dresselhaus, Z. Ren, G. Chen, D. Vashaee, Physical Review B **80**, 155327 (2009).

[90] M. Haras, V. Lacatena, T. M. Bah, S. Didenko, J.-F. Robillard, S. Monfray, T. Skotnicki, E. Dubois, IEEE Electron Device Letters **37**, 1358 (2016).

[91] S. LeBlanc, S. K. Yee, M. L. Scullin, C. Dames, K. E. Goodson, Renewable and Sustainable Energy Reviews **32**, 313 (2014).

[92] J. P. Taborda, J. Romero, B. Abad, M. Muñoz-Rojo, A. Mello, F. Briones, M. M. Gonzalez, Nanotechnology **27**, 175401 (2016).

[93] A. A. Wilson, M. M. Rojo, B. Abad, J. A. Perez, J. Maiz, J. Schomacker, M. Martín-Gonzalez, D.-A. Borca-Tasciuc, T. Borca-Tasciuc, Nanoscale **7**, 15404 (2015).

[94] M. Hu, X. Zhang, D. Poulikakos, Physical Review B **87**, 195417 (2013).

[95] Y. D. Kuang, L. Lindsay, S. Q. Shi, G. Zheng, Nanoscale **8**, 3760 (2016).

[96] X. Gu, R. Yang, Journal of Applied Physics **117**, 025102 (2015).

[97] X. Zhang, H. Xie, M. Hu, H. Bao, S. Yue, G. Qin, G. Su, Physical Review B **89**, 054310 (2014).

[98] J. Reparaz, I. Marcus, A. Goñi, M. Garriga, M. Alonso, Journal of Applied Physics **112**, 023512 (2012).

[99] P. Ferrando-Villalba et al., Nano Research **8**, 2833 (2015).

[100] J. A. Perez-Taborda, M. Muñoz Rojo, J. Maiz, N. Neophytou, M. Martin-Gonzalez, Scientific Reports **6**, 32778 (2016).

[101] J.-K. Yu, S. Mitrovic, D. Tham, J. Varghese, J. R. Heath, Nature Nanotechnology **5**, 718 (2010).

[102] J. Tang, H.-T. Wang, D. H. Lee, M. Fardy, Z. Huo, T. P. Russell, P. Yang, Nano Letters **10**, 4279 (2010).

[103] E. Dechaumphai, D. Lu, J. J. Kan, J. Moon, E. E. Fullerton, Z. Liu, R. Chen, Nano Letters **14**, 2448 (2014).

[104] M. C. Wingert, S. Kwon, M. Hu, D. Poulikakos, J. Xiang, R. Chen, Nano Letters **15**, 2605 (2015).

[105] J. Chen, G. Zhang, B. Li, Applied Physics Letters **95**, 073117 (2009).

[106] J. J. Urban, Nature Nanotechnology **10**, 997 (2015).

[107] P. E. Hopkins, J. C. Duda, C. W. Petz, J. A. Floro, Physical Review B **84**, 035438 (2011).

[108] G. Chen, M. Dresselhaus, G. Dresselhaus, J.-P. Fleurial, T. Caillat, International Materials Reviews (2013).

[109] J. Wei, H. Liu, X. Tan, L. Cheng, J. Zhang, D. Fan, J. Shi, X. Tang, RSC Advances **4**, 53037 (2014).

[110] E. V. Jelenkovic, K. Tong, W. Cheung, S. Wong, B. Shi, G. Pang, Solid State Electron. **50**, 199 (2006).

[111] I. Nakamura, T. Ajiki, H. Abe, D. Hoshi, M. Isomura, Vacuum **80**, 712 (2006).

[112] W. Qin, D. G. Ast, T. I. Kamins, Journal of Applied Physics. **92**, 168 (2002).

[113] M. Lindorf, H. Rohrmann, G. Span, S. Raoux, J. Jordan-Sweet, M. Albrecht, Journal of Applied Physics **120**, 205304 (2016).

[114] M. Lindorf, H. Rohrmann, G. L. Katona, D. L. Beke, H. F. Pernau, M. Albrecht, Materials Today: Proceedings **2**, 557 (2015).

[115] M. Seibt, S. Buschbaum, U. Gnauert, W. Schröter, D. Oelgeschläger, Physical Review Letters **80**, 774 (1998).

[116] T. J. Konno, R. Sinclair, Philosophical Magazine B **71**, 163 (1995).

[117] Z. Wang, J. Wang, L. Jeurgens, E. Mittemeijer, Physical Review B **77**, 045424 (2008).

[118] K. U. M. Kumar, R. Brahma, M. G. Krishna, A. K. Bhatnagar, G. Dalba, Journal of Physics: Condensed Matter **19**, 496208 (2007).

[119] K. Kumar, M. G. Krishna, Journal of Nanomaterials **2008**, 57 (2008).

[120] S. Raoux, H.-Y. Cheng, J. L. Jordan-Sweet, B. Muñoz, M. Hitzbleck, Applied Physics Letters **94**, 183114 (2009).

[121] G. Gerlach, W. Dotzel, *Introduction to Microsystem Technology: A Guide for Students* (Wiley Microsystem and Nanotechnology) (Wiley Publishing, 2008).

Chapter 9

Silicon Growth Technologies for PV Applications

Guilherme Manuel Morais Gaspar,
Antoine Autruffe and José Mário Pó

Additional information is available at the end of the chapter

http://dx.doi.org/10.5772/intechopen.68351

Abstract

Crystalline silicon is the most used semiconductor material for solar cell applications accounting for more than 90% of the market share. Nowadays, multicrystalline and monocrystalline silicon are mainly produced from directional solidification and Czochralski method, respectively. Solar cells made of these two types of material have shown efficiencies below the theoretical limit due to the presence of impurities in the silicon feedstock, challenges during the solidification process and issues in the different solar cell flowchart steps. This book chapter focuses on the solidification of the silicon photovoltaic value chain and corresponding growth technologies. A detailed description of the directional solidification and Czochralski method apparatus is initially presented. The several types of defects generated during the solidification as well as the source of impurities incorporation into the ingots are described in detail. Among different defects, dislocations and grain boundaries are presented for directional solidification and voids and oxygen-related defects for the Czochralski method. At the end, alternative methods to grow silicon substrates from both liquid and gaseous phase at lower cost and moderate qualities compared to standard processes are presented. These methods are still not considered to be cost-effective compared to more traditional ones due to the higher defect density.

Keywords: silicon, growth, monocrystalline, multicrystalline, Czochralski method, directional solidification, ribbons, impurities, defects, solar cells

1. Introduction

Crystalline silicon has been the dominant semiconductor material used in the fabrication of solar cells with a market share of over 90%. Silicon is abundant in earth crust and presents unique physical and chemical properties. Associated with the strong technological developments of its purification and solidification as well as advanced solar cell concepts, silicon is an interesting candidate to remain the foremost photoactive substrate for photovoltaic (PV) applications over the next decades. In 2016, multicrystalline silicon and monocrystalline silicon solar cells have reached maximum efficiencies of 21.3 and 25%, respectively [1]. These values are still far from the theoretical efficiency limit of about 31% for homojunction solar cells. It gives rise to the fact that optimization of the first steps of the PV value chain is needed. **Figure 1** shows the flowchart of the PV value chain where it can be divided into six main categories, from silicon feedstock to the final system installed outdoors. In order to obtain a silicon ingot from the feedstock, several growth processes can be used, namely Czochralski process for monocrystalline silicon (Cz-Si) and directional solidification for multicrystalline silicon (mc-Si). The silicon feedstock is initially molten at 1414°C in a crucible and subsequently solidified according to the corresponding method. The solidification step determines the quality of the ingot, where both impurities can be incorporated and defects generated and will impact the solar cell performance. Therefore, this step of the silicon PV value chain is of high importance to the cost effectiveness of the technology and needs to be carefully studied and optimized.

The production of mc-Si solar cells dominates the market share, as mc-Si technologies accounted over 60% of all module produced in 2014 [2]. Mc-Si materials are mainly formed by directional solidification, achieved in Bridgman or vertical gradient freeze (VGF) type of furnaces. The material produced in this way is relatively cheaper than monocrystalline silicon. It has, however, much higher densities of crystalline defects compared to singe crystals, and consequently, lower energy conversion efficiencies are obtained. Mc-Si ingots contain a variety of impurities, in dissolved and precipitated form. Some of them are intentionally added to the melt (dopants) to increase and control the conductivity of the silicon material. N- and p-type ingots are usually doped with phosphorus and boron, respectively. Other impurities are unintentionally incorporated into the crystal and drastically affect the material electrical properties. These are referred as contaminants and originate from the silicon feedstock, the crucible and coating, and the furnace environment. Among them, metallic impurities (e.g., iron and chromium) and light elements (e.g., carbon) are usually found.

In 2014, Cz-Si market share was well below the mc-Si one, accounting for approximately 34% [2]. It is, however, expected that this technology will have a significant increase over the next years due to expected improvements and enhanced productivity. Although with lower amount of defects compared to mc-Si solar cells, Cz-Si is still facing important challenges with the incorporation of impurities and their interaction with defects. The main source of unintentionally added impurities is the crucible and feedstock. The silica crucible dissolves during the process feedstock melting and subsequent steps of the process. Oxygen is released into the melt and transported toward the solid-liquid interface, which together with point defects generated at the interface, form oxygen-related defects. These are nowadays the type of defects that have the most detrimental impact on solar cell efficiency [3]. Carbon is normally found in much lower amounts compared to oxygen, but still with high contribution to the efficiency degradation even at low concentrations.

These processes waste very high amounts of material during sawing with important impact on the final module cost. Wires with nearly the same thickness of the desired wafer thickness are used, and thus, a large amount of material is wasted (commonly referred as kerf losses). A possible solution will be to develop low cost techniques by using lower energy budgets compared to standard processes and avoid the wafering step. At the same time, a reasonable

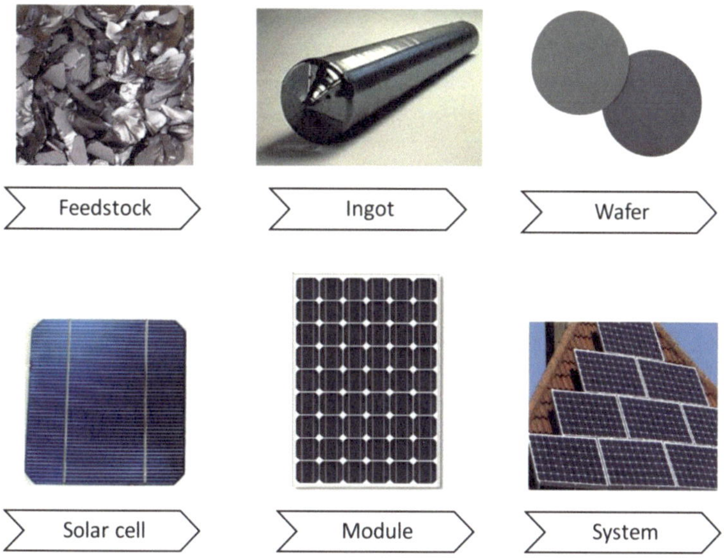

Figure 1. Flowchart of the silicon PV value chain.

performance of these materials is expected in order for the technique to become cost effective. Alternative techniques that short-circuit the ingot and wafering step have been developed and used to produce thin films and ribbons [4].

Next section describes the apparatus of standard processes for the production of silicon for solar cells, namely directional solidification for mc-Si and Cz method for monocrystalline silicon. The main impurities and their incorporation mechanisms as well as the most common defects found in both type of materials are described. In addition, alternative methods to produce and grow Si substrate materials from both liquid and gaseous phase feedstock are presented.

2. Growth techniques

2.1. Directional solidification

2.1.1. Process details

Silicon feedstock is placed in silicon nitride coated silica crucible. The coating layer prevents the silicon to stick to the crucible wall. Sticking points generate stresses in the ingot leading to plastic deformation and therefore to the development of crystalline defects such as dislocations. It also increases the level of oxygen contamination. The silicon nitride coating is applied to the inner walls by spraying a water-based slurry. It is then dried and oxidized by firing at 1100°C. This oxidation step is necessary to avoid the silicon melt to wet the coating layer. The silica crucible is a consumable as it cracks during the cooling of the ingot.

The feedstock is then melted in a controlled argon atmosphere. A near vertical temperature gradient is established in the hot zone, and directional solidification occurs from bottom to top.

The temperature ramp-down is achieved while maintaining a relatively high vertical temperature gradient of approximately 500–1000 K/m in order to (1) keep a near-planar solidification interface and (2) obtain good mixing conditions in the melt. The importance of point (2) will be developed in Section 3. Cooling can technically be performed either by:

- Controlled heat extraction from the bottom of a crucible. In such a case, the crucible is immobile and the process is often referred to as "Bridgman."
- Downward movement of the crucible through the vertical temperature gradient. The process is then often referred to as vertical gradient freeze (VGF).
- A combination of the two first methods.

The melt surface is flushed with argon during the entire process to remove the oxygen evaporating from the silicon melt. The chosen growth rates are low (typically 1 cm/h) in order to maintain favorable solidification conditions—that is, an effective mixing of the melt, and a near-planar solidification interface. Once the solidification step is over, the ingot is cooled down progressively to limit the development of stresses.

Industrial mc-Si ingots solidified in Bridgman/VGF furnaces have square horizontal cross sections and weigh up to 1 ton. The larger ingots are referred as "G6," as they will later in the solar cell process be cut into 6 × 6 bricks of 120 × 120 mm^2 top surfaces. The trend is toward larger ingot size, which gives a higher yield and lowers the price of the produced material.

2.1.2. Defects in mc-Si ingots

As mentioned in the previous section, mc-Si ingots crystallized by directional solidification methods contain more crystalline imperfections than single-crystalline ingots. Extended defects—as opposed to point defects—constitute the most important category, as they are found in high densities in mc-Si materials and have a considerable impact on solar cell properties. They are also responsible for the occurrence of *internal gettering*, where mobile impurities diffuse in the material and precipitate at defects. This phenomenon occurs when the ingot cools down during the directional solidification process and continues during the high-temperature steps of the solar cell process. Extended defects can be separated in two categories:

- *Surface defects*—that is, mostly *grain boundaries*—correspond to the interfaces separating grains of different crystalline orientations;
- *Line defects*—that is, *dislocations*—correspond to a line in the crystal lattice around which atoms are misaligned;

2.1.2.1. Grain boundaries

Multicrystalline silicon ingots are typically composed of elongated grains extending through its height, with diameters ranging from a few millimeters to a few centimeters. This multicrystalline structure is the result of the initial multiplicity of nucleation points on the silicon nitride coating layer. The grains composing an ingot have different crystalline orientations and are separated by interfaces referred as *grain boundaries*. Depending on the misorientation between the grains, the boundaries can have different degrees of coherency and consequently different impacts on the output solar cell performances. A boundary with a higher degree of coherency is generally considered to have a lower impact on the material electrical properties, and a lower ability to internally getter impurities by precipitation. Grain boundaries can be classified into two main categories, according to their misorientation angle and the arrangement of atoms at their interface:

- A specific type of grain boundaries with defined misorientations shows low interfacial energies and high degrees of coherency. These boundaries correspond to particular atom arrangements, where the two adjacent grains share coincident sites. They are referred as *coincident site lattice (CSL) grain boundaries*. A Σ-value defines the degree of fit, corresponding to the reciprocal density of coincident sites. For example, two grains separated by a perfect Σ27 grain boundary share one coincident site each 27 atoms. Σ3 grain boundaries—or *twins*—are a particular case of CSL boundary. They have a very high degree of coherency and an extremely low interfacial energy. Σ3 boundaries are in their majority formed during the growth of the ingot by a phenomenon called *twinning*. CSL boundaries are very common in silicon, in particular Σ3 boundaries. Twins of higher degrees $\Sigma 3^n$—Σ9, Σ27, for example—are also often found in mc-Si materials.
- Any deviation from the ideal coincident site lattice misorientations is accommodated by the introduction of line defects—or *intrinsic dislocations*—in the CSL grain boundary. *Random grain boundaries* are boundaries deviating considerably from the CSL configurations. They consequently have low degrees of coherency and high interfacial energies.

2.1.2.2. Dislocations

One of the main drawbacks of directional solidification when compared to single-crystal solidification methods is the invariable incorporation of numerous dislocations in the ingot. Dislocations are formed during the crystal growth and the cooling of the ingot and are found in densities ranging from 10^4 to 10^8 cm^{-2}. As opposed to grain boundaries, dislocations are mobile in silicon material. This means that once generated, a dislocation will have the ability to migrate and potentially multiply, annihilate or rearrange in energetically favorable structures. They have drastic impacts on the material electrical properties and heavily getter mobile impurities. Research has been conducted in order to limit the development of dislocations in the ingot, and multiplication mechanisms and dislocation interaction processes are necessary for this purpose.

As mentioned in the previous section, a dislocation is a linear defect. A line vector therefore defines each dislocation. A displacement vector—also referred as burgers vector **b**—characterizes in addition the nature of the dislocation. The edge and screw dislocations are special cases where the angle between the line direction and **b** is 90° and 0°, respectively (see **Figure 2**). All other types of dislocations are mixed cases, with both edge and screw components. In silicon, dislocations tend to align along the <1 1 0> dense direction, and the most common Burgers vector is ½ <110> **a**, being **a** the lattice vector.

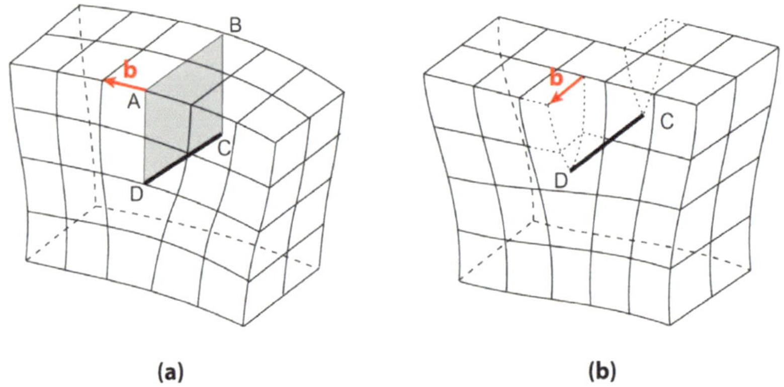

Figure 2. (a) Model of an edge dislocation (line DC), perpendicular to the Burgers vector (b) Model of a screw dislocation (Line DC), parallel to the Burgers vector **b**. Modified from Ref. [5].

In mc-Si, dislocations are mostly generated at grain boundaries, where the energy barrier for plastic deformation and dislocation emission is lowered. Grain boundary topological imperfections particularly trigger dislocation emission, as they act as stress concentrators. Semicoherent grain boundaries—or near-CSL boundaries—have the highest ability to generate dislocations, where incoherent boundaries tend to accommodate for stresses developing in the ingot. Precipitates and inclusions also generate dislocations, mainly on the bottom, sides and top of the ingot. In silicon, dislocations move only at temperatures close to the melting point. When the motion occurs within a surface that contains both the dislocation line and **b**, it is referred as *gliding*. Dislocations glide in the so-called slip planes, and along specific slip directions. In silicon, the most favorable slip planes are the {1 1 1} densely packed planes, and the <1 1 0> dense directions are the slip directions. As each of the four {1 1 1} planes contains three distinct <1 1 0> vectors, 12 slip systems are available for dislocations in silicon. When the motion occurs within a plane that does not contain **b**, it is referred as *climbing*. This mechanism is controlled by the presence of point defects. Multiplication phenomena are responsible for the formation of so-called *dislocation clusters*, containing very high densities of dislocations. The multiplication mechanisms are not yet clear, but it has been demonstrated that clusters are formed during the growth of the crystal. The dislocations generated during the cooling of the ingot are sparser as they do not have the same ability to move in the crystal. Dislocations in silicon tend to rearrange by alignment in arrays perpendicular to the slip planes, by a recovery mechanism called polygonization. A high fraction of the dislocations present in clusters is polygonized. These dislocation arrays form low-energy structures and define low-angle grain boundaries.

2.1.2.3. Structure control

Much effort is currently put on research focusing on techniques to control the structure of the mc-Si ingots solidified by directional solidification. The objective is to produce material with lower densities of extended defect. Two main directions are pursued:

- One of them focuses on obtaining ingots with larger grains, the objective being to eliminate grain boundaries and to approach single crystalline materials. The most common approach is to control the structure by seeded-growth: monocrystalline seeds are preplaced on the bottom of the crucible and partially melt before solidification. The orientation of the growing crystal is imposed by this mean, and large grains grow from bottom to top. This technology faces several technical challenges, as (1) parasite grains can tend to grow from the side-walls of the crucible, (2) dislocations originating from the seed-joints spread into the crystals, and (3) a domain of bad electrical properties is observed right above the seeds. The recent development has been, however, very successful to hinder these problems, and results on the solar cell level show a great potential;

- The second direction focuses on lowering the average dislocation density. The objective is to obtain a more random structure in order to favor the development of random grain boundaries, less likely to generate dislocations. This random structure is achieved by a better control of the nucleation step, or by seeding, using a bed of silicon particles. Mc-Si solidified using this method is typically composed of smaller grains and contains almost no dislocation cluster. This material shows also great results on the solar cell level and is now widely implemented industrially;

2.1.3. *Incorporation of Impurities*

2.1.3.1. *Contaminants*

Although carbon and oxygen can be found in mc-Si ingots in concentrations ranging from 5 to 10 ppma, metallic impurities are the contaminants with the most dramatic effects on the mc-Si electrical properties. Metallic impurities are typically found at ppb levels in the middle of mc-Si ingots, which is often enough to extensively limit the solar cell performances. Dissolved metallic impurities are often mobile in silicon and tend therefore to be internally gettered by precipitation at extended defects. Metallic impurities originate from several sources: The silicon feedstock contains invariably metallic impurities that will be later partially incorporated in the crystal. The feedstock purity is, however, very high, up to 6N for polysilicon. In addition, directional solidification being a very effective purification process (see Section 2.1.3.2), the amount of metallic impurities typically found in mc-Si ingots, cannot be solely attributed to the silicon feedstock. The silica crucible and the silicon nitride coating contain both typically 10–50 ppm of metallic impurities and are therefore important contamination sources. Impurities originating from the crucible/coating are incorporated in the melt during the growth of the crystal, and in the ingot, once solidification is finished. The effect of the solid contamination is visible on vertical ingot cuts when mapping minority carrier lifetime. A region of lower performance—often referred as red zone—is found on the bottom and the sides, close to the crucible. This region extends typically 2–5 cm in the ingot and corresponds to the diffusion of metallic impurities from the crucible/coating, in the solidified ingot (see **Figure 3**).

2.1.3.2. *Impurity segregation*

The solid solubility of impurities in silicon is higher than their liquid solubility. It implies that foreign elements are segregated at the solid-liquid interface during the silicon solidification. A

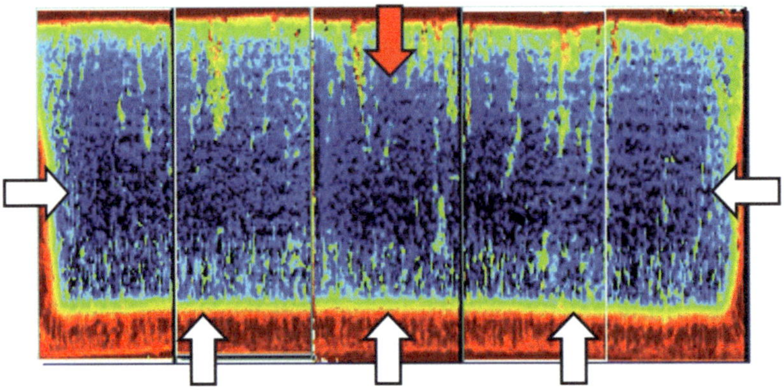

Figure 3. Minority carrier lifetime map of a vertical cross-section of a mc-Si ingot. The region of lower performance appears close to the crucible wall as a result of the in-diffusion of impurities from the crucible and coating. The top part of the ingot is influenced by the back-diffusion of segregated impurities. Modified from [6].

thermodynamic partition ratio k_0 is defined for each element and characterizes its ability to be incorporated in the growing crystal:

$$k_0 = C_s/C_l \tag{1}$$

C_s is the solid solubility and C_l is the liquid solubility. Equilibrium partition ratios of selected impurities are listed in **Table 1**. Metallic elements typically have very low segregation ratio, which makes directional solidification an efficient purification process. Carbon, oxygen and doping elements are less segregated as their equilibrium partition ratios are closer to unity.

Once the solidification is finished, the impurity distribution over the height of the ingot is commonly described by the Scheil-Gulliver equation:

$$C_s = k_0 \times C_0(1 - z/H)^{k_0 - 1} \tag{2}$$

where C_S is the impurity concentration at the vertical position z and H is the total ingot height. Examples of segregation profiles are given in **Figure 4**. The Scheil-Gulliver equation assumes no diffusion in the solid and a homogeneous liquid. In real cases, this last assumption does not hold, as all elements have a limited diffusivity in liquid silicon, and the stirring driven by natural or forced convection cannot lead to a perfectly homogeneous liquid. Impurities rejected in the liquid during solidification have therefore a tendency to build up and form a boundary layer of higher concentration close to the solid-liquid interface. This phenomenon is accounted for by the introduction of a so-called *effective partition ratio* k, where $k > k_0$.

The segregation of impurities with low partition ratios results in higher concentration in the top of the ingot (see **Figure 4**). During the ingot cooling, the impurity will then tend to back-

http://dx.doi.org/10.5772/intechopen.68351

Element	Equilibrium partition ratio, k_0
Fe	8×10^{-6}
Ti	3.6×10^{-4}
Cr	1.1×10^{-5}
Cu	4×10^{-4}
Ni	8×10^{-6}
B	0.8
P	0.35
O	0.25^{-1}
C	0.07

Table 1. Equilibrium partition ratio of selected impurities in silicon. From Ref. [7].

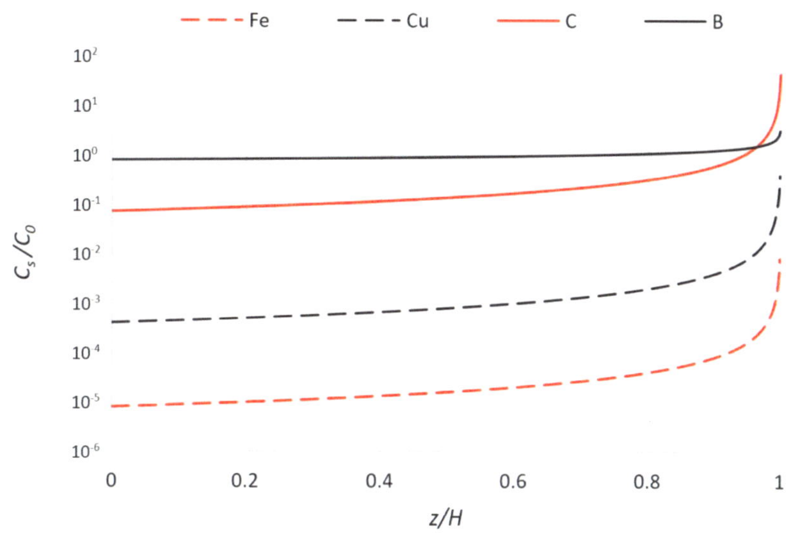

Figure 4. Examples of segregation profiles, as given by the Scheil-Gulliver equation using the equilibrium partition ratio k_0.

diffuse in the material, while it is still warm. This phenomenon is not captured by the Scheil-Gulliver equation due to the "no solid-diffusion assumption," but is visible on minority carrier lifetime maps as it affects the quality of the ingot over distances of several centimeters. The Scheil-Gulliver equation finally assumes that there is no outer-source of impurity—that is, the final concentration profile is the only result of the segregation of impurities initially present in the silicon feedstock. In the case of directional solidification of silicon, this assumption does not hold, as the crucible and the coating are important source of impurities contaminating the melt continually during the ingot solidification. In order to account for the flux of impurities $J_c(t)$

originating from the crucible/coating interior walls, an impurity mass balance in the liquid is made at a given time t during solidification

$$\frac{dC_l}{dt} = \frac{1}{1 - f_s(t)} \left[\frac{A(t)}{V} J_c(t) + (1-k)R(t)C_l \right] \tag{3}$$

With C_l the concentration in the liquid, $f_c(t)$ the solid fraction, $A(t)$ the lateral area of contact between the liquid and the crucible, V the total silicon volume (solid + liquid), and $R(t)$ the normalized growth rate ($R(t) = \frac{df_s}{dt}$). A modification of Eq. (2) can be then analytically derived from Eq. (3), assuming constant normalized growth rate R and impurity flux J_c from the crucible/coating:

$$C_s = k[C_0 + \beta(z)] \times (1 - z/H)^{k-1} \tag{4}$$

with

$$\beta(z) = \frac{4LH\,J_c}{VR(1-k)} \left[1 + \frac{1}{k-2} \right] [1 - (1 - z/H)^{2-k}] \tag{5}$$

where L is the width of the crucible. $\beta(z)$ is the height-dependent correction index of concentration unit accounting for the contamination from the crucible/coating. An application example of Eq. (4) is given here for the case of iron, using different values of J_c. An initial concentration of iron in the melt of 1.5×10^{13} cm^{-3} is considered. The chosen effective partition coefficient of iron is 2×10^{-5}, the growth rate is 3×10^{-3} cm/s, the crucible width is 70 cm, and the silicon is filled up to 40 cm height. The results are plotted in **Figure 5**.

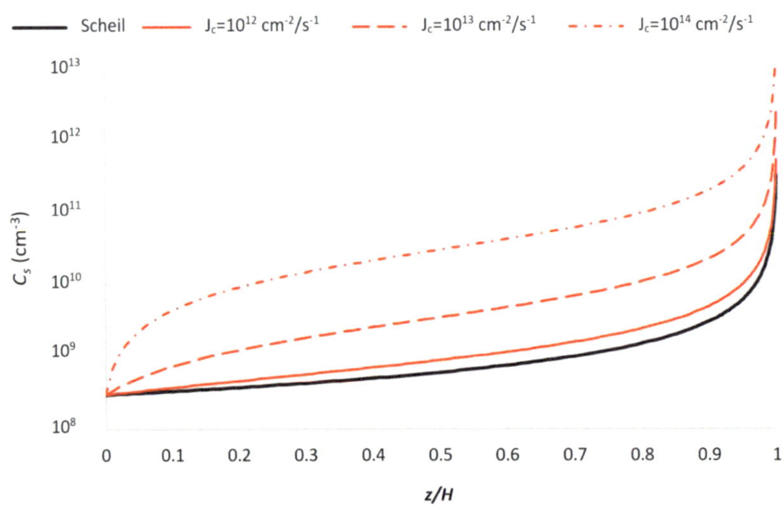

Figure 5. Application example from Eq. (4). Concentration profiles over the height of the ingot are plotted for different values of J_c.

http://dx.doi.org/10.5772/intechopen.68351

2.2. Czochralski growth

2.2.1. Process details

2.2.1.1. Apparatus

The Cz method apparatus includes several individual components to grown crystals under clean conditions and achieved the solidification of high-quality monocrystalline ingots. The main components are the hot zone, the pulling and rotating systems and the process control system. The hot zone is considered the most important part of the whole apparatus as it determines most of the thermal growth conditions of the crystal. This includes the graphite susceptor, giving mechanical support to the silica crucible, the graphite heater, responsible for providing controlled amounts of heat through the entire process, and the thermal shield, normally used for (1) conditioning the thermal field close to the solid-liquid interface and (2) controlling the gas flow near the melt-gas interface.

The Cz method offers several degrees of freedom. Both crystal and crucible can be rotated and lifted at variable speeds which allow process optimization according to the customer requirements. At the industrial-scale, the process is automated and follows predefined growth conditions. However, the continuous monitoring of process parameters and growth conditions is needed to suppress natural fluctuations of the growth conditions. Among them, the most important is the meniscus inspection by an IR camera. The pulling rate of the seed and the temperature of the heaters are continuously updated according to the feedback received from the monitoring system. The crystal diameter is maintained approximately contestant by this means.

Before the crystal starts to grow, the furnace chambers are evacuated, and inert gas is purged in.

2.2.1.2. Process steps

The growth of a Cz silicon crystal starts with the stacking of high purity polysilicon feedstock in the crucible, where either solar-grade or electronic grade silicon are normally used. The silicon chunks are placed strategically in order to (1) avoid movements during subsequent feedstock melting and (2) minimize the contact between the crucible and the silicon to limit the incorporation of oxygen. The feedstock is then molten by rising the temperature of the heaters above the silicon melting point. The stabilization of the melt is achieved over the next few hours, which is fundamental to obtain a more stable melt temperature and homogeneous solute distribution. A dislocation-free silicon seed of a specific crystallographic orientation (normally <1 0 0>) is finally dipped into the melt, and a meniscus is formed at the seed end. Due to the thermal shock between the melt and the cooler seed, dislocations are generated and need to be eliminated prior full crystal growth. A millimeter-range diameter neck is then grown at high pulling rates, ranging from 2 to 4 mm/min, to promote dislocation diffusion toward the crystal surface and thus stimulate their eradication. It needs to be long enough to ensure dislocation-free growth of the subsequent crystal. The total length of the neck is typically of tens of centimeters. To reach the desired crystal body diameter, two additional steps are needed, that is, crown and shoulder. The crown is grown at constant pull speed, usually lower than the crystal body, while the heater temperature is decreased. It leads to a constant enlargement of the crystal diameter at a rate depending on both parameters. The shoulder corresponds to a sharp transition at the stage close to the nominal body diameter and is needed to avoid further diameter increase. A sharp increase of the pull speed followed by a decrease of the nominal body pull speed over a short distance will result in a vertical growth of the crystal. The crystal body is finally grown at constant speed, with small adjustment over the entire process in order to adjust for fluctuations of the diameter. The active monitoring system inspecting the meniscus communicates with the controlling system responsible for the regulation of the pulling speed. Nowadays, the crystal body is grown at a nominal pulling rate of approximately 1 mm/min and diameters between 150 and 200 mm for PV applications. The pulling speed and other parameters may vary upon further process optimization. The process ends with the growth of the tail. A gradual decease of the diameter takes place at this stage, and the final part of the ingots presents thus a conical shape. This last step may be optimized in order to have a moderate cooling rate and avoid the formation of high density of oxygen-related defects and thermal stress. Finally, the ingot is brought up to a receiving chamber where it is cooled down to room temperature. A schematic of the different parts of a Cz silicon ingot is shown in **Figure 6**

2.2.2. Incorporation of impurities

In addition to the doping impurities intentionally added to the feedstock, oxygen and carbon are normally found in relatively large amounts in Cz-Si ingots, ranging from 15 to 21 ppma and 0 to 1 ppma, respectively. Nowadays, the control of the oxygen transport in the melt and consequent incorporation into the crystal is one of the main challenges to achieve high-quality crystals. Oxygen originates from the dissolution of the silica crucible, and its incorporation into the melt depends on boundary conditions between crucible and melt, for example, the crucible wall temperature. The dissolved oxygen is transported towards the solid-liquid interface through melt convection. Despite an equilibrium partition ratio close to unity, only a small amount is incorporated into the crystal, that is, less than 1%. Due to the high vapor pressure of oxygen, the majority of the oxygen is indeed evaporated at the melt-free surface, where the evaporation rate depends mostly on the free melt surface area in contact with purging gas. The oxygen transport over large distances and eventually incorporated at Cz crystal depends on an interplay between crucible dissolution rate, melt convection and evaporation at the melt-free surface.

At the industrial scale, Cz ingots are grown in large crucibles, depending on the crystal diameter. To achieve a homogeneous distribution of temperature and impurities in the melt

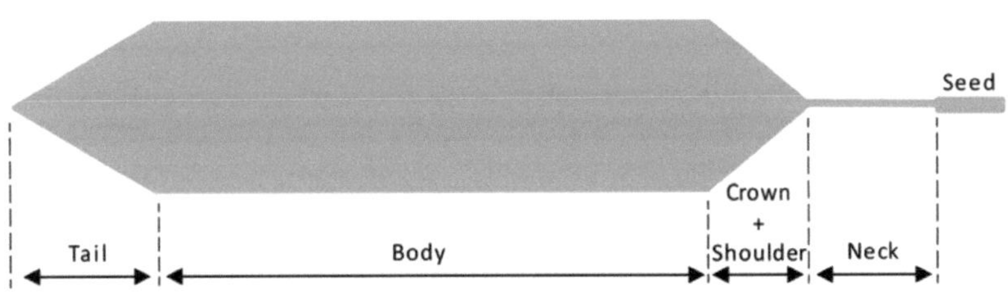

Figure 6. Cross section of a Cz-Si crystal and its different parts after the growth process.

as well as control the solid-liquid interface shape, both crystal and crucible are rotated in opposite directions. It results in forced melt convection which together with other mechanisms such as buoyancy driven flows and Marangoni forces, results in the development of several individual convection cells. Melt convection has also an important role in the solid-liquid interface shape, thermal field in the melt and impurities distribution. Buoyancy driven convection is a result of the temperature gradient between the growing crystal and the heated melt, intentionally created to drive the growth of crystal. Given the nature of the buoyance forces and their strong impact on the melt flow, these can deliver variable amounts of heat and impurities (e.g., oxygen) to the solid-liquid interface. The rotation of the crystal results in liquid areas of variable angular velocities below the solid-liquid interface. The dominant one is the so-called Taylor-Proudman cell and occupies most of the central area below the interface. The centrifugal forces pushing liquid outwards due to crystal rotation compete against the buoyancy flows. The interaction between the two can be controlled by the crystal rotation rate, where higher crystal rotations result on a larger extension of the central force convection cell. The enlargement of this cell may increase the oxygen transport from the crucible walls toward the crystal and govern its axial and radial distribution. Marangoni forces enhance the buoyancy-driven convection. They arise from variations of surface tension along the melt-gas surface, as a result of radial temperature gradients. This leads to the formation of a flow driven from the hot crucible walls to the solid-liquid interface and may influence the crystal growth conditions. Lower oxygen concentration is typically measured at the edge of the crystals in comparison with the center due to the effect of the buoyancy and Marangoni flows.

The evaporated oxygen at the melt surface can react with the graphite parts of the hot-zone and form carbon monoxide (CO) gas. The control of the argon flow in the furnace is then crucial to remove the CO from the furnace and prevent it to dissolve back in the molten silicon, which constitutes the main origin for carbon contamination. Carbon and oxygen contamination are thus closely related. The carbon in the melt is mostly segregated into the crystal after it reaches the solid-liquid interface. Attempts are made to use crucibles made of other material than silica to prevent oxygen incorporation into the melt. In addition to the optimization of the argon flow conditions, alternative materials and coated graphite parts are pathways to minimize carbon contamination of the melt. These new technologies are still facing challenges, and silica and graphite remain the materials of choice for crucibles and crucible holders and heaters, respectively.

2.2.3. Defects in Cz-Si crystals

2.2.3.1. Point defects

During Cz silicon growth, a large amount of native point defects are generated at the interface and are subsequently incorporated in the crystal, the so-called vacancies and self-interstitials. Vacancies correspond to missing atoms in the crystal lattice while self-interstitials to extra ones occupying empty sites in the lattice. Impurity atoms are extrinsic point defects. Substitutional impurities take over silicon atoms, and interstitial ones occupy empty sites in the crystal lattice. Carbon dissolves substantially in the silicon crystal lattice and oxygen interstitially.

The incorporation of native point defects depends on both the growth conditions at the solid-liquid interface and the crystal cooling. The growth rate together with the defects diffusion into the crystal plays a crucial role on the dominant type of point defects and corresponding density near the solid-liquid interface. At lower growth rates, the diffusion of point defects into the crystal dominates, and the crystal becomes interstitial rich due to their higher diffusivity compared to vacancies. Voronkov and his collaborators have successfully described the incorporation of point defects at the solid-liquid interface during Cz silicon growth [8, 9]. Based on the ratio between growth rate (*V*) and radial thermal gradient (*G*), it is possible to determine whether vacancies or interstitials are generated in excess. The transition between the two regimes is defined by the critical *V*/*G* ratio, and values ranging from 0.12 and 0.2 mm^2/min K are found in literature [10–12]. Given that *G* increases toward the crystal edge and assuming that the crystal is grown at a steady-state rate, *V*/*G* also changes over the crystal radius, as shown in **Figure 7**(left). Therefore, the density of point defects incorporated at the interface also change accordingly. The critical ratio may also be affected by the presence of impurities at the interface. Species such as oxygen can result in a lower critical value and thus promote the incorporation of vacancies. Nowadays, industrial-scale crystals are grown in the vacancy-rich regime due to the relatively high growth rates and large crystals used to increase their productivity.

2.2.3.2. Voids

Voids correspond to agglomerates of vacancies and, due to their nature, belong to the category of volume-type defects. They are formed due to their agglomeration of vacancies during crystal cooling at the temperature range of 1150–1000°C [13]. At temperatures close to 1000°C, vacancies preferentially bond with oxygen which results on dramatic reduction of their the size. Void sizes range from 100 to 300 nm and present a twin or triple structure with facets oriented along the {1 1 1} crystallographic orientation [14]. In addition to TEM, voids can be visualized with several other characterization techniques, namely etching methods, where the etch pits designation differs according to the used method. Secco etching can be used to delineate voids that show up as D-defects at the tip of a wedge-shaped etch pattern. They can further be visualized with optical techniques. Given their small sizes, the formed patterns are the fingerprint of an existing void and make the void density possible to be determined.

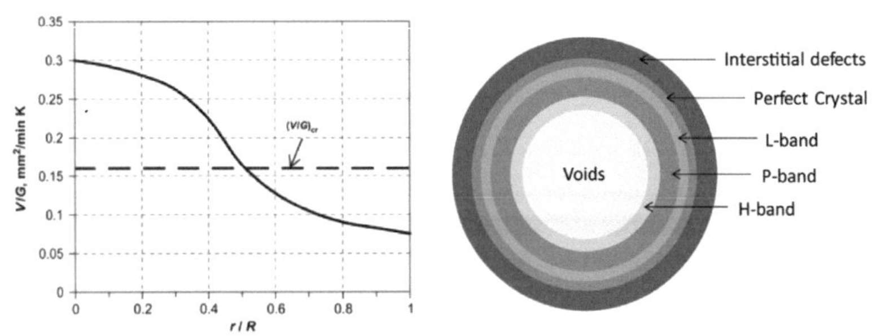

Figure 7. (Left) *V*/*G* distribution along the radial direction of a Cz crystal grown under steady-state conditions [8]. (Right) Defect rings distribution for variable *V*/*G* along the crystal radial direction, modified from Ref. [16].

Industrial-scale Cz silicon crystals for PV applications tend to be grown at higher growth rates and decreasing oxygen concentrations due to continuous process optimization. Therefore, it is expected to find crystals of higher vacancy concentrations and consequently higher amount of voids in the future.

2.2.3.3. *Defect bands*

Cz silicon crystals may be grown at different growth rates and dimensions according to the application and pulling conditions. Both the diameter and the growth rate influence the type and amount of defects incorporated, and given that the gradient at the interface changes with the radial position, several types of defects may be found within the same wafer. Defects are distributed as circular bands through the radius according to their nature. This can happen mainly at the top of the ingot, where large variations of the crystal diameter are needed to reach the body width. As such, the V/G ratio may vary, and all sorts of defects can be generated from interstitials to voids. Due to the lower G at the center of the ingot, this ratio is higher at this region, and it is thus more likely to find vacancy-related defects close to the crystal center. Between the two extreme cases and close to the critical V/G ratio, several bands can be found from interstitial defects to voids regions in the following order: L-band, P-band and H-band. In addition, there is a narrow region between the interstitial region and the L-band, the so-called *perfect crystal*, where interstitial and vacancies are of very low densities, and nucleation of other defects is not promoted. The P-band is found at areas of slightly higher V/G ratio than the critical one and corresponds to the maximum residual vacancy concentration after the nucleation of voids [15]. Oxide particles are found in densities in the order of 10^8 cm^{-3} and can be visualized as oxygen-induced stacking faults (OiSF) after thermal oxidation and etching. This corresponds to an area of high recombination activity in as-grown crystals and must be avoided. Process optimization is of upmost importance to accommodate this band at the very top of the crystal, for example, in the crown area. A narrow band of smaller particles and higher density compared to the P-band (in the range of 10^{10} cm^{-3}) is formed at the periphery of the last one. It corresponds to the so-called L-band, where only oxide particles can be formed due to the low initial vacancy concentration. On the other hand, particles and voids coexist in the H-band due to the higher initial vacancy concentration compared to the previous two. Due to the higher density of oxide particles, higher recombination activity is found at the L- and H-band compared to P-band after sufficient thermal budget to growth them above a certain size. The different bands that can coexist in a crystal radial cut are schematically shown in **Figure 7**(right)

2.2.3.4. *Thermal donors*

Interstitial oxygen atoms are not electrically active by themselves, but the same criterion is not applied to their agglomerates. One important class of oxygen related defects is the so-called thermal donors (TD). They are formed during crystal cooling at temperatures close to 450°C [17], and their size and density depend on the cooling rate of the crystal and the interstitial oxygen concentration. Their denomination arises from the fact that they are a source of electrons which results on a direct impact in the electrical properties. Their presence changes the resistivity of the material leading to its reduction in n-type and increases in p-type silicon.

Nevertheless, it leads to an increase of the recombination activity of minority carriers and results on poor material performances. It is possible to completely dissolve TD by applying a rapid thermal annealing for few seconds at 800°C [18]. Other class of TD can be found in Cz silicon that grows at higher temperatures than the previously described. These are formed at temperatures ranging from 650 to 850°C and are mostly developing in silicon with higher concentration of carbon.

2.3. Alternative methods

2.3.1. *Liquid phase growth*

Ribbon growth processes lead to the fabrication of substrates from the melt directly into a planar one where a crystalline structure suitable for solar cell applications is obtained. The ribbon growth processes can be divided into either vertical or horizontal growth, depending on the crystal pulling direction. The most mature methods of growing ribbons vertically are edge-defined film-fed growth (EFG) and string ribbon (SR), and ribbon growth on a substrate (RGS) for horizontally grown substrates. In the EFG method (see **Figure 8A**), the lower part of the meniscus is formed and shaped by a graphite die, through which an octagonal silicon pipe is directly pulled from the melt at rates approximate to 1.7 cm/min. The pipe is subsequently cut into slices by laser so that it can be processed into solar cells. Due to the use of the graphite die, a high carbon contamination is present in these ribbons [19]. In the SR method, strings are fed through the molten Si to provide edge support for the growing ribbon (**Figure 8B**). It is then used to pull the ribbon from the melt. Si substrates obtained using SR method can be grown at 1–2 cm/min and exhibit low oxygen concentration, but still with high carbon concentrations [20]. The RGS technique (**Figure 8C**) allows the fastest growth rates among these techniques (600 cm/min), and consequently, a high throughput can be achieved [21]. However, it has the main limitation of using a substrate to support the crystal growth. Furthermore, crystals produced with this technique exhibit high carbon concentrations [22], which compromise the solar cells quality.

Another interesting vertical ribbon growth method is the ribbon of a sacrificial template (RST) [23]. This method combines two of the previous concepts; the vertical growth direction is combined with the use of a substrate (**Figure 8D**). It allows faster vertical growth rates, but with the disadvantages of the vertical and horizontal techniques in terms of contamination.

2.3.2. *Gaseous phase growth*

The substrate material for the solar cell can be grown directly from silane or chlorosilane gases in a low-cost substrate. For such procedure, chemical vapor deposition (CVD) is used for depositing silicon on a substrate at high temperatures. Using a gaseous mixture of H_2 and the precursor SiH_4 or SiH, it is thermally decomposed at the hot surface of the substrate. Si layers of a few microns are deposited on a substrate that is prepared so that the deposited film is releasable onto a carrier substrate [25], as shown in **Figure 9**. Such technique is applied to produce thin films with an amorphous structure, which is a limitation toward the fabrication of high efficiency solar cells.

http://dx.doi.org/10.5772/intechopen.68351

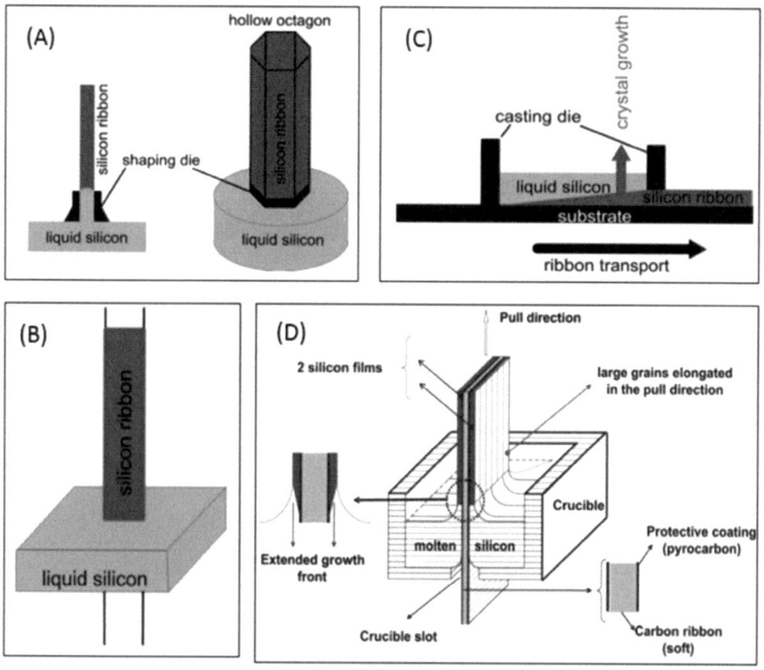

Figure 8. Ribbon growth techniques schematics: (A) edge-defined film fed growth; (B) string ribbon growth; (C) ribbon growth on a substrate; (D) ribbon on a sacrificial template growth schematics. Adapted from Ref. [24].

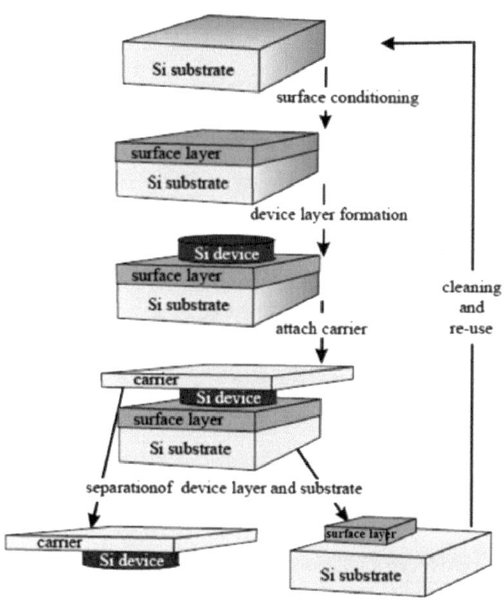

Figure 9. Gas-phase film growth by CVD onto a releasable substrate. Adapted from Ref. [26].

An interesting method to grow silicon ribbons for solar cells directly from a gaseous feedstock is the so-called silicon on dust substrate (SDS) process [27–29]. In this process, a self-supporting silicon sheet is produced directly from a silicon gaseous feedstock by a fast CVD step using gaseous SiH_4. With this method, it is possible to achieve high deposition rate at low temperature (approximately 800°C), and at atmospheric pressure. These characteristics make it a fast deposition process with low energy budget. The SiH_4 is decomposed by the follow pyrolysis reaction,

$$SiH_{4(g)} \rightarrow 2H_{2(g)} + Si_{(s)} \tag{6}$$

The deposition of Si occurs on top of a thin silicon dust layer that is placed on a quartz plate. This deposition process guaranties high homogeneous nucleation rates of silicon nanoparticles of high purity. The resulting film has a nanoporous structure, which after being detached from the dust layer has a thicknesses ranging from 300 to 400 μm. The remaining detached silicon dust acts as a sacrificial layer that can then be reused for a new deposition. The drawback of this method is that the resulting Si sheet is porous Si and of a very low nanocrystalline quality, which makes it unsuitable for solar cell applications. Therefore, the sheet goes through a crystallization process by float zone melting. A localized molten zone is created by a focused radiation in the form of a thin line which goes through the film crystallizing it. After the crystallization step, the resulting sheet has a multicrystalline structure that is suitable to be used as a substrate for solar cells. For this process, a deposition rate of 20 μm/min and constant advance speed of 10 mm/min can be achieved. With low impurity sources, this technique is mainly limited by the high dislocation density, which is produced due to high thermal gradients during crystallization. Recent developments in this technique foresee solar cells of efficiencies of 14%, which is relatively low when compared to standard and more mature processes [30].

Author details

Guilherme Manuel Morais Gaspar[1]*, Antoine Autruffe[2] and José Mário Pó[3]

*Address all correspondence to: guilherme.morais.gaspar@gmail.com

1 Tyndall National Institute, University College Cork, Cork, Ireland

2 Elkem AS, Kristiansand, Norway

3 Faculty of Sciences, University of Lisbon, Campo Grande, Lisbon, Portugal

References

[1] PHYC.ORG. Efficiency Records Chart [Internet]. Available from: https://phys.org/news/2016-02-solar-cell-efficiency-nrel.html [Accessed: Feb. 2017]

[2] International Technology Roadmap for Photovoltaic (ITRPV). 2014

[3] Gaspar G, et al. Identification of defects causing performance degradation of high temperature n-type Czochralski silicon bifacial solar cells. Solar Energy Materials and Solar Cells. 2016;**153**:31–43

[4] G. Hahn, et al. New crystalline silicon ribbon materials for photovoltaics. Journal of Physics: Condensed Matter. 2004;**16**:R1615–R1648.

[5] Autruffe A. Silicon directional solidification: Impurity segregation and defects [PhD thesis]. Trondheim (Norway): Norwegian University of Science and Technology; 2014.

[6] C. W. Lan, et al. Engineering silicon crystals for photovoltaics. CrystEngComm. 2016;**18**: 1474–1485.

[7] R. Hull (editor). Properties of Crystalline Silicon. Emis Series, The Institution of Engineering and Technology; 1999.

[8] VoronkovVV. Grown-in defects in silicon produced by agglomeration of vacancies and self-interstitials. Journal of Crystal Growth. 2008;**310**:1307–1314

[9] FalsterR, et al. On the properties of the intrinsic point defects in silicon: a perspective from crystal growth and wafer processing. Physica Status Solidi (b). 2000;**222**:219–244

[10] Voronkov VV, et al. Vacancy-type microdefects formation in Czochralski silicon. Journal of Crystal Growth. 1998;**194**:76–88

[11] von Ammon W, et al. The dependence of bulk defects on the axial temperature gradient of silicon crystals during Czochralski growth. Journal of Crystal Growth. 1995;**151**: 273–277

[12] Abe T, et al. Swirl defects in float-zoned silicon crystals. Physica B+C. 1983;**116**:139–147

[13] Takano T, et al. Relationship between grown-in defects and thermal history during CZ Si crystal growth. Journal of Crystal Growth. 1997;**180**:363–371

[14] Itsumi M. Octahedral void defects in Czochralski silicon. Journal of Crystal Growth. 2002;**237–239**:1773–1778

[15] Voronkov VV, et al. Grown-in microdefects, residual vacancies and oxygen precipitation bands in Czochralski silicon. Journal of Crystal Growth. 1999;**204**:462–474

[16] Gaspar G. N-type Czochralski silicon solidification: oxygen- and copper-related defects formation [PhD thesis]. Trondheim (Norway): Norwegian University of Science and Technology; 2016.

[17] Kaiser W, et al. Mechanism of the formation of donor states in heat-treated silicon. Physical Review. 1958;**112**:1546–1554

[18] Tokuda Y, et al. Thermal donor annihilation and defects production in n-type silicon by rapid thermal annealing. Journal of Applied Physics. 1989;**66**:3651–3655

[19] Wald FW, editor. Crystals: Growth Properties and Applications. Berlin (Germany): Springer; 1981

[20] Hanoka JI. Continuos, Automated Manufacturing of String Ribbon Si PV Modules, Final Report. NREL/SR-520-30622. 2011

[21] A. Schonecker, et al. Ribbon-Growth-On-Substrate: Progress in High-Speed Crystalline Silicon Wafer Manufacturing. 29th IEEE Photovoltaic Specialists Conference; 20–24 May; New Orleans (USA). 2002.

[22] Lange H, et al. Ribbon Growth on Substrates (RGS) – A new approach to high speed growth of silicon ribbons for photovoltaics. Journal of Crystal Growth. 1990;**104**:108

[23] Belouet C. Growth of silicon ribbons by the RAD process. Journal of Crystal Growth. 1987;**82**:110

[24] Hahn G, et al. Review on ribbon silicon techniques for cost reduction in PV. In: Proceedings of 4th World Conference on Photovoltaic Energy Conversion; 07–12 May; Waikoloa (USA). 2006.

[25] F. R. Faller, et al. High temperature CVD for crystalline silicon thin film solar cells. IEEE Transactions on electron devices. 1999;**46**:2048–2054.

[26] Pó JM. High quality silicon ribbons for solar cells [thesis]. Lisbon (Portugal): University of Lisbon; 2017.

[27] Pinto CR, et al. Zone melting recrystallization of self-supported silicon ribbons obtained by fast CD from silane. In Proceedings of the 21st EUPVSEC; Dresden (Germany). 2006

[28] Augusto A, et al. First multicrystalline silicon ribbons using the continuous SDS process. In 37th IEEE Photovoltaic Specialists Conference (PVSC); 19–24 Jun; Seattle (USA). 2011

[29] Serra JM, et al. The silicon on dust substrate path to make solar cells directly from a gaseous feedstock. Semiconductor Science and Technology. 2009;**24**:045002

[30] Augusto A, et al. Residual stress and dislocations density in silicon ribbons grown via optical zone melting. Journal of Applied Physics. 2013;**113**:083510

Chapter 10

Guiding On-Chip Optical Beams Without Diffraction in a Rod-Type Silicon Photonic Crystal

Wei Li, Hao Li, Ming Li, Haiyang Huang and Fuwan Gan

Additional information is available at the end of the chapter

http://dx.doi.org/10.5772/67949

Abstract

Guiding on-chip optical beams without diffraction is very important in the future's all-photonic circuits. Herein, both theoretically and experimentally, we study an all-angle quasi-self-collimation phenomenon occurring in photonic crystals composed of silicon nanorods. When the all-angle quasi-self-collimation phenomenon occurs, the optical beams can be incident onto such photonic crystals from directions covering a wide range (extremely close to all-angle) of incident angles direction and become highly localized along even a single array of rods, which finally achieve results in the narrow-beam propagation without divergence. The propagation length is expected to be 1000 times larger than the wavelength of light. Theoretically, it is shown that such all-angle quasi-self-collimation phenomenon is owing to the symmetry change of the lattice of photonic crystals. By changing the symmetry of a photonic crystal to straighten the isofrequency contours, the photonic crystal shows the all-angle quasi-self-collimation effect. Experimentally, we show the observation of all-angle quasi-self-collimation phenomenon occurring in a rod-type silicon photonic crystal fabricated on by patterning a silicon-on-insulator (SOI) wafer. The experimentally observed propagation length is more than 0.4 mm over the telecom wavelength range, even though at large angle of incidence, which is a relatively large length scale for on-chip optical interconnection.

Keywords: without diffraction, silicon photonic crystal, all-angle quasi-self-collimation, SOI

1. Introduction

By using an index gradient, a well-designed waveguide array [1], or nonlinear optical effects [2, 3], beam spreading arising from geometrical origin can be manipulated in many ways. Photonic crystals (PCs) provide an alternative to beam steering based on photonic band-gap (PBG) formation and related dispersion phenomena such as negative refraction, super-prisms, and slow light. Self-collimation (SC) as a dispersion-related phenomenon is worth studying because it has important implications for subwavelength focusing, channel-less waveguiding and diffraction suppression [4–7], etc. In a SC PC, the light beams can propagate without diffraction, since propagation directions are paralleled to the group velocity, i.e., $V_g = \nabla_k \omega(k)$, where ω is the optical frequency for a determined wave vector k. Thus, we can determine that the SC effect is ascribed to the flat part of the equifrequency contours (EFCs) [8, 9]. Based on the SC effect, it can be used to design novel SC-based photonic devices, such as devices for diffraction inhibition, channel-less waveguides, subwavelength focusing or imaging [4, 5, 10, 11].

In recent years, the development of fabrication techniques has enabled the SC PCs to function at optical frequencies. In experiments, it has been recognized that SC phenomenon exists in many different kinds of structures, containing pillar-type and hole-type PCs [4, 12, 13], and quasi-zero-average-index structures [14], or even three-dimensional PCs [15]. Lately, wide-angle SC phenomena have been reported in PCs with square lattices by the composition of elliptical air holes [16]. Technically speaking, contrary to the hole-type PC, there are a great potential for active components with the pillar-type structure, benefiting from the possibility of the heat dissipation capability and electrical contraction. In addition, the light interacts with the medium (to be sensed) which more strongly surrounds the rods, and in optofluidics, the rod structure allows for fluid penetration better than the structure composed of air holes in a dielectric slab, therefore, the rods are more suitable for sensing applications.

However, there are still some problems to be solved. For example, the SC effect may be limited by the angle of incidence of light, which makes SC-based super-integrated devices difficult to manufacture. In this paper, the model is clarified for a full-angle quasi-SC PC, and the lower reflections for our quasi-SC PC by utilizing the destructive interference-based method. This all-angle quasi-SC is related to a flat equifrequency contour (EFC) across the entire Brillouin zone. In contrast to the conventional self-collimation, the all-angle quasi-SC shown in this chapter exhibits two unique properties. First, the light waves can be injected from any angle, even up to 90° [17]. Second, the electromagnetic energy can be achieved for high degree of localization in the case of nondiffraction [5]. For hole-type PCs, light waves are confined between the pores owing to the full-angle quasi-SC phenomenon, whereas for pillar-type PC, the electromagnetic energy can be positioned highly along a narrow path just as in a single nanorod array, as presented in this paper. Moreover, we will also show our recent experimental work on the all-angle quasi-SC in PCs composed of silicon nanorods. We believe this work motivates the current research on silicon photonics.

http://dx.doi.org/10.5772/67949

2. Theoretical model and analysis

First, the two-dimensional (2D) rod-shaped silicon PC is considered with a rectangular lattice in the air, as shown on the inset in **Figure 1(a)**. The breadth and length of the rectangular lattice are denoted by a and b, respectively, and the radius of the rod is $r = 0.3a$. Assume that the silicon rod is lossless and nondispersive near the telecommunication frequency and exhibit a refractive index of 3.5. Then, EFCs can be calculated by plane-wave expansion method. To simplify the model throughout this chapter, only one single frequency $f = 0.2\ c/a$ is considered for the transverse electric (TE) modes. This frequency is located in the first photonic band, and the corresponding isofrequency contour plot is shown in **Figure 1**.

2.1. Straightness of the EFCs and collimation ability of the beams

As shown in **Figure 1(a)–(d)**, the large angle SC effect in PC can be obtained by changing the symmetry of the PC, that is, by increasing the aspect ratio of the length to breadth "$\beta = b/a$". As shown in **Figure 1(d)**, the large angle SC effect is indicated by the straightness of the EFC

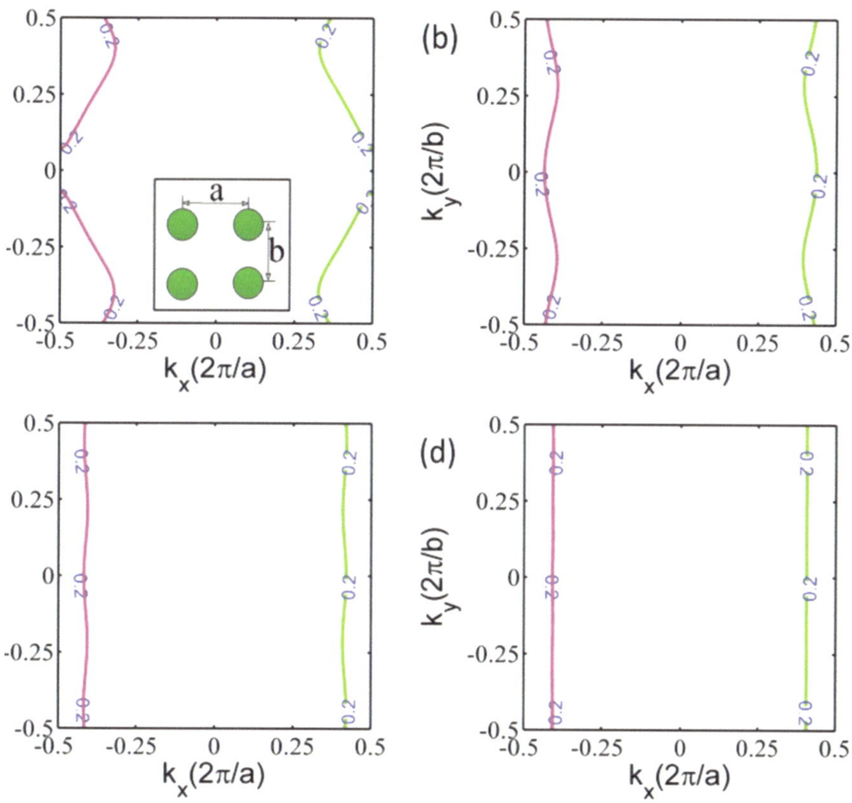

Figure 1. EFCs for the normalized frequency $f = 0.2c/a$, fixed a, $r=0.3a$, and different lattice length-breadth ratios β: (a) β = 1.25, (b) β = 1.75, (c) β = 2.3, and (d) β = 3.5. Inset image in (a): schematic illustration of the 2D PC.

plotted over the entire Brillouin zone. The light beams can reduce beam divergence along with the increase of β value, resulting in self-collimation propagation along the ΓX direction. Therefore, for the purpose of improving the collimation ability, a higher EFC straightness must be obtained. As described in **Figure 1**, the straightness linearity of the EFCs is determined by β.

In order to study the correlation between length-breadth ratio "β" and EFC straightness, the least-squares method [17] is used to quantify EFC straightness as shown in **Figure 1**. According to this method, a quasistraight EFC whose curve can be illustrated by a function "Y = F(X)" (Y representing ky and X representing kx). Assuming that a straight line with the equation

$$\bar{Y} = AX + B \tag{1}$$

can be applied to fit the quasistraight EFC, where A and B are undetermined coefficients. For the purpose of obtaining A and B, $\varepsilon = \sum_{i=1}^{n}\left[Y_i - \left(AX_i + B\right)\right]^2$ is defined, where i is the number of sampling points, then, by utilizing the minimum condition, i.e., ∂ε/∂A, ∂ε/∂B, we obtain

$$A\sum X_i^2 + B\sum X_i = \sum X_i Y_i \tag{2}$$

and

$$A\sum X_i + nB = \sum Y_i \tag{3}$$

From Eqs. (2) and (3), A and B can be calculated as follows:

$$A = \frac{n\sum X_i Y_i - \sum X_i \sum Y_i}{n\sum X_i^2 - \left(\sum X_i\right)^2} \quad B = \frac{\sum Y_i \sum X_i^2 - \sum X_i \sum X_i Y_i}{n\sum X_i^2 - \left(\sum X_i\right)^2} \tag{4}$$

Finally, we can quantify the straightness quality of the EFCs through the straightness factor

$$L = \Delta L_{max} - \Delta L_{min} \tag{5}$$

Where $\Delta L_{max} = [Y - \bar{Y}]_{max} = [Y - AX - B]_{max}$ and $\Delta L_{min} = [Y - \bar{Y}]_{min} = [Y - AX - B]_{min}$ are the maximum and the minimum convexities, respectively.

According to the straightness factor L is defined by Eq. (5), the averaged deviation angle of the propagation direction can be illustrated for a quasi-collimated light beam. Furthermore, we note that the beam width will become broader while the beam propagates through the quasi-SC PC, with determining the beam waist by $W(D) \approx W_0 + L\frac{\lambda D}{\pi W_0}$ where λ is the wavelength in the air, W_0 and D (we assume $D \gg \lambda$, $D \gg W_0$) describe the initial waist and propagation distance of the beam, respectively. The smaller L corresponds to a straighter EFC and better self-collimation behavior. Just in case L = 0, a perfectly straight EFC corresponding to a strict self-collimation without beam divergence is obtained, $W(D) = W_0$ which the D can take any value.

For most practical applications, the condition L = 0 is too strict. A sufficiently small L is usually acceptable. We proposed $L_0 = 0.01$ as the critical value for the straightness factor (corresponding to the pink dash line in **Figure 2**). In this case, the quasi-collimated beam shows almost no diffraction if $L \le L_0$, for example, relative to a typical Gaussian beam with an initial waist $W_0 = 10\lambda$, the

http://dx.doi.org/10.5772/67949

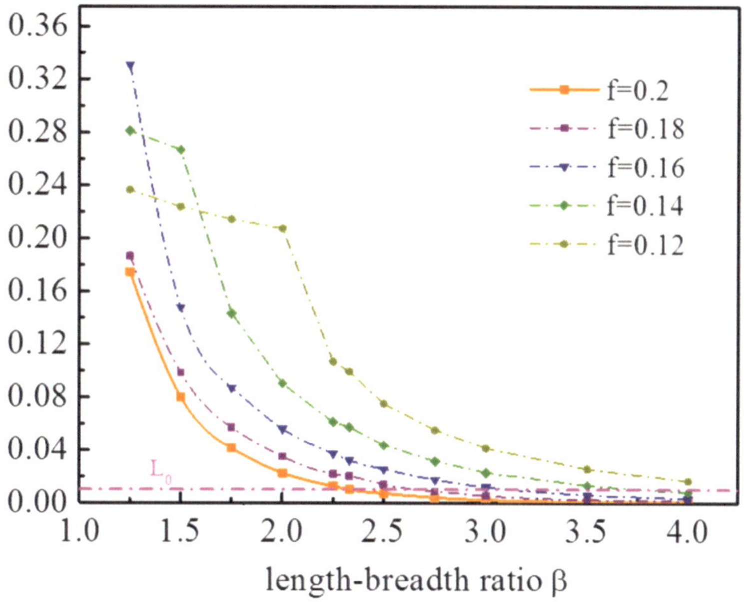

Figure 2. Straightness factor L plotted as a function of the length–breadth ratio β of the rectangular lattice. The normalized frequencies are in unit of c/a.

beam broadening is less than 1% as the beam propagates at a distance scale of 100λ. We consider this quasi-collimated beam to be sufficient for most practical applications with self-collimation phenomenon.

Using Eq. (5), the straightness factor L of the EFCs can be calculated. **Figure 2** shows the relationship between the straightness factor of the EFCs and the rectangular lattice structure. As β increases, the straightness factor L decreases obviously. This result is consistent with the result in **Figure 1**. As shown in **Figure 1(a)–(d)**, we can measure that the corresponding straightness factors are $L = 0.1743$, $L = 0.014$, $L = 0.01$, and $L = 8.09 \times 10^{-4}$, respectively. For $f = 0.2c/a$, which is the normalized frequency, the condition $L = L_0$ corresponds to $\beta = 2.3$, with the corresponding EFC shown in **Figure 1(c)**. In order to provide usable operating frequency quasi-SC effect on additional information, L in the other four frequencies as a function of β is also shown in **Figure 2**, from which they can be seen with $f = 0.2c/a$ very similar. Furthermore, bandwidth of quasi-SC effect becomes broader and broader by increasing β, because more and more EFCs have their straightness factor L smaller than L_0.

As shown in **Figure 2**, it is possible to realize a better and better SC by increasing β. For the case of $\beta \gg 1$, the system will become separated from each other, so in this case, we can investigate only one of the row of rods. We find that the SC effect is still valid in this single row of rods. However, there are many differences between their single row, and also there are many differences between a single row and a PC having a rectangular lattice. For the sake of simplicity, we limit ourselves to only explore the PC shown in **Figure 1**(with $\beta = 2.3$) at a single frequency $f = 0.2c/a$.

In order to study the whole-angle quasi-collimation phenomenon of the PC at the frequency $f = 0.2c/a$, with $\beta = 2.3$, $a = 0.4$ μm, and $r = 0.3a$, a Gaussian beam with the beam width 2.3 μm (a little bit larger than one wavelength 2 μm) is incident to the PC at any incident angle. **Figure 3** shows the calculation model, which can be applied to calculate the electromagnetic field by a perfectly matched layer (PML) finite-difference-time-domain (FDTD) method [10, 11] with a grid size 20 nm. The PC's right, 50 × 71 rods array (as shown in the blue bars in **Figure 3**) are manually set to be absorbed (a small imaginary part is added to the refractive index of rod, i.e., the index becomes 3.5 + 0.02i) which set to avoid the size-dependent Fabry-Perot effect. A PML boundary condition is acted by absorption rods array for the PC. Thus, the structure may be available to simulate an infinite-length PC. As shown in **Figure 4** and the subsequent figures, the field distribution is presented. However, only the field which is located in the nonabsorption region is shown. When a Gaussian beam propagates with a different incident angle in air (top panels) and PC (lower panels), the field distribution in this case is depicted in **Figure 4(a)–(c)**. We can see that the beam does not diverge in the PC which shown in **Figure 4**. On the contrary, the electron beam has great divergence in the air.

2.2. Coupling analysis

We investigate the optical coupling efficiency of the SC PC in this section. Owing to strong reflection, there are the large coupling losses (about 50%), which are shown in **Figure 4**. For improving the coupling efficiency, we propose a method based on destructive interference. **Figure 5** depicts an antireflection layer (ARL), which is used in front of the quasi SC PC. The ARL is also constituted of the silicon rods array, which a lattice constant $b = 2.3a$ is equal to the length of the rectangular lattice of the quasi-SC PC. The silicon rods radius in the ARL is

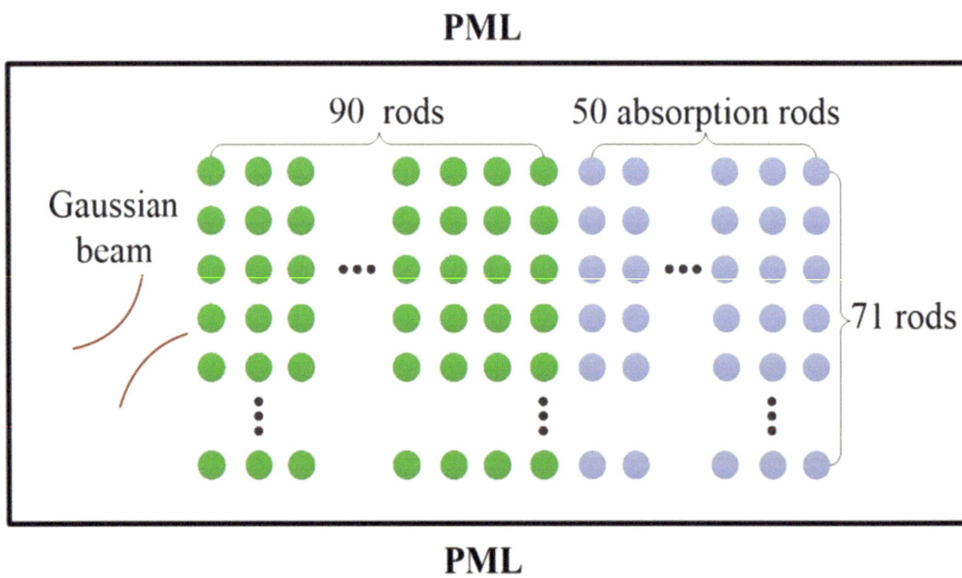

Figure 3. FDTD calculation model.

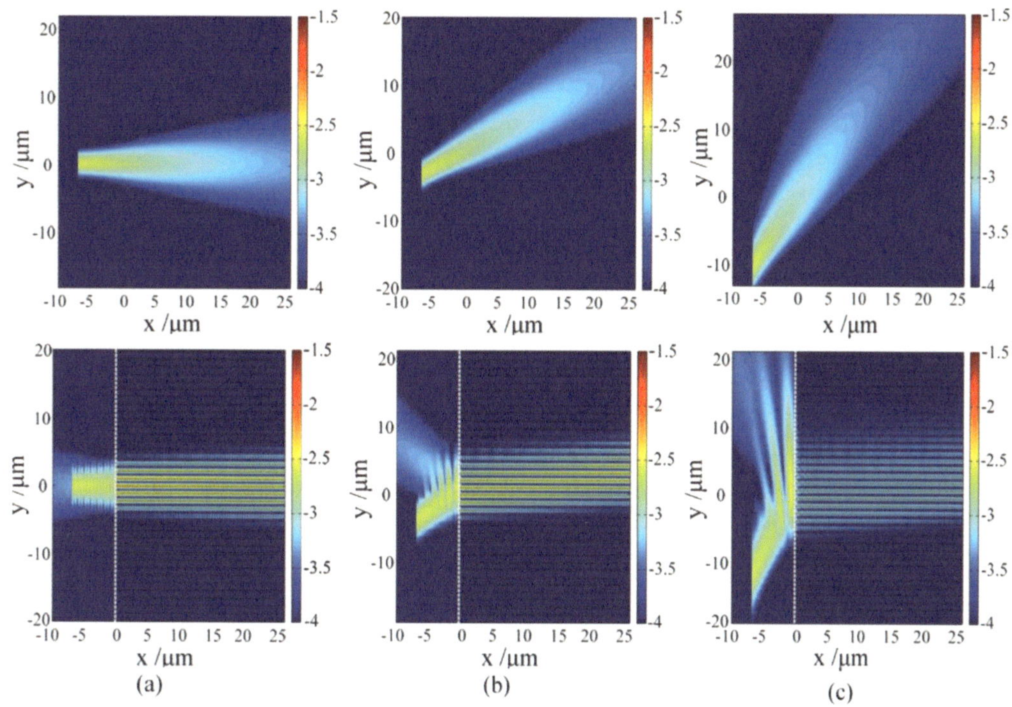

Figure 4. The field distribution when Gaussian beam irradiated to the air (top panel) and PC (lower panel) having different incident angles θ: (a) θ = 0°, (b) θ = 30°, and (c) θ = 30°. The white dashed line presents the interface between the air and PC, and the field distribution is plotted using a logarithmic color map .

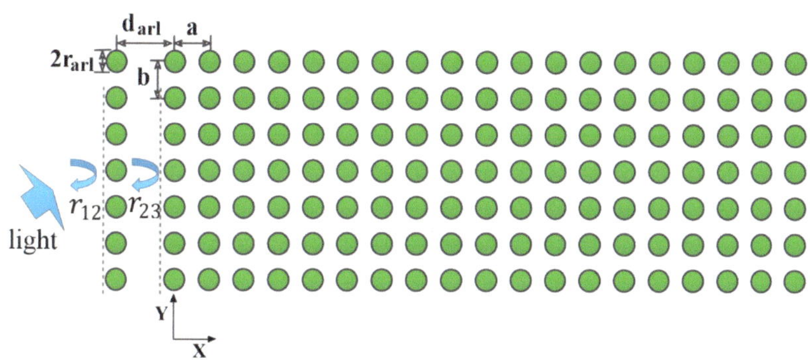

Figure 5. Schematic illustration of the ARL placed in front of the quasi-SC PC. The variable ARL parameters are the distance d_{arl} and the radius r_{arl} .

r_{arl}, and the distance between the ARL and the quasi-SC PC is d_{arl}. The parameters d_{arl} and r_{arl} should be well-designed to improve the coupling efficiency.

The coupling efficiency $\kappa = 1 - |r_{tot}|^2$ is directly determined by the total reflection coefficient r_{tot}, which can be written as:

$$r_{tot} = \frac{r_{12} + r_{23} e^{i2\beta}}{1 + r_{12} r_{23} e^{i2\beta}} \tag{6}$$

where $\beta = k_0 \cdot (d_{arl} - r - r_{arl})$ is the phase shift of the light beam as it crosses the ARL (k_0 denotes the wave vector in the air), r_{12} is the reflection coefficient of the ARL and r_{23} is the reflection coefficient of the semi-infinite quasi-SC PC in the air when the light beam is striking the PC surface from air. By applying the multiple scattering method [18], both r_{12} and r_{23} can be calculated.

We try to find the appropriate r_{arl} and d_{arl} values to reduce total reflection $|r_{tot}|^2$. First, we calculate the coupling efficiency for an initial value of r_{arl} with $r_{arl} = r$, and the resulting color map is shown in **Figure 6(a)**. In this figure, the red area corresponds to high coupling efficiency. As shown in **Figure 5**, there are two regions, which allow for high coupling efficiency over a relatively wide incidence angle. We can select the appropriate d_{arl} according to **Figure 1**. As shown in **Figure 5**, for instance, 3.28a for an incident angle of 0°–20°, or 3.44a for an incident angle of 20°–30°. Then, for the optimization of d_{arl}, r_{arl} can be further optimized by improving the coupling efficiency, as shown in **Figure 6(b)** and **(c)**. The d_{arl} values are fixed to be 3.28a and 3.44a, respectively. From **Figure 6(d)**, the enhanced coupling efficiency is described versus the

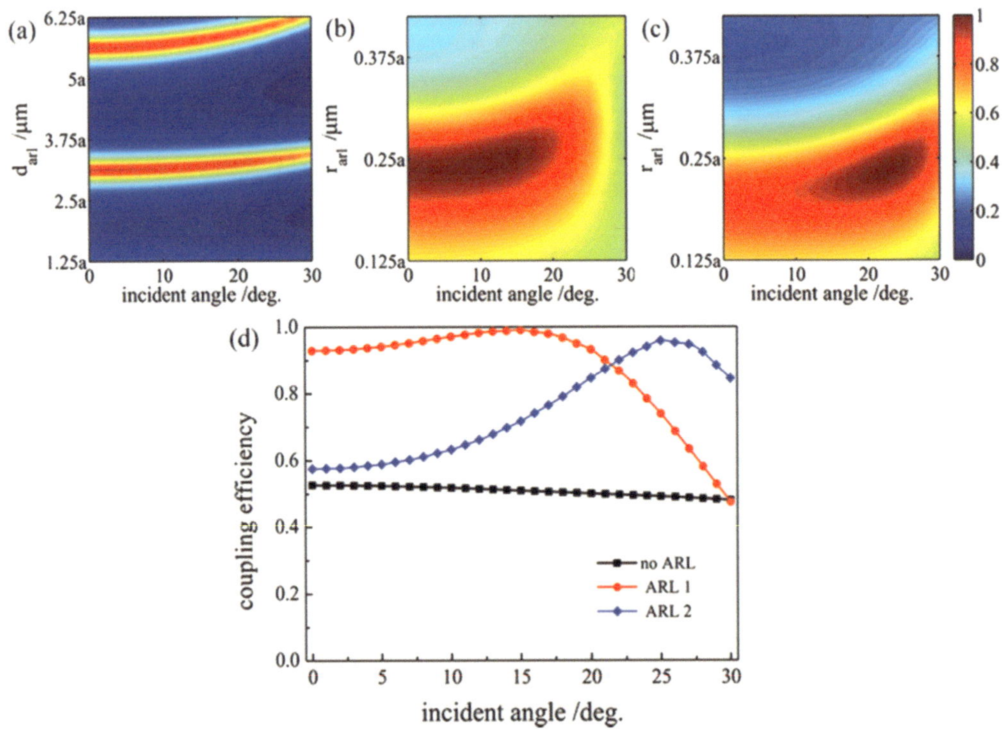

Figure 6. (a) The coupling efficiency as a function of d_{arl} and the incident angle, which d_{arl} and r_{arl} are fixed at 0.3a. (b) The coupling efficiency as a function of r_{arl} and the incident angle θ, which an optimized d_{arl} is fixed at *3.28a*. (c) The coupling efficiency as a function of r_{arl} and an incident angle θ which an optimized d_{arl} is fixed at *3.44a*. (d) Coupling efficiency as a function of the incident angle without and with ARL. The structure parameters of ARL 1 and ARL 2 are $d_{arl} = 3.28a$, $r_{arl} = 0.26a$ and $d_{arl} = 3.44a$, $r_{arl} = 0.26a$, respectively.

http://dx.doi.org/10.5772/67949

incident angle. The coupling efficiency is greatly improved over a wide incidence angle range by using the well-designed ARL.

In order to further improve the coupling efficiency, graded multiple ARLs are provided with same lattice constant (b = 2.3a) as shown in **Figure 7(a)**. They locate in front of the photonic crystal, which the distance is a. There are 20 layers in the graded multiple ARLs, and the rods radius is varying from the left to the right slowly, i.e., for the ith layer, the radius of the rods is $r_{ARLs}(i) = i(r/20)$, where i = 1, 2,..., 20. By using the multiple ARLs, the coupling efficiency is improved for nearly all incident angles, as shown in **Figure 7(b)**.

In order to verify the improvement of the coupling efficiency, the field distribution is calculated for the Gaussian beam at four angles of incidence emitted at a PC with multiple ARLs, and the resulting field distribution is described in **Figure 8(a)–(d)**, respectively. For multiple ARLs structure, the Gaussian beams are almost completely coupled into the PC with little reflection. Comparing **Figure 8** with **Figure 4**, it is indicated that the multiple ARLs can suppress the reflection.

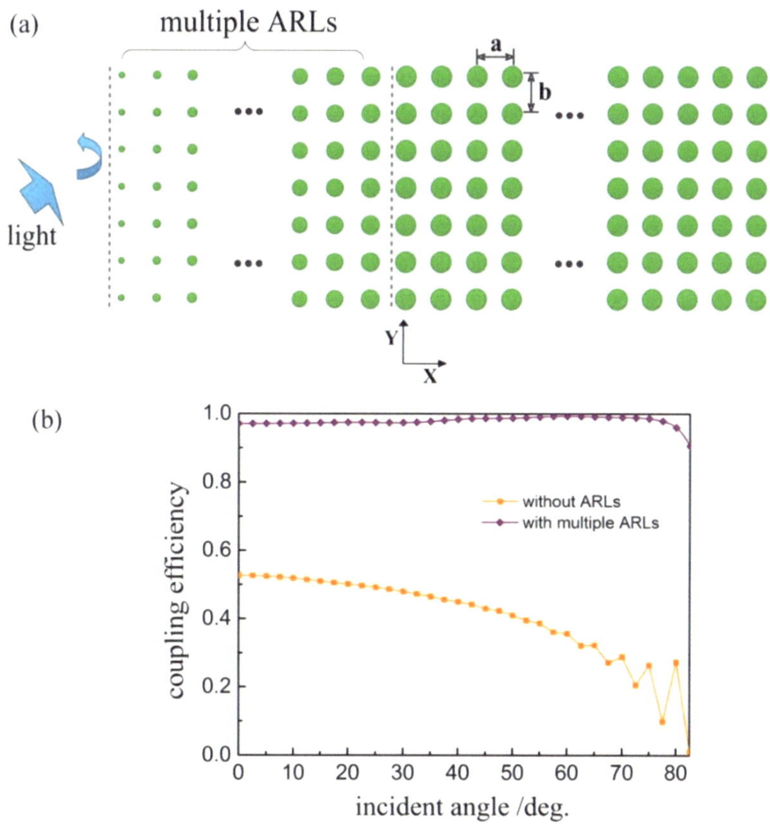

Figure 7. (a) Schematic illustration of the multiple ARLs. (b) Coupling efficiency as a function of the incident angle without and with multiple ARLs.

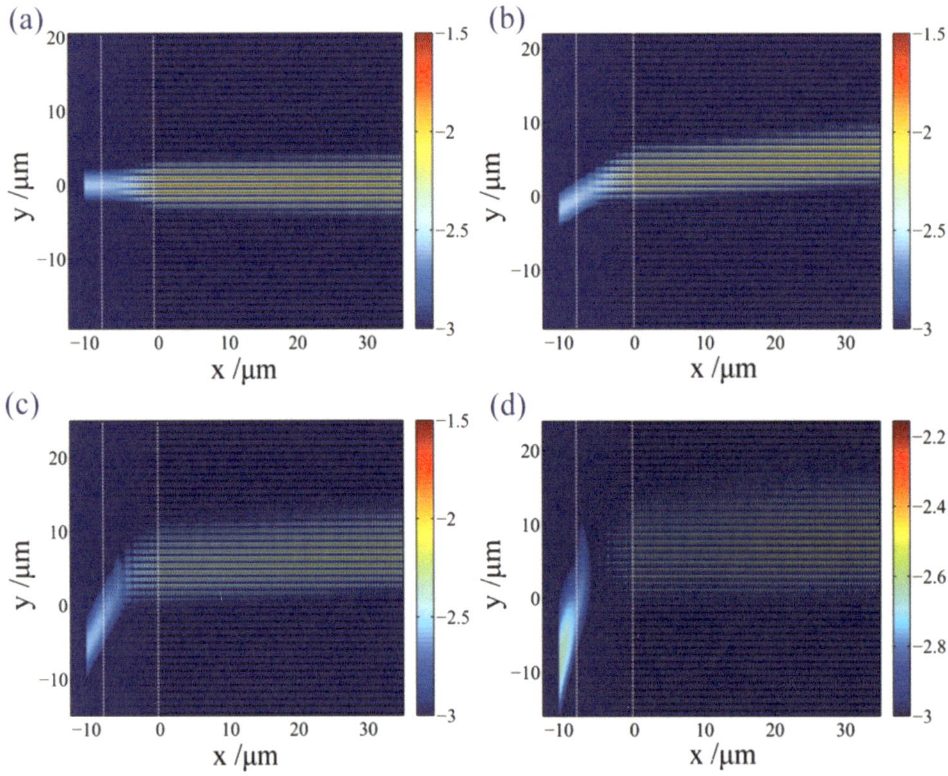

Figure 8. Field distribution when Gaussian beam emitting at the PC with multiple ARLs at different incident angles θ: (a) θ = 0°, (b) θ = 30°, (c) θ = 60°, and (d) θ = 75°. The white dashed line presents the interface, and the field distribution is described using a logarithmic color map.

3. Experimental model, results and discussion

We present the demonstration of a full-angle quasi-SC phenomenon based on pillar-type PCs in this section. Light waves are received into these pillar-type PCs from a wide range of incidence angles and can be positioned highly along the array of rods. We observed that the propagation length was 0.4 mm in the wavelength range of 1540–1570 nm. For on-chip photonic applications, this value might be sufficient propagation distance.

3.1. Experimental model

In order to increase the angular collimation range, a simple and feasible approach is to reduce the symmetry of the photonic construct by using a rectangular lattice [19]. Here, the 2D pillar-type PC composed of a rectangular lattice fabricated by patterning silicon on insulator (SOI) wafer, as presented in **Figure 9(a)**. By increasing a rectangular lattice of side ratio, a flat EFC across the entire Brillouin zone can be obtained. However, due to the small filling factor of the dielectric material, an excess ratio will result in leakage of the light wave in the other direction

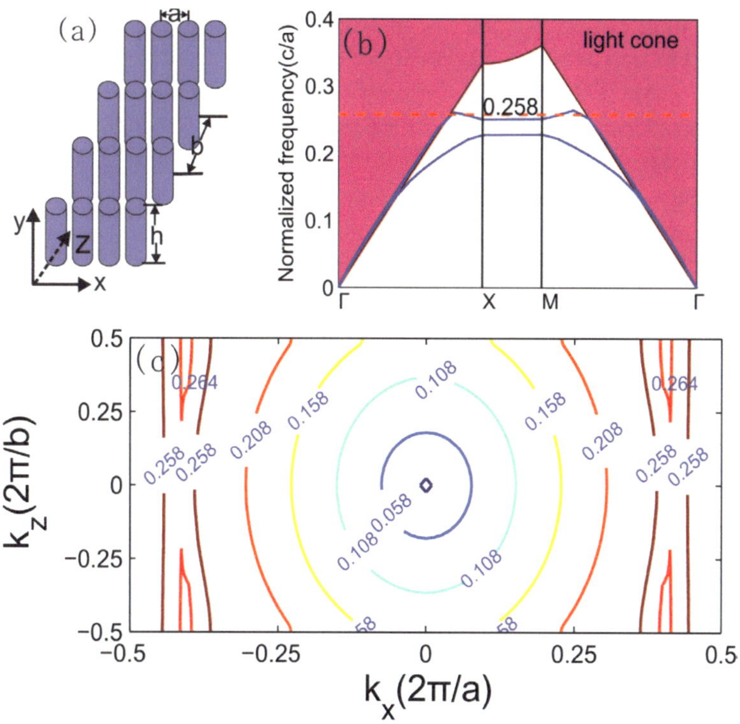

Figure 9. (a) Schematic diagram of a 2D pillar-type PC, which consists of a rectangular lattice formed by patterning the SOI wafer. (b) Band diagram of rectangular lattice PC. The dashed line indicates the frequency as 0.258c/a. The colored area describes a light cone. (c) An equifrequency contours of the second frequency band.

(z-direction). As a compromise between the full-angle quasi-self-collimation effect and the collimating efficiency, we choose a PC structure with $b/a = 2.4$. A top silicon layer with a height $h = 2{:}025a$ was used as a board layer. The buried oxide (BOX) layer forms a high refractive index-contrast structure with the air as the cladding. According to this particular structure, out-of-plane light confinement is achieved owing to total internal reflection. For the silicon and BOX, the refractive indices are $n_{Si} = 3.5$ and $n_{SiO2} = 1.5$, respectively. Using plane wave expansion (PWE), we can obtain the band diagrams for the TM-like polarization (with the E-field along the y-direction), as shown in **Figure 9(b)**. The colored area is a light cone, below which are guided mode to be positioned in the plate [20, 21].

The light propagation direction in PCs depends entirely on the gradient direction of the corresponding EFC, since the propagation direction is the same as the group velocity $v_g = \nabla k\omega$ (k), where ω is the optical frequency corresponding the wave vector k [22]. Thus, we can consider that the self-collimation effect is derived from the flat portion of the EFC. However, the non-diffractive propagation is confined owing to the use of a relatively broad beam of light, which the wave vectors are limited to the local flat portion of the EFC. For a narrow beam including a large range of k vectors, for instance, to obtain all-angle quasi-self-collimation, we require a flat EFC across the entire first Brillouin zone (as we discussed in the above section). Then, we calculate the corresponding EFC of the second frequency band as shown in **Figure 9(c)**.

It is noteworthy that, at the f = 0.258 c/a, the flat EFC is flat across the first Brillouin zone, which is expected to lead to a large angle self-collimation. To validate our prediction, 3D finite difference time domain (FDTD) method is used to perform the simulation. A Gaussian beam having a width of 0.25b at f = 0.258 c/a is emitted into the designed PC. The calculated electric field intensity is drawn in **Figure 10(a)**. The self-calibration is achieved even though the small beam circumference, which the light source contains a wide range of k vectors. The E_y field intensity is presented along the z-direction at x = 40a in **Figure 10(b)**. It is noteworthy that the EM energy is highly positioned along the array of a single nanorod. The E_y field intensity is only $1/10^5$ which located in the third row to each one side of this single nanorod array, for instance, a very weak electric field. It will ensure good confinement for light waves when placing of four rows of nanorods on each side of the central single nanorod array.

3.2. Experimental results and discussion

The parameters of the dielectric nanorod chain design is a = 400 nm, r = 160 nm, and h = 810 nm for which these operate at frequencies of optical communications. The PC is fabricated on a platform of silicon-on-insulator (SOI) wafers by four main processing steps. First, we require that thermal oxidation is needed to form an oxide layer, which serves as a hard mask for the purpose of transferring the pattern into the Si layer by using reactive ion etching (RIE). Second, we apply electron beam (E-beam) lithography to transfer this pattern to the high-sensitivity and high-resolution E-beam Resist ZEP-520. Third, the hard mask is etched using RIE in a CHF_3 plasma. Finally, after using the inductively coupled plasma (ICP) etching process, which a thermally grown SiO_2 layer is used as a hard mask, the resulting dielectric rods are about 810 nm. **Figure 11** shows the scanning electron microscope (SEM) image of the nanorods, the rods have sidewall profiles approximately 90° to the substrate; the screw thread is caused by deep etching with using the Bosch process.

To show the large angle of self-collimation effect, three different samples with tilted PC areas were designed in the experiment. We introduce light waves to irradiate the PCs by a waveguide, which a width is 3 m as shown in left panels (a), (b), and (c) of **Figure 12**, it is designed that the tilt angles with respect to the incident waveguide are set to be 0°, 45°, and 75°, respectively. Theoretically, note that self-collimation can even be observed for an incident angle of almost 90°. However, due to impedance mismatch, larger incident angle causes more

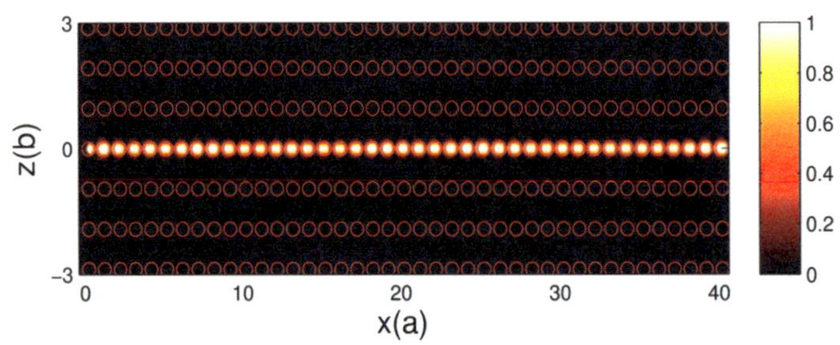

Figure 10. (a) The field intensity of the electric field component Ey (TM polarization) calculated using the FDTD method. (b) The Ey field intensity along the z-direction at x = 40a.

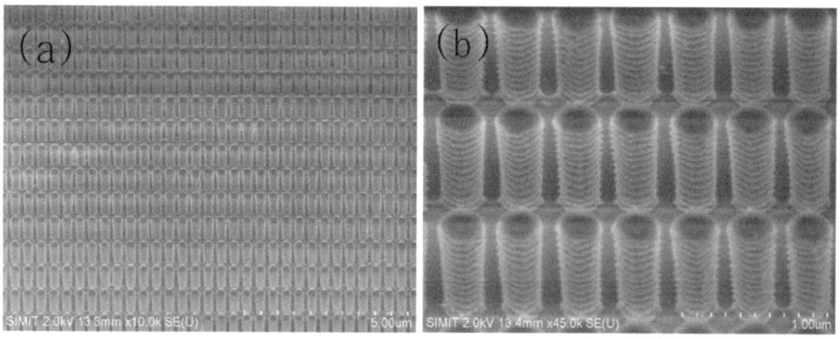

Figure 11. SEM images of the fabricated device: (a) Rectangle lattice pillar-type PCs with a = 400 nm, b = 2.4a, and h = 2.025a. (b) Magnified image.

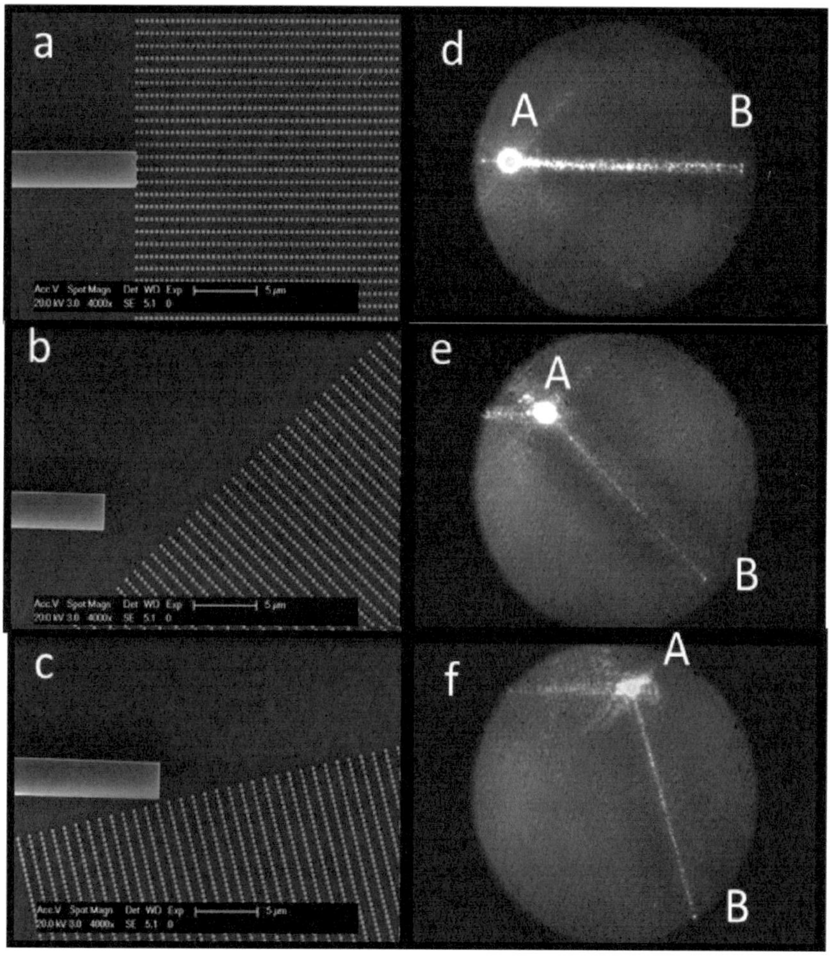

Figure 12. The left part of FIG: SEM images of PC samples with different tilt angle (a) 0°, (b) 45°, and (c) 75°. The right part of FIG: (d), (e), and (f): a top view of the scattered light, which indicates the ray trace of the light beam captured by the infrared camera. A and B are marks of the light spots generated by strong scattering at the interface.

reflection losses. In our experiments, the maximum incident angle is set at 75°, so that our camera can be used to capture a sufficient amount of light scattering due to structural defects. The TM polarized light from the tunable laser source is introduced into the lensed fiber. In order to visualize the EM energy transfer in the PC, we use an infrared camera to capture the ray trajectory of the EM energy (through the scattered light). Light traces were observed clearly in the 1540–1570 nm wavelength range. Coupled resonator waveguides (CROWs) based on weak coupling between high-Q cavities can also achieve functions similar to those of our guided-wave PC structures. Unfortunately, their operating frequencies are limited to the resonant frequencies of the whispering gallery modes of individual rods [23]. In contrast, our SC effect is based on the dispersion relation of the entire PC, operating over a bandwidth wider than the resonant frequency of a single rod. **Figure 12(d)–(f)** corresponds to capturing the results at 1550 nm wavelength with incident angles of 0°, 45°, and 75°, respectively. At the point of A, this light spot is contributed by strong scattering which located on the interface of the waveguide surface and the PCs. At the point of B, this light spot which situated at the opposite end of the PC region, also provides an interface for optical scattering. There are 1000 lattice periods in the routing from A and B, corresponding to the propagation length value of 0.4 mm. For the on-chip optical interconnection, it is a very considerable length scale. For the large-angle self-collimation effect, previous experiments shown that the light propagating had at least 100 lattice periods without significant divergence [24]. In our experiment, the propagation length can be as high as 1000 lattice periods, and ray trajectory of the light is clearer than the previous.

For a detailed description of the quantitative results, the field intensity of the nanorod array determined from the FDTD calculations [25] is plotted in **Figure 13(a)**. The experimentally determined propagation loss of the self-collimated beam is measured by plotting the relative intensity along the nanorods in **Figure 12(d)**. The experimental propagation loss is determined to be 17.6 dB/mm, which is higher than the simulation result as shown in **Figure 13**. It is believed that the extra propagation loss is mainly attributed to the

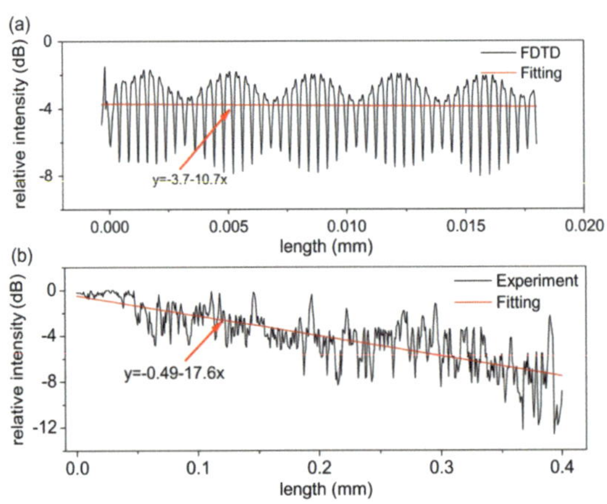

Figure 13. (a) Theoretical propagation loss from the 3D FDTD simulation. (b) Experimental propagation loss of the self-collimated beam determined by plotting the relative intensity (in logarithmic scale) along the nanorods in **Figure 11(d)**.

http://dx.doi.org/10.5772/67949

roughness of the pillars, but we can reduce this roughness by improving the manufacturing process. Our simulation shows that the coupling efficiency can be as high as 92, 90, and 45% corresponding to the incident angles of 0°, 45°, and 75°, respectively. Furthermore, through modifying the interface [26, 27], the coupling efficiency can be enhanced by suppressing scattering at the point A. Therefore, it is believed that the millimeter-scale propagation length can be easily observed with the rod structure based on the above-mentioned improvements.

4. Conclusion

In conclusion, an all-angle quasi-self-collimation effect can be achieved by changing the rectangular-lattice symmetry of photonic crystal. The quasi-self-collimation effect is identified and quantified by a straightness factor L, which is based on the least-squares method. With the decrease of straightness factor L, the photonic crystal possesses a more powerful self-collimation effect. Besides, the efficiency of light coupling to the quasi-self-collimation photonic crystal is investigated and is greatly improved by applying a carefully designed antireflection structure. Experimentally, we have designed, fabricated, and characterized a structure based on silicon nanorods that exhibits the all-angle quasi-SC phenomenon. The millimeter-scale propagation length and broad wavelength range exhibited by our structure may be sufficient for on-chip photonic applications. Our recent study shows such all-angle quasi-self-collimation effect can also be used to design a compact silicon nanoantenna array for high-efficiency vertical light emission [28]. Such a large-angle SC phenomenon resulting from the nanorod structure may also find applications in sensing and optofluidics.

Acknowledgments

This research was sponsored by the National Natural Science Foundation of China (Grant Nos. 61475180, and 61275112), and the Science and Technology Commission of Shanghai Municipality (Grant Nos. 14JC1407601, and 14JC1407602). The figures in this chapter were extracted from our previous papers (H. Li et al, IEEE Photon. J. 5, 2201306 (2013) and M. Li et al., IEEE Photon. J. 7, 4700108 (2015)).

Author details

Wei Li*, Hao Li, Ming Li, Haiyang Huang and Fuwan Gan

*Address all correspondence to: waylee@mail.sim.ac.cn

State Key laboratory of Functional Materials for Informatics, Shanghai Institute of Microsystem and Information Technology, Chinese Academy of Sciences, China

References

[1] H. S. Eisenberg, Y. Silberberg, R. Morandotti, and J. S. Aitchison, "Diffraction management," Phys. Rev. Lett., Vol. 85, pp. 1863–1866, Aug. 2000.

[2] W. Chen, and D. L. Mills, "Gap solitons and the nonlinear optical response of superlattices," Phys. Rev. Lett., Vol. 58, pp. 160–163, Jan. 1987.

[3] B. J. Eggleton, R. E. Slusher, C. M. de Sterke, P. A. Krug, and J. E. Sipe, "Bragg grating solitons," Phys. Rev. Lett., Vol. 76, pp. 1627–1630, Mar. 1996.

[4] H. Kosaka, T. Kawashima, A. Tomita, M. Notomi, T. Tamamura, T. Sato, and S. Kawakami, "Self-collimating phenomena in photonic crystals," Appl. Phys. Lett., Vol. 74, no. 9, pp. 1212–1214, 1999.

[5] L. L. Zhang, Q. W. Zhan, J. Y. Zhang, and Y. P. Cui, "Diffraction inhibition in two-dimensional photonic crystals," Opt. Lett., Vol. 36, no. 5, pp. 651–653, Mar. 2011.

[6] Y. Jin, and S. He, "Canalization for subwavelength focusing by a slab of dielectric photonic crystal," Phys. Rev. B., Vol. 75, p. 195126, May. 2007.

[7] P. A. Belov, C. R. Simovski, and P. Ikonen, "Canalization of subwavelength images by electromagnetic crystals," Phys. Rev. B., Vol. 71, p. 193105, May. 2005.

[8] R. S. Chu, and T. Tamir, "Group velocity in space-time periodic media," Electron. Lett., Vol. 7, no. 14, pp. 410–412, 1971.

[9] T. Yamashita, and C. J. Summers, "Evaluation of self-collimated beams in photonic crystals for optical interconnect," IEEE J. Sel. Area. Comm., Vol. 23, no. 7, pp. 1341–1347, Jul. 2005.

[10] C. Lv, W. Li, X. Y. Jiang, and J. C. Cao, "Far-field super-resolution imaging with a planar hyperbolic metamaterial lens," Europhys. lett., Vol. 105, no. 2, Jan. 2014.

[11] H. Li, A. Wu, W. Li, X. Lin, C. Qiu, Z. Sheng, X. Wang, S. Zou, and F. Gan, "Millimeter-scale and large-angle self-collimation in a photonic crystal composed of silicon nanorods," IEEE Photonics J., Vol. 5, no. 2, pp. 2201306–2201306, Apr. 2013.

[12] P. T. Rakich, M. S. Dahlem, S. Tandon, M. Ibanescu, M. Soljacic, G. S. Petrich, J. D. Joannopoulos, L. A. Kolodziejski, and E. P. Ippen, "Achieving centimetre-scale supercollimation in a large-area two-dimensional photonic crystal," Nat. Mater., Vol. 5, pp. 93–96, Feb. 2006.

[13] T. Shih, A. Kurs, M. Dahlem, G. Petrich, M. SoljaCic, E. Ippen, L. Kolodziejski, K. Hall, and M. Kesler, "Supercollimation in photonic crystals composed of silicon rods," Appl. Phys. Lett., Vol. 93, no. 13, p. 131111, 2008.

[14] V. Mocella, S. Cabrini, A. S. P. Chang, P. Dardano, L. Moretti, I. Rendina, D. Olynick, B. Harteneck, and S. Dhuey, "Self-collimation of light over millimeter-scale distance in a quasi-zero-average-index metamaterial," Phys. Rev. Lett., Vol. 102, p. 133902, Apr. 2009.

[15] M. Qi, E. Lidorikis, P. T. Rakich, S. G. Johnson, J. Joannopoulos, E. P. Ippen, and H. I. Smith, "A three-dimensional optical photonic crystal with designed point defects," Nature, Vol. 429, no. 6991, pp. 538–542, Jun. 2004.

[16] Z. Lu, S. Shi, J. A. Murakowski, G. J. Schneider, C. A. Schuetz, and D. W. Prather, "Experimental demonstration of self-collimation inside a three-dimensional photonic crystal," Phys. Rev. Lett., Vol. 96, p. 173902, May 2006.

[17] H. Abdi, "The method of least squares," in Encyclopedia of research design. Thousand Oaks, CA, USA: Sage. pp. 705–708, 2007.

[18] H. Li, W. Li, J.-J. Du, A.-M. Wu, C. Qiu, Z. Sheng, X. Wang, S.-C. Zou, and F.-W. Gan, "Optical total reflection and transmission with mode control in a dielectric subwavelength nanorod chain," Chin. Phys. B., Vol. 22, no. 11, p. 117807, Nov. 2013.

[19] Y. Xu, X. Chen, S. Lan, Q. Guo, W. Hu, and L. Wu, "The all-angle self-collimating phenomenon in photonic crystals with rectangular symmetry," J. Opt. A: Pure Appl. Op., Vol. 10, no. 8, p. 085201, 2008.

[20] J. Witzens, M. Loncar, and A. Scherer, "Self-collimation in planar photonic crystals," IEEE J. Sel. Top. Quantum Electron., Vol. 8, no. 6, pp. 1246–1257, nov/dec. 2002.

[21] Li. Ming, Li. Wei, Haiyang Huang, Jing Wang, Li. You, Wu. Aimin, Zhen Sheng, Xi Wang, Shichang Zou, Fuwan Gan, "All-angle quasi-self-collimation effect in a rod-type silicon photonic crystal," IEEE Photonic J., Vol. 7, no. 1, p. 4700108, Feb. 2015.

[22] J. D. Joannopoulos, S. G. Johnson, J. N. Winn, and R. D. Meade, "Photonic crystals: molding the flow of light (Second Edition)," 2nd ed. Princeton University Press, Princeton, USA. Feb. 2008.

[23] A. Yariv, Y. Xu, R. K. Lee, and A. Scherer, "Coupled-resonator optical waveguide: a proposal and analysis," Opt. Lett., Vol. 24, no. 11, pp. 711–713, Jun. 1999.

[24] L. Gan, F. Qin, and Z. Li, "Broadband large-angle self-collimation in two-dimensional silicon photonic crystal," Opt. Lett., Vol. 37, no. 12, pp. 2412–2414, Jun. 2012.

[25] W. Li, Z. Liu, X. Zhang, and X. Jiang, "Switchable hyperbolic metamaterials with magnetic control," Appl. Phys. Lett., Vol. 100, no. 16, pp. 16108-1–16108-5, Apr. 2012.

[26] V. Mocella, P. Dardano, L. Moretti, and I. Rendina, "Influence of surface termination on negative reflection by photonic crystals," Opt. Exp., Vol. 15, no. 11, pp. 6605–6611, May. 2007.

[27] L. Lin and Z. Li, "Sensitivity to termination morphology of light coupling in photonic-crystal waveguides," Phys. Rev. B., Vol. 69, no. 19, pp. 193103-1–193103-4, May. 2004.

[28] Haiyang Huang, Li. Hao, Li. Wei, Wu. Aimin, Xin Chen, Xuefeng Zhu, Zhen Sheng, Shichang Zou, Xi Wang, and Fuwan Gan, "High-efficiency vertical light emission through a compact silicon nanoantenna array," ACS Photonics, Vol. 3, no. 3, pp. 324–328, Feb. 2016.

Chapter 11

Recent Advances in Silicon Photodetectors Based on the Internal Photoemission Effect

Maurizio Casalino

Additional information is available at the end of the chapter

http://dx.doi.org/10.5772/67720

Abstract

Silicon technologies provide an excellent platform in order to realize microsystems where photonic and microelectronic functionalities are monolithically integrated on the same substrate. In recent years, a lot of passive and active silicon photonic devices have been optimized to work at telecom wavelengths where, unfortunately, silicon has a neglectable optical absorption due to its bandgap of 1.12 eV. Although silicon cannot detect wavelengths above 1.1 μm, in recent years, tremendous advances have been made in order to make it suitable for operation in the near-infrared spectrum. One of the approaches is to take advantage of the internal photoemission effect through a Schottky junction where a metal absorbs the incoming radiation and emits hot carriers into silicon making sub-bandgap detection possible. The present chapter describes the more recent advances in the field of the silicon photodetectors based on the internal photoemission effect showing as devices based on new emerging materials and complex nanostructure are leading this family of device to compare favorably with the well-established technologies commonly used for telecom wavelengths based on germanium and III–V semiconductors.

Keywords: silicon, internal photoemission effect, photodetectors, near-infrared, graphene

1. Introduction

Silicon (Si) photonics is a discipline of paramount importance in the field of integrated optics, and, nowadays, new Si-based commercial products are already available on the market [1, 2]. Although Si photodiodes are commonly realized to operate at visible wavelengths, their development at wavelengths of interest for telecommunications is not a trivial task to reach because Si is transparent at wavelengths above 1.1 μm. Conventional near-infrared Si-based PDs are based on the integration with III–V compound semiconductors as: InGaAs [3, 4], Ge [5], and SiGe [6]. Concerning InGaAs, a hybrid approach [7] is commonly followed because of the high lattice mismatch between InGaAs and Si (8.1%) making the monolithic integration with Si very difficult [8]. On the other hand, the growth of Ge on a Si substrate can be traditionally obtained with a two-step epitaxial growth technique [8, 9] in order to realize a buffer layer where a pure crystalline Ge can be epitaxially grown on Si substrates [10]. However, the requirement for a buffer layer that causes problems in both thermal budget and planarity [11, 12] prevents its monolithic integration on Si. Indeed, even if Si four-channel optical receivers based on GePDs have been successfully realized by both Intel [13] and Luxtera [14], it is worth noting that they are only flip-chip mounted to a Si electronic circuitry. In alternative, there has been progress in fabricating SiGe-based PDs. SiGe has been considered as semiconductor because it is still implemented in the CMOS process flow [6], but, unfortunately, the Ge content of the available layer is estimated to be 25–35% [15], and consequently, the detection wavelength is lower than 1200 nm [16]. In order to circumvent all these drawbacks, many approaches have been followed to realize silicon compatible PDs [17]. In particular, very promising is the approach based on the internal photoemission effect (IPE), that is, the exploitation of photon-assisted transmission of hot carriers across a potential barrier at metal-semiconductor interfaces. In the last decade, much effort has been focused on this field, and both impressive results and new structures have been reported. Indeed, IPE has been combined with nanoscale metallic structures, including metal stripes supporting surface plasmon polaritons (SPPs) [18, 19], Si nanoparticles (NPs) [20], metallic gratings [21], and antennas [22]. In addition, IPE has been combined with new structures based on two-dimensional materials (like graphene) able to replace metal in the Schottky junction [23]. After these research processes, IPE-based Si PDs show the potentialities to compare favorably with Ge-based devices, while also offering new advantageous characteristics. Indeed, IPE-based PDs are very fast thanks to the unipolar nature of the Schottky junction, and they have already shown the capability to be monolithically integrated with Si-based charge coupled devices for infrared applications [24]. IPE-based Si PDs were already summarized in a previous work of some years ago [25].

In the present chapter, the huge advances made in this field, are reported. In the first section, IPE theory will be elucidated in detail taking advantage of the new recent developments. Then, the main structures reported in the literature, and the most significant results obtained in recent years will be reviewed and discussed, comparing the performance of devices based on the different approaches. In particular, the second section will illustrate the state-of-the-art of the main both surface-illuminated and waveguide IPE-based Si PDs reported in literature, while the third section will be dedicated on devices able to combine IPE with plasmonic effects. Finally, in the fourth section, the potential of new devices taking advantage of both newly emerging materials and smart structures will be addressed.

2. Internal photoemission theory

IPE is the optical excitation of electrons in the metal to energy levels above the Schottky barrier and then the transport of these electrons to the energy bands of the semiconductor. A band diagram for a metal/p-Si junction is sketched **Figure 1**.

It is well-known that IPE is typically a very weak effect due to many factors: (1) the low absorption due to high reflectivity of the metal layer, (2) the conservation of momentum during carrier emission over the potential barrier which lowers the carriers emission probability into semiconductor, (3) the excitation of carriers lying in states far below the Fermi energy, which get very low probability to overcome the Schottky barrier. The result is that IPE-based devices are characterized by low responsivity R (i.e., the ratio between the photogenerated current I_{ph} and the incoming optical power P_{inc}):

$$R = \frac{I_{ph}}{P_{inc}} = \frac{\lambda\,[\mathrm{nm}]}{1242} \cdot \eta_e \qquad (1)$$

where λ is the wavelength of the incident photon expressed in nm, and η_e is the external quantum efficiency of the device, that is, the number of charge carriers collected per incident photon. Alternatively, the number of carriers collected per absorbed photons is called the internal quantum efficiency η_i.

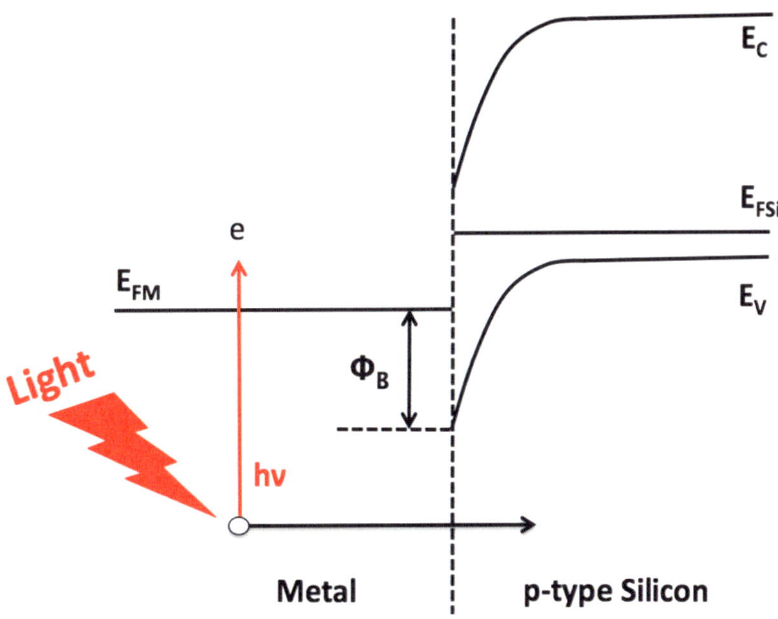

Figure 1. Energy band diagram for a metal/p-type Si junction (E_{FM} and E_{FSi} are the metal and Si Fermi level. E_C and E_V are the conduction and valence band energy of Si and ϕ_B is the Schottky barrier of the metal/p-Si junction.

Fowler described the first electron model concerning electron photoemission from metal into vacuum in the 1931 [26]. Subsequently, in the 1960s, Cohen et al. [27] modified the Fowler's model in order to take into account the carrier photoemission from a metal into a semiconductor, elaborating the commonly used formula:

$$\eta_{\mathrm{i}} = C \cdot \frac{(h\nu - \phi_{\mathrm{B}})^2}{h\nu} \tag{2}$$

where ϕ_B is the Schottky barrier, $h\nu$ is the energy photon, $C = 1/8(E_F + \phi_B)$ is named the quantum efficiency coefficient, and E_F is the metal Fermi level. In this context, it is worth noting that Eq. (2) was obtained under the zero temperature approximation and for thick metal films.

Subsequently Elabd and Kosonocky reviewed the IPE model in order to obtain better agreement with the experimental data [28] and, always under the zero temperature approximation, they obtained approximately the same Eq. (2) but characterized by a different quantum efficiency coefficient that results $C = 1/8\phi_B$. In addition, the authors extended the photoemission model to the case of thin metal films by introducing a multiplicative gain factor arising from the increased escape probability of the hot carriers into the metal due to the scattering with metal surfaces [28]. In the last years, many other authors have investigated IPE theory and new physical models more and more complex have been proposed [29–31]. Recently, Scales and Berini extended the Elabd and Kosonocky's theory in order to take into account of the escape probability through a double Schottky barrier [32]. A very intriguing approach was proposed by Vickers, who derived a theoretical model in which the estimation of the internal quantum efficiency η_i is given by the product of two factors [33]:

$$\eta_i = F(\phi_{\mathrm{B}}) \cdot P(d) \tag{3}$$

where the Fowler factor F is the fraction of the excited carriers having appropriate momentum and energy to overcome the potential barrier, P is a scattering term taking into account the probability that one of these excited (hot) carriers will be emitted over the potential barrier after scattering by cold carriers and the metal boundary surfaces, d is the metal thickness, and ϕ_B is the junction potential barrier. The main advantage of the Vickers model is that the zero temperature approximation is removed making it valid at any temperature. Indeed, the factor F incorporating the temperature dependence is shown in the following formula:

$$F = \frac{1}{4 E_F h\nu}\left(\frac{(h\nu - \phi_{\mathrm{B}})^2}{2} + (k_{\mathrm{B}} T)^2\left\{\frac{\pi^2}{6} + \sum_{i=1}^{\infty}\frac{1}{i^2}\left(-e^{-\frac{h\nu-\phi_{\mathrm{B}}}{k_{\mathrm{B}}T}}\right)^i\right\}\right) \tag{4}$$

where k_B ant T are the Boltzmann constant and the absolute temperature, respectively. It is worth noting that, in the limit $T \rightarrow 0$, Eq. (4) reduces to the well-known Eq. (2). Recently, Casalino [34] showed that by making a change of variables $m = (h\nu - \phi_B)/k_BT$ and putting in evidence the factor $(k_BT)^2$ in (4), it is possible to define a F^* factor as:

$$F^* = \frac{4 E_{\mathrm{F}} h\nu}{(k_{\mathrm{B}} T)^2} F = \left(\frac{m^2}{2} + \frac{\pi^2}{6} + \sum_{i=1}^{\infty}\frac{1}{i^2}\left(-e^{-m}\right)^i\right) \tag{5}$$

Three terms of Eq. (5): $m^2/2$, $\pi^2/6$ and the polylogarithm function $\sum_{i=1}^{\infty}\frac{1}{i^2}(-e^{-m})^i$ have been plotted in **Figure 2** by varying m between 10^{-6} and 2×10^5 in order to understand the weight that any term plays into (5). It is possible to estimate that F^* approaches $m^2/2$ within 10% for $m \geq 5.43$, within 5% for $m \geq 7.91$ and within 1% for $m \geq 18$.

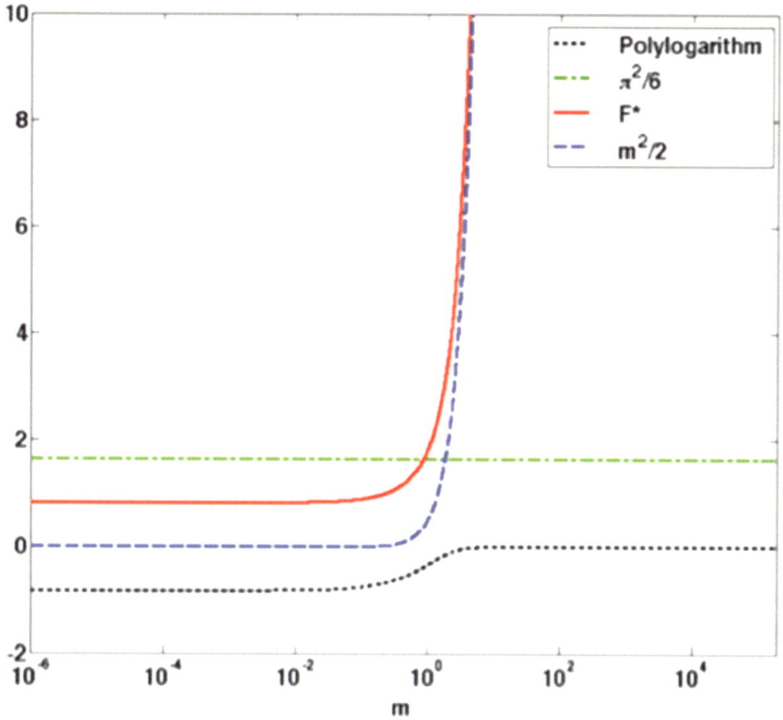

Figure 2. Plot of the F^* factor as a function of m. Three terms forming F^*: $m^2/2$, $\pi^2/6$ and the polylogarithm function are reported, too. X-axis is plotted on log scale.

Most important, Casalino found that F^* approaches $m^2/2 + \pi^2/6$, that is, it is possible to neglect the polylogarithm term and to reduce Eq. (4) to the following Eq. (6) [34]:

$$F = \frac{1}{8\, E_F\, h\nu}\left((h\nu - \phi_B)^2 + \frac{(k_B\, T\pi)^2}{3} \right) \tag{6}$$

within 10% when $h\,\nu - \phi_B \geq 0.035$ eV, within 5% when $h\,\nu - \phi_B \geq 0.046$ eV, and within 1% when $h\,\nu - \phi_B \geq 0.074$ eV. It is worth noting that Eq. (6) is a slight modification of Eq. (2) but can be used also at room temperature in the limit of the aforementioned errors.

On the other hand with reference to Eq. (3), the scattering term P allows extending IPE theory to a thin film where the escape probability of carriers excited in the metal is increased. Because the P term derived by Vickers is further complex, he proposed the following analytical approximation, very simple but with no physical meaning [33]:

$$P = \frac{L^*}{d} \cdot \sqrt{1 - e^{-\frac{d}{L^*}}} \tag{7}$$

where L^* is the mean free path in the metal, and d is the metal thickness. Eq. (7) is valid for thin films but in the limit $d/L^* > 0.2$. By following the same line of reasoning, more recently Casalino introduced the following Eq. (8) able to extend the operating range in the limit $d/L^* > 0.002$ [34]:

$$P = \frac{L^*}{d} \cdot \left(\sqrt{1 - e^{-\frac{d}{L^*}}} + 0.1 \cdot e^{-4.1\frac{d}{L^*}}\right) \tag{8}$$

Because the lower the metal thickness the higher the scattering term P, there is always an advantage to work with very thin metal thickness to the point that it was recently to proposed to replace metal with two-dimensional material such as graphene able to form Schottky junction with Si [23]. Finally, it should be mentioned that device efficiency can also be increased by applying a reverse bias to the Schottky junction, and this increase is due to the lowering of the Schottky barrier ϕ_B when a reverse bias is applied to the junction (image-force effect [35]).

It is worth noting that in this section we have focused only on the device efficiency of the PDs because it is the Achilles' heel of all IPE-based devices. However, many others figures of merit are useful in order to compare different PDs: bandwidth, noise equivalent power (NEP), and voltage operation are the main specifications of any commercial PD datasheet, and their definitions can be found everywhere [36].

3. Surface-illuminated and waveguide Schottky PDs

Schottky surface-illuminated PDs are typically less responsive than Schottky waveguide PDs where the optical power is confined close to the metal-semiconductor interface and can be absorbed along its propagation. However, in some cases, the surface-illuminated structures are the only option: for instance, in imaging applications where the vision can be improved in critical conditions (such as smoke and fog) thanks to reduced scattering at NIR wavelengths [37] or in reflectography applications where the transparency of most pigments to NIR wavelengths has been used to investigate ancient paintings [38]. Historically, surface-illuminated Schottky PDs have been used in the field of Schottky-barrier infrared focal-plane array (FPA) technology [24]. PtSi/pSi PD is the most popular surface-illuminated Schottky-barrier device, commonly used for detection in the 3–5 μm spectral range [39]. However, this device is characterized by two main drawbacks: low quantum efficiency (about 1%) and low operating temperature requirements (77 K or below). In order to overcome these drawbacks, in 2006, Casalino et al. proposed to incorporate a Schottky junction inside a Fabry-Perot optical microcavity in order to enhance the device efficiency at both near-infrared wavelengths and room temperature [40, 41]. Subsequently, in 2012, after a first experimental proof-of-concept demonstration [42, 43], the same authors fabricated and characterized a new surface-illuminated Schottky device based on a resonant cavity Fabry-Perot structure. The device was formed by a dielectric bottom mirror, a metallic top mirror and, in the middle, a silicon cavity [44]. The dielectric bottom mirror was a distributed Bragg reflector (DBR) formed by alternating $\lambda/4$-layers of amorphous hydrogenated silicon (a–Si:H) and silicon nitride (Si_3N_4), while the top mirror was realized by means of a copper (Cu) layer able to work as both an absorber and an optical mirror at the same time. It was demonstrated that, when the DBR mirror reflectivity approaches the reflectivity of the metallic top mirror, that is, when critical coupling conditions are fulfilled, the maximum responsivity can be obtained at the cavity resonance wavelengths. The critically coupled Cu/pSi Fabry-Perot PD exhibited a maximum responsivity of 0.063 mA/W around

1550 nm and measurements of junction capacitance in the pF range encouraged the pursuit of greater bandwidth making possible to operate at several GHz. On the other hand, the dark current density is reported to be as high as 28 mA/cm^2 at −1 V. More recently, Desiatov et al [45]. have demonstrated surface-illuminated Aluminum(Al)/Si Schottky PDs at near-IR wavelengths based on pyramidally shaped devices created in Si by potassium hydroxide (KOH) anisotropic etching. The advantage of KOH etching is that it is possible to fabricate plasmonic devices with a nanometric active area, without requiring sophisticated equipments such as electron beam lithography (EBL) or focused ion-beam (FIB). SEM micrographs of the fabricated device are shown in **Figure 3(a)** and **(b)**. The Si pyramids arranged in an array structure work as efficient and broadband light concentrators able to collect the light from a large area and to confine it into a small active pixel area, thereby providing high responsivity and low dark current at the same time. The responsivity of the device at −0.1 V was found to be 5, 12, and 30 mA/W for incident optical wavelengths of 1550 nm, 1300 nm, and 1064 nm, respectively. Moreover, the device showed a dark current of 80 nA at −0.1 V. The authors claim in this work that the efficiency is enhanced, not only by the increased absorption, but also by the nanoscale apex of the pyramid able to increase the escape probability from Al into Si.

Concerning waveguide Schottky PDs, in 2008, Zhu et al. [46] described the first nickel silicide (NiSi) Schottky PD integrated into a silicon-on-insulator (SOI) waveguide working at both NIR wavelengths and at room temperature. In this case, the author reports a responsivity, bandwidth, and dark current of 4.6 mA/W, 2 GHz and 3 nA at −1 V of reverse bias applied,

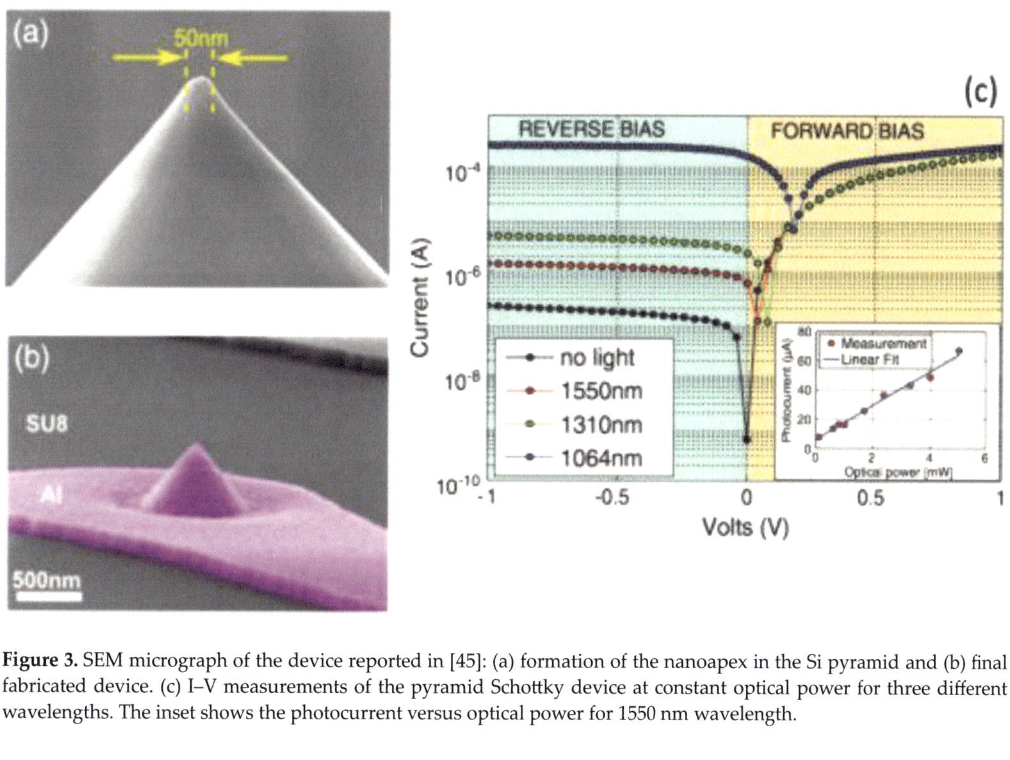

Figure 3. SEM micrograph of the device reported in [45]: (a) formation of the nanoapex in the Si pyramid and (b) final fabricated device. (c) I–V measurements of the pyramid Schottky device at constant optical power for three different wavelengths. The inset shows the photocurrent versus optical power for 1550 nm wavelength.

respectively. A similar device from the same research group, but based on a metal-semiconductor-metal (MSM) configuration, shows both higher responsivity and larger dark current [47]. It is worth noting that in the aforementioned guiding structures, the absorbing metal is always deposited along the direction of the propagating light, but another possibility is that the active metal layer is placed on the vertical exit surface of the output waveguide, that is, normal to the propagating beam.

This proposed PD design is reported in [48], and it is based on an asymmetric MSM junction integrated onto a SOI substrate. The active metal is copper (Cu) that results in contact with Si only on the vertical exit wall of the optical waveguide. In practice, Cu works as a mirror (as well as an active absorbing layer), enabling the possibility of fabricating an integrated cavity if a second mirror is realized on the waveguide (for instance by means of deep trenches) in order to get a substantial shrinkage of the footprint together with an increase in the device performance. The integrated PD was characterized by a responsivity value at 1550 nm of 0.08 mA/W and a dark current of 10 nA at −1 V. In 2013, Casalino et al. proposed an optimized version of this device showing the possibility of tackling the typical responsivity/dark current trade-off [49, 50]. Indeed, in [50], it was shown that, by taking advantage of a small contact area of about 3 μm^2, it was possible to increase the reverse bias applied even up to 21 V while maintaining a limited dark current value of only 2.2 nA. The increase in reverse voltage allows reducing the Schottky barrier height (due to the image force effect [35] mentioned in Section 2), increasing the responsivity up to 4.5 mA/W. In addition, an experimental bandwidth of 1 GHz was demonstrated. Finally, the Schottky PD proposed shows the potentialities to work at longer wavelengths than NIR (i.e., wavelengths in the range 2–3 μm) [50].

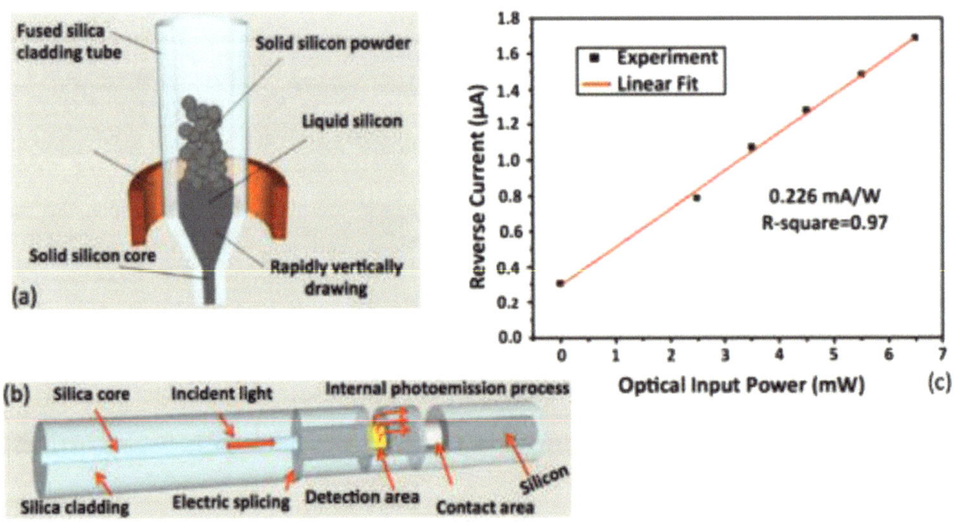

Figure 4. (a) A schematic illustration showing the fabrication process for making the Si-cored fiber and (b) the Schottky PD directly integrated on the Si-cored fiber described in [53]. (c) Measured photocurrent versus optical power from a 1550 nm laser.

In this section, we believe that other two structures proposed in the current literature deserve to be mentioned. The first concerns the possibility to combine Schottky PDs with microring resonators, and these structures have been investigated theoretically in the current literature [51, 52] providing encouraging results; however, no experimental validation has been carried out to so far. The second is reported in [53] where a very intriguing in-line IPE-based Si PD was fabricated directly on an optical fiber. In particular, an Au Schottky layer was placed directly on the Si core of a non-conventional optical fiber fabricated by starting from polycrystalline n-type Si powder, firstly packed in a fused silica tube, and then melted at an appropriate temperature as shown in **Figure 4(a)** and **(b)** [53]. The authors report a responsivity and dark current combination of 0.226 mA/W and 0.3 μA, respectively, at −0.45 V for a wavelength of 1550 nm as shown in **Figure 4(c)**. This device could become very interesting in the field of Lab-on-Fiber technology.

4. Surface-plasmon Schottky PDs

Surface-plasmon Schottky PDs allow combining IPE with the excitation of surface plasmon polaritons (SPPs) in a waveguide provided of a proper metal layer [54]. In other words, when the SPP is excited, it will be absorbed during its propagation along the metal stripe deposited on the Si waveguide. Absorbed photons will generate photoexcited carriers able to be emitted through the Schottky junction into Si making the sub-bandgap detection possible also at NIR wavelengths in agreement with the internal photoemission mechanism. The P. Berini's group has theoretically [55] and experimentally [18] investigated the performance of Schottky PDs integrated with Si-based waveguides supporting SPPs.

A first proposed structure is based on an Au stripe deposited on p-Si to form a Schottky contact [18, 19]. The authors demonstrate that the plasmonic mode excitation, localized at the Au/p-Si interface, occurs at NIR wavelengths under both end-facet and top illumination. In particular, under end-facet illumination, the measured responsivities in an asymmetric 1.5-μm-wide, 40-μm-long, and 40-nm-thick Au stripe waveguide were 0.942 and 0.941 mA/W at 1310 nm and 1550 nm, respectively, and at −0.1 V of reverse bias applied. On the contrary, under top illumination, the measured responsivity in an asymmetric 6.5-μm-wide, 40-μm-long, and 40-nm-thick Au stripe waveguide was 0.559 mA/W at the wavelength of both 1550 nm and −0.1 V of reverse bias applied. Moreover, the device dark current was in the μA range. Previously, the same authors reported on a similar device working under a strong reverse bias of −210 V close to the breakdown condition. The 2.5-μm-wide, 75-μm-long, and 135-nm-thick Au stripe on n-Si shows a responsivity of 2.35 mA/W at 1550 nm [56].

It is worth mentioning that the aforementioned devices are all based on asymmetric cladding configurations, and on the other hand, a symmetric cladding configuration, where the Au stripe is buried in Si [57, 58], gives the advantage of increased efficiency thanks to the emission through two Schottky junctions [32]. Even if no experimental validation has been performed in symmetric cladding SPP Si PDs so far, Knight et al. [59] proved increased responsivity for plasmonic nanostructures embedded in Si. Indeed, Ti (2 nm)/Au (35 nm) planar nanowires

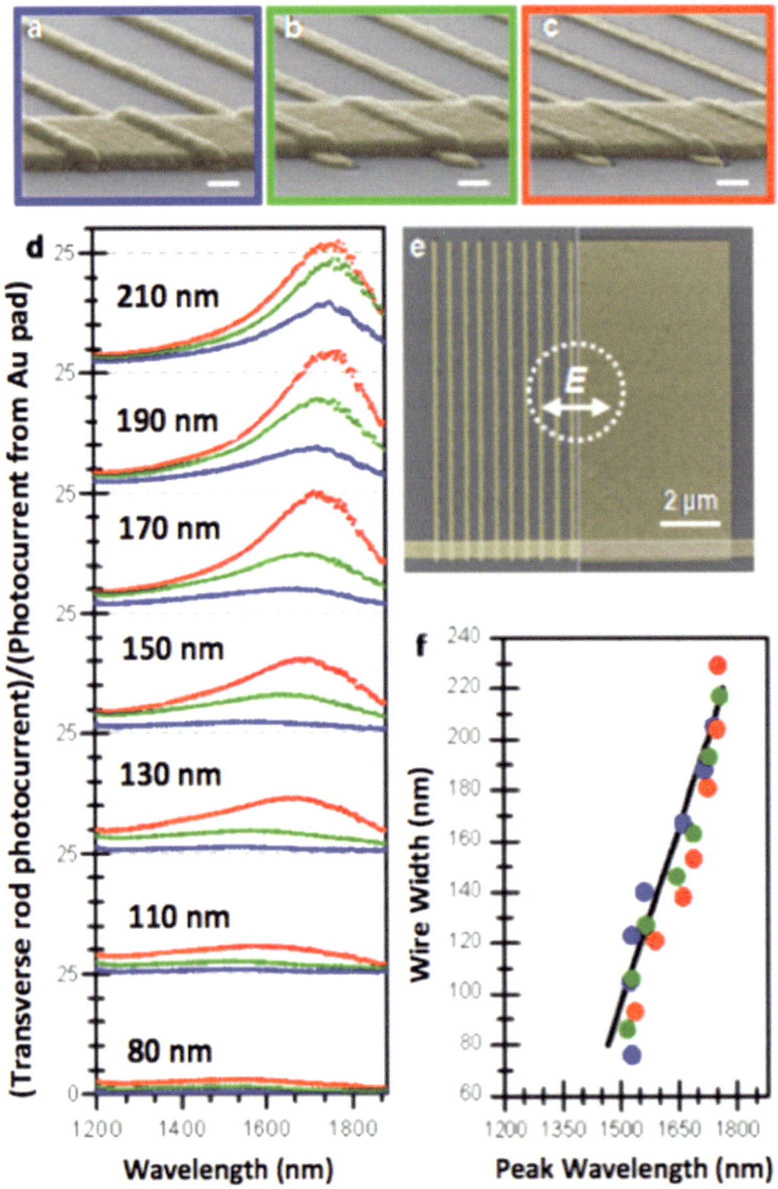

Figure 5. (a–c) Three representative SEM images of devices with widths of 120 ± 10 nm embedded about: 5 nm (blue), 15 nm (green), and 25 nm (red) into the silicon substrate. Scale bars are 100 nm. (d) Measured photocurrent spectra for increasing widths, where each spectrum is the photocurrent from (e) a nanowire array normalized to the response from a solid Au pad. The dotted white circle indicates the laser spot FWHM of 3 μm. Scale bar is 2 μm. (f) Calculated absorption peak wavelengths (black line) agree closely with experimentally observed enhancement peaks.

arranged into a 10 μm × 10 μm array were embedded in Si to a depth of 5 nm, 15 nm, and 25 nm in such a way that not only the flat surfaces but also the lateral surfaces of the planar nanowires are able to emit photoexcited carriers through the Schottky junctions (**Figure 5**).

The authors prove that an increased photocurrent can be achieved with respect to nonembedded plasmonic elements, in other words the higher the embedding depth, the higher the responsivity, and this trend was most clearly observable for the nanowires with the widest transverse dimensions. Indeed, devices based on the highest width to thickness ratio exhibited a photocurrent of about 25 times greater than the non-embedded structures. For incident light polarized transverse to the length of the nanowire, the maximum reported responsivity was 65 μA/W at 1550 nm. The authors attribute the responsivity enhancements to the increased carrier probability emission occurring through the three metal/Si interfaces. The embedded Schottky junctions and the main results reported in [59] are reported in **Figure 5**. In the 2012, Goykhman et al. demonstrated a NIR Al/Si Schottky PD integrated with a submicrometer Si waveguide fabricated taking advantage of the standard microelectronic LOCOS (Local Oxidation of Silicon) technique [60]. The 320 nm × 1000 nm Al active area in contact with the Si forms a Schottky PD with responsivity of about 12.5 mA/W at 1550 nm and −0.1 V bias, while the leakage current is only 30nA. The measured responsivity is about two orders of magnitude higher than that published two years earlier by the same authors for a similar device [61], and they attribute the enhanced responsivity to the presence of plasmonic effects due to the surface roughness at the boundary between the Al and the Si. In the same year, Zhu et al. demonstrated PDs based on nickel silicide nanoparticles embedded in the space charge region of a waveguide p-n Si junction [20]. While the idea of enhancing the IPE by using nanoparticles (NPs) was already explored in the past [62, 63], it is worth noting that the previously reported structures were able to operate only at cryogenic temperatures (77 K) and at wavelengths shorter than those required for telecommunications [64, 65]. The enhanced responsivity of PDs based on NPs is ascribed both to the increased light absorption due to the excitation of localized surface plasmon resonance (LPSR) and to the increased emission probability of the excited carriers due to the spherical shape of the NPs [66]. On the other hand, the device proposed in [20] is able to work at both room temperature and 1550 nm and its fabrication is very interesting: the junction was realized by depositing a ~1-nm-thick titanium (Ti) film and a ~3-nm-thick nickel (Ni) film on p-Si, and a subsequent rapid thermal annealing process at 200°C for 30 s produces the nickel silicide (NiSi). After removing the un-reacted metal using a proper etching solution (Piranha at 90°C), a 200-nm-thick film of amorphous Si was deposited, followed by phosphorus implantation and a rapid thermal annealing at 700°C for 5 min in order to get the dopant activation. It is worth noting that during this rapid thermal process, the Ni-silicide agglomerated to form NPs. At the same time, the amorphous Si crystallizes to form polycrystalline Si. The measured responsivity at room temperature depends on both wavelength and polarization, and the maximum value is 30 mA/W for −5 V and TE polarization. Finally, a dark current density and bandwidth of 2.84 A/cm^2 and 6 GHz at −5 V were reported, respectively. In [67] are reported Schottky PDs based on Au nanorods randomly distributed on a 30-nm-thick Au film deposited on Si. Various nanorods lengths ranging from 50 to 100 nm were considered, while their diameter was 10 nm. The authors claimed that the Au nanorods are capable of inducing SPR excitation under illumination at 1300 nm and 1550 nm and an enhanced photocurrent with respect to the same devices without nanorods was demonstrated at zero bias. In 2011, Knight et al. [22] reported on strong optical absorption in a small active metal region due to the excitation of resonant plasmons in metallic nanoantenna. The PD consisted of three hundred nanoantennas arranged in a 15 × 20 array. Each nanoantenna used 30-nm-high and 50-nm-wide rectangular Au nanorods with lengths ranged from 110 to 158 nm. Unfortunately, the PD responsivity depends on the complex optimization of several factors (the properties of the materials involved, the geometry, and the efficiency of the uppermost indium-tin-oxide electrical contact layer) and a very low responsivity of 10 μA/W and 3 μA/W at 1250 nm and 1550 nm were reported without reverse bias applied, respectively.

Another approach in order to obtain hot electron generation induced by plasmon modes was proposed by A. Sobhani et al. [21] who investigated the phenomenon of extraordinary optical transmission (EOT) in metallic gratings. The same plasmon modes that give rise to EOT also give rise to the excitation of electrons that can be emitted above the Schottky barrier into a semiconductor substrate as shown in **Figure 6**.

Gratings based on Au with different geometries were fabricated on n-type Si by using an interfacial layer of titanium in order to promote adhesion with Au. A measured maximum responsivity of 0.6 mA/W and 0.47 mA/W at 1300 nm and 1500 nm, for zero bias voltage, was demonstrated, respectively. Device measurements showed a drastic increase in responsivity

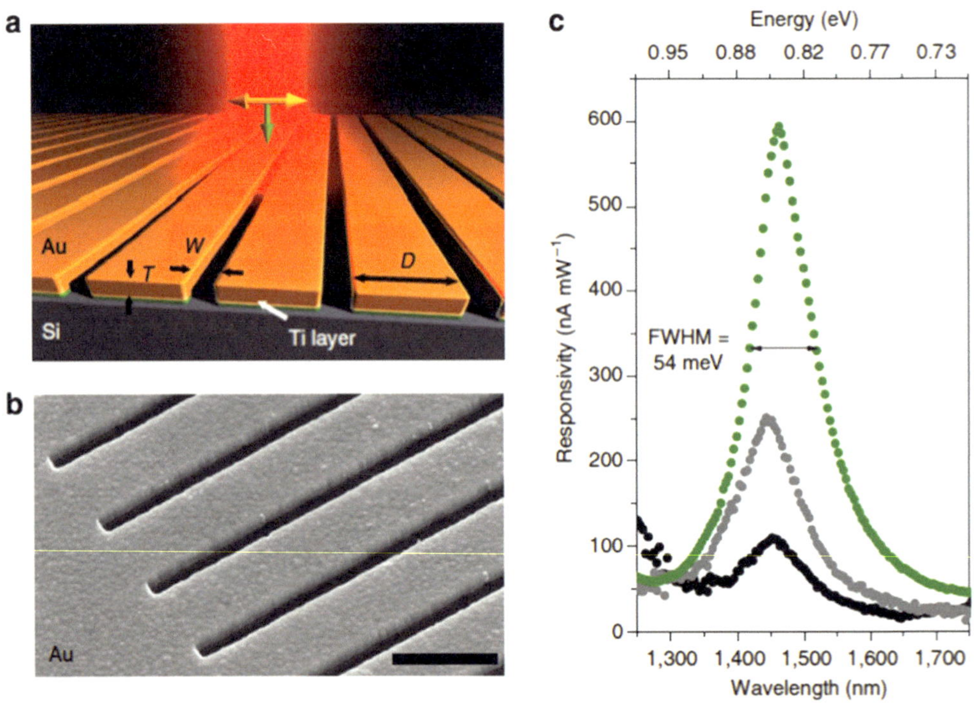

Figure 6. (a) Schematic of a gold grating on an n-type silicon substrate with a 2-nm Ti adhesion layer, oriented transverse to the laser polarization. Polarization of the incident laser and its k vector are represented with horizontal (yellow) and vertical (green) arrows, respectively. (b) Scanning electron microscopy image of gold grating structure with grating thickness (*T*)=200 nm, interslit distance (*D*)=950nm and slit width (*W*)=250 nm. For all structures the array measured 12 x 12 mm. The scale bar is 1 mm. (c) Photocurrent responsivities of grating-based photodetectors for three different gold layer thicknesses, T=93 nm (lower black curve), 170nm (grey curve in the middle) and 200nm (higher green curve), showing a strong intensity dependence on grating thickness.

with respect to structures based on nanoantennas previously mentioned. This increase mainly arises from the possibility to design the grating geometry in such a way to generate hot electrons primarily near the Schottky interface. Finally, this approach allows tuning the responsivity peak over a broad wavelength regime ranging from 1295 to 1635 nm, simply by changing the grating geometry.

Concerning the possibility to detect in a wide range of NIR wavelengths, in 2014 Nazirzadeh et al. [68] demonstrated Si Schottky PDs, operating in the 1200–1600 nm spectrum, based on Au nanoislands able to plasmonically enhance the sub-bandgap photon generation and collection. In this work, randomly distributed Au nanoislands on Si surface were realized by annealing a continuous thin Au layer at temperatures of 300°C, 450°C, and 600°C, respectively. The structure annealed at 450°C showed the highest responsivity of 2 mA/W at 1300 nm, while a dark current of about 30 μA at −1 V was reported. On the contrary, a device annealed at 300°C showed a responsivity of 0.6 mA/W at 1550 nm, while a dark current of about 100 μA at −1 V was demonstrated. The main advantage of the proposed PD is that all nanostructures were realized without requiring high-resolution electron beam lithography (EBL). Very recently, in 2016, Muehlbrandt et al. reported a novel plasmonic internal photoemission detector (that authors name with the PIPED acronym) characterized by a Si waveguide where the two sidewalls are metallized by two different metals: gold (Au) and titanium (Ti) [69]. The Au-Si-Ti waveguide is able to guide SPPs dissipating their energy mainly at Ti/Si interface because Ti is characterized by a larger imaginary part of the complex refractive index with respect to Au. In this work, authors demonstrate that when an external positive DC-bias voltage is applied (positive polarity from gold to titanium), the band diagram of the Au-Si-Ti structure changes in such a way that IPE at 1550 nm can occur over the Ti/Si Schottky barrier. The structure based on asymmetric Au-Si-Ti waveguides with a width of 75 nm and a length of 5 μm shows a responsivity at 1550 nm and dark current of 0.126 A/W and ~10 μA, respectively, at a bias voltage of 3.25 V. Finally, the detector exhibits optoelectronic bandwidths of at least 40 GHz.

5. Graphene/Si Schottky PDs

Deep investigations have been realized on graphene since its discovery in 2004 [70]. This is because electrons in graphene behave as massless two-dimensional particles leading to a wide absorption over wavelengths from the ultraviolet to the infrared thanks to both inter-band and intra-band transitions [71, 72]. In the field of the photodetection, this property is of fundamental importance because the semiconductors used for realizing PDs are characterized by a limited absorption range. This is because graphene-based PDs are nowadays one of the most studied photonic devices over a broadband range of wavelengths, in particular at NIR. It is well-known that graphene forms a Schottky junction with Si [73], and it results a good candidate for the realization of IPE-based devices due to its very low thickness that promises to increase the emission probability of the photoexcited carriers over the Schottky junction. In 2013, Amirmazlaghani et al. reported on an exfoliated graphene/Si Schottky PD operating at 1550 nm [23] showing a maximum experimental responsivity of 9.9 mA/W at both 1550 nm and −16 V. On the other hand, the measured dark current was 2.4 μA. A key point of the work is that the proposed device shows an experimental responsivity higher than that predicted by the traditional IPE theory. The authors explain the discrepancy by claiming that IPE theory should be totally revised when two-dimensional materials are considered, and they propose an alternative model in a good agreement with the experimental results. It has been shown that the insertion of a graphene monolayer in between the metal and the semiconductor provides an IPE enhancement [74] and very recently Levy et al. have proposed a phenomenological model able to explain physics behind this enhancement [75].

Recently Goykhman et al. demonstrated a NIR Schottky PD integrated with a SOI waveguide based on silicon-graphene junctions where graphene is grown by chemical vapor deposition CVD system [74]. Device is shown in **Figure 7**.

In particular, the rib waveguide was coupled to a single layer graphene (SLG)/Au contact able both to form a Schottky junction with Si and to support surface plasmonic modes that confined the optical beam to the SLG/Si interface. The PD length was ~5 μm, while the Si waveguide width was 310 nm. A responsivity of 0.085 A/W, at both, 1 V and 1550 nm, with respect to a dark current of 20 nA, was reported. Finally, the authors show that responsivity increases up to 0.37 A/W at a higher voltage of—3 V thanks to the combined effect of two processes: tunneling through the graphene/Si Schottky junction (thermionic-field emission, TFE [76]) and avalanche multiplication of within the Si depletion region.

Finally, Vabbina et al. have fabricated and characterized a very interesting two-dimensional Schottky PD working at 1440 nm [77]. The Schottky junction is realized by putting in contact the p-type molybdenum disulfide (MoS_2) with graphene, and the absorption mechanism is based on IPE: under illumination, photo-generated holes travel from the graphene into MoS_2 over the Schottky barrier. MoS_2 is a two-dimensional material attracting much interest due to its direct bandgap of 1.80 eV giving to the material semiconduc-

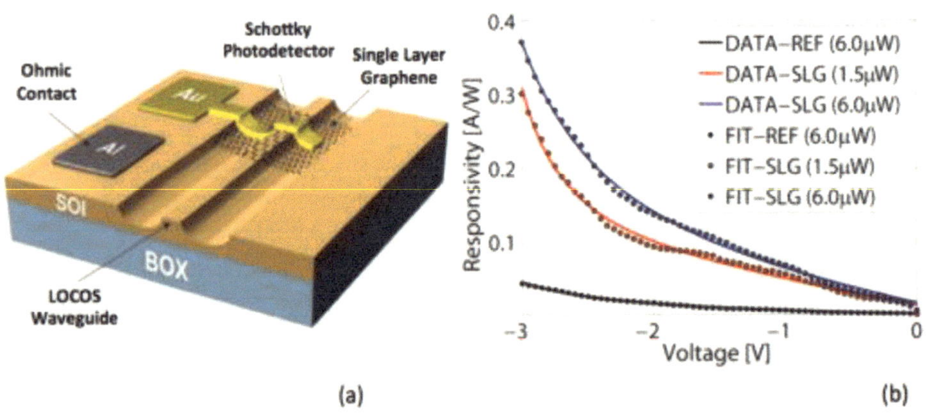

Figure 7. (a) Schematic metal-SLG-Si Schottky PD proposed in [74]. (b) Responsivity of metal-SLG-Si and reference metal-Si PDs for a reverse bias ranging from 0 to –3 V (reference metal-Si device is used to demonstrate the key role of graphene). Solid lines show a fit of the bias dependent responsivity based on combined TFE and avalanche multiplication processes.

tor characteristic. The thin film graphene/MoS_2 junction was deposited on an oxidized silicon substrate. Few layers of MoS_2 were deposited by a combined sputtering-CVD technique, while graphene was synthesized by a CVD process. The authors show a Schottky barrier of 0.139 eV, and the spectral responsivity when the energy photon is higher ($h\nu > E_{gMoS2}$) and lower ($h\nu < E_{gMoS2}$) than MoS_2 bandgap (E_{gMoS2}) was experimentally measured at −2 V of reverse bias applied. In particular, author shows that when $h\nu > E_{gMoS2}$ a maximum responsivity of 0.52 A/W is observed at 590 nm, on the other hand, beyond 670 nm where $h\nu < E_{gMoS2}$, the photo-current is due to IPE from the graphene to the MoS_2. In this last case, a maximum responsivity and noise equivalent power (NEP) of 1.26 A/W and 7.8×10^{-12} W/√Hz were respectively reported at 1440 nm and −2 V. Measured dark current is about 300 μA at −2 V.

6. Conclusions

In this work, an overview of recent advances in the field of the near-infrared silicon photodetectors based on the internal photoemission effect has been presented. Firstly, we have described the detection mechanisms of the devices especially in light of the models proposed in the recent literature. Thereafter, a quantitative comparison of the state-of-the-art IPE-based silicon PDs including Si nanoparticles, metal stripes supporting SPPs, antennas, metallic gratings, and two-dimensional materials has been given and summarized in **Table 1**.

Device	Device type	Size	Responsivity	Bandwidth	Dark current/dark current density
Casalino et al. [44]	Surface-illuminated Fabry-Perot microcavity	100-μm-thick silicon cavity	0.063 mA/W @1550 nm (−0.1 V)	GHz range estimated	3.5 mA (−1 V)
Desiatov et al. [45]	Surface-illuminated Si pyramids	Pyramid apex ~50 nm	5 mA/W @ 1550 nm 12 mA/W @ 1300 nm 30 mA/W @ 1064 nm (−0.1 V)	–	80 nA (−0.1 V)
Casalino et al [50]	Waveguide	Active area ~3 μm^2	4.5 mA/W @ 1550 nm (−21 V)	1 GHz	2.2 nA (−21 V)
Berini et al. [18]	Au strip supporting SPPs	Asymmetric 1.5-μm-wide 40-μm-long 40-nm-thick strip	0.942 mA/W @ 1310 nm 0.941 mA/W @ 1550 nm (−0.1 V)	–	0.3 μA (−0.45 V)
Goykhman et al. [60]	Nanoscale bus waveguide	Al active area 0.32 μm × 1 μm	12.5 mA W @ 1550 nm (−0.1 V)	–	30 nA (−0.1 V)

Device	Device type	Size	Responsivity	Bandwidth	Dark current/dark current density
Zhu et al. [20]	NPs supporting LPSR	Few nm NPs	30 mA/W @ 1550 nm (−5 V for TE polarization)	6 GHz	2.84 A/cm^2 (−5 V)
Knight et al. [22]	15 × 20 matrix of Au nanoantennas	30-nm-height 50-nm-width nanoantenna	10 μA/W @ 1250 nm 3 μA/W @ 1550 nm (No bias)	–	–
Sobhani et al. [21]	Au grating based on EOT	Different grating geometries (few hundreds of nanometers)	0.6 mA/W @ 1300 nm 0.47 mA/W @ 1500 (No bias)	–	–
Nazirzadeh et al. [68]	Au nanoisland	–	2 mA/W @ 1300 nm (450°C) 0.6 mA/W @ 1500 nm (300°C)	–	30 μA (450°C) 100 μA (−300°C) both −1 V
Muehlbr et al. [69]	Si waveguide with metallizedsidewalls (PIPED)	Asymmetric Au-Si-Ti 75-nm-wide 5-μm-long 300-nm-thick	0.126 A/W @ 1550 nm (3.25 V)	40 GHz (200 nm–wide 20 μm-long)	
Amirmaz et al. [23]	Surface-illuminated graphene/Si	–	9.9 mA/W @ 1550 nm (−16 V)	–	2.4 μA (−16 V)
Goykhman et al. [74]	Graphene integrated with a rib SOI waveguide	PD length ~5 μm Si waveguide width ~310nm	0.37 A/W @ 1550 nm (−3 V)	–	20 nA (−1 V)
Vabbina et al. [77]	Graphene/MoS_2 Schottky PD on oxidized Si	–	1.26 A/W @ ~1550 nm (−2 V)	–	300 μA (−2 V)

Table 1. Summary of selected IPE-based Si PDs at NIR wavelengths published since 2010.

The most part of IPE-based PDs have been realized for operation at telecom wavelengths, in particular at 1550 nm, and are characterized by a large GHz bandwidth thanks to the unipolar nature of the Schottky junction. In addition, a low dark current to the nA range has been obtained thanks to the use of plasmonic structures allowing strong absorption in a small metal layer in contact with silicon.

Historically, the main drawback of the IPE-based PDs is their external responsivity limited to the mA/W range at room temperature; this work put in evidence that taking advantage of both the integration of two-dimensional materials with silicon and plasmonic structures, a responsivity in the A/W range can be obtained. These values are comparable with the well-established near-infrared technologies based on germanium and III–V semiconductors and open the path to the investigation of more complex structures that can make the IPE-based Si Schottky PDs very promising to play a key role for telecommunications.

http://dx.doi.org/10.5772/67720

Author details

Maurizio Casalino

Address all correspondence to: maurizio.casalino@na.imm.cnr.it

Institute for Microelectronics and Microsystems (IMM) of the Italian National Research Council, Naples, Italy

References

[1] Jalali B, Fathpour S: Silicon photonics. J. Lightwave Technol. 2006;**24**:4600–4615.

[2] Paniccia M: Integrating silicon photonics. Nat. Photon. 2010;**4**:498–499.

[3] Kang Y, Mages P, Clawson AR, Yu PKL, Bitter M, Pan Z, Pauchard A, Hummel S, Lo YH: Fused InGaAs–Si Avalanche photodiodes with low noise performances. IEEE Photon. Tech. Lett. 2002;**14**:1593–1595.

[4] BelingA, Campbell J: InP-based high-speed photodetectors. J. Lightwave Technol. 2009; **27**:343–355.

[5] Koester SJ, Schaub JD, Dehlinger G, Chu, JO: Germanium-on-SOI infrared detectors for integrated photonic applications. IEEE J. Sel. Top. Quantum Electron. 2006;**12**: 1489–1502.

[6] Harame DL, Koester SJ, Freeman G, Cottrel P, Rim K, Dehlinger G, Ahlgren D, Dunn JS, Greenberg D, Joseph A, Anderson F, Rieh JS, Onge SAST, Coolbaugh D, Ramachandran V, Cressler JD, Subbanna S: The revolution in SiGe: impact on device electronics. Appl. Surf. Sci. 2004;**224**:9–17.

[7] Jones R, Park HD, Fang AW, Bowers JE, Cohen O, Raday O, Paniccia MJ: Hybrid silicon integration. J. Mater. Sci. Mater. Electron. 2009;**20**:3–9.

[8] Eng PC, Sung S, Ping B: State-of-the-art photodetectors for optoelectronic integration at telecommunication wavelength. Nanophotonics. 2015;**4**:277–302.

[9] Michel J, Liu J, Kimerling LC: High-performance Ge-on-Si photodetectors. Nat. Photon. 2010;4:527–534.

[10] Kang Y, Liu H-D, Morse M, Paniccia MJ, Zadka M, Litski S, Sarid G, Pauchard A, Kuo Y-H, Chen H-W, Zaoui WS, Bowers JE, Beling A, McIntosh DC, Zheng X, Campbell JC: Monolithic germanium/silicon avalanche photodiodes with 340GHz gain–bandwidth product. Nat. Photon. 2009;3:59–63.

[11] Vivien L, Osmond J, Fédéli J-M, Marris-Morini D, Crozat P, Damlencourt J-F, Cassan E, Lecunff Y, Laval S: 42 GHz p.i.n Germanium photodetector integrated in a silicon-on-insulator waveguide. Opt. Express. 2008;**17**:6252–6257.

[12] Wang J, Lee S: Ge-photodetectors for Si-based optoelectronic integration. Sensors. 2011;11:696–718.

[13] Alduino A, Liao L, Jones R, Morse M, Kim B, Lo W-Z, Basak J, Koch B, Liu H-F, Rong H, Sysak M, Krause C, Saba R, Lazar D, Horwitz L, Bar R, Litski S, Liu A, Sullivan K, Dosunmu O, Na N, Yin T, Haubensack F, Hsieh I-W, Heck J, Beatty R, Park H, Bovington J, Lee S, Nguyen H, Au H, Nguyen K, Merani P, Hakami M, Paniccia MJ: Demonstration of a high speed 4-channel integrated silicon photonics WDM link with hybrid silicon lasers. Hot Chips. 2010;**22**:Session 3.

[14] Narasimha A, Analui B, Balmater E, Clark A, Gal T, Guckenberger D, Gutierrez S, Harrison M, Ingram R, Koumans R, Kucharski D, Leap K, Liang Y, Mekis A, Mirsaidi S, Peterson M, Pham T, Pinguet T, Rines D, Sadagopan V, Sleboda TJ, Song D, Wang Y, Welch B, Witzens J, Abdalla S, Gloeckner S, De Dobbelaere P: A 40-Gb/s QSFP optoelectronic transceiver in a 0.13 μm CMOS silicon-on-insulator technology. In: Proceeding of the Optical Fiber Communications Conference (OFC); 2008; OMK7.

[15] Alloatti L, Srinivasan SA, Orcutt JS, Ram RJ: Waveguide-coupled detector in zero-change complementary metal-oxide-semiconductor. Appl. Phys. Lett. 2015;**107**:041104.

[16] Meng H, Atabaki A, Orcutt JS, Ram RJ: Sub-bandgap polysilicon photodetector in zero-change CMOS process for telecommunication wavelength. Opt. Express. 2015;**23**: 32643–32653.

[17] Casalino M, Coppola G, De La Rue RM, Logan DF: State-of-the-art all-silicon sub-bandgap photodetectors at telecom and datacom wavelengths. Laser Photon. Rev. 2016:**10**;895–921.

[18] Berini P, Olivieri S, Chen C: Thin Au surface plasmon waveguide Schottky detectors on p-Si. Nanotechnology. 2012;**23**:444011.

[19] Akbari A, Tait RN, Berini P: Surface plasmon waveguide Schottky detector. Opt. Express. 2010;**18**:8505–8514.

[20] Zhu S, Chu HS, Lo GQ, Bai P, Kwong DL: Waveguide-integrated near-infrared detector with self-assembled metal silicide nanoparticles embedded in a silicon p-n junction. Appl. Phys. Lett. 2012;**100**:061109.

[21] Sobhani A, Knight MW, Wang Y, Zheng B, King NS, Brown LV, Fang Z, Nordlander P, Halas NJ: Narrowband photodetection in the near-infrared with a plasmon-induced hot electron device. Nat. Commun. 2013;**4**:1643.

[22] Knight MW, Sobhani H, Nordlander P, Halas NJ: Photodetection with active optical antennas. Science. 2011;**332**:702–704.

[23] Amirmazlaghani M, Raissi F, Habibpour O, Vukusic J, Stake J: Graphene-Si Schottky IR detector. IEEE J. Quant. Electron. 2013;**49**:589–594.

[24] Kimata M, Ueno M, Yagi H, Shiraishi T, Kawai M, Endo K, Kosasayama Y, Sone T, Ozeki T, Tsubouchi N: PtSi Schottky-barrier infrared focal plane arrays. Opto-Electron. Rev. 1998;**6**:1–10.

[25] Casalino M, Sirleto L, Iodice M, Coppola G. Silicon photodetectors: the challenge of detecting near-infrared light. In: Gateva S editor. Photodetectors. InTech; 2012. p. 51–76. DOI: 10.5772/36014..

[26] Fowler RH: The analysis of photoelectric sensitivity curves for clean metals at various temperatures. Phys. Rev. 1931;38:45–56.

[27] Cohen J, Vilms J, Archer RJ: Investigation of Semi-conductor Schottky Barriers for Optical Detection and Cathodic Emission. Air Force Cambridge Research Labs. 1968; Report No. 68–0651.

[28] Elabd H, Kosonocky WF: Theory and measurements of photoresponse for thin film Pd2Si and PtSi infrared Schottky-barrier detectors with optical cavity. RCA Rev. 1982;**43**: 569–589.

[29] Stuart R, Wooten F, Spicer WE: Monte Carlo calculations pertaining to the transport of hot electrons in metals. Phys. Rev. 1964;**135**:A495–A505.

[30] Czernik A, Palm H, Cabanski W, Schulz M, Suckow U: Infrared photoemission of holes from ultrathin (3–20 nm) Pt/Ir-compound silicide films into silicon. Appl. Phys. 1992;**55**:180–191.

[31] Sellai A, Dawson P: Monte Carlo calculations of quantum yield in inhomogeneous PtSi/p-Si Schottky barriers. Semicond. Sci. Technol. 1998;**13**:700–704.

[32] Scales C, Berini P: Thin-film Schottky barrier photodetector models. IEEE J. Quant. Electron. 2010;**46**:633–643.

[33] Vickers VE: Model of Schottky barrier hot-electron-mode photodetection. Appl. Opt. 1971;**10**:2190–2192.

[34] Casalino M: Internal photoemission: theory revisited and theoretical limitations on the performance of near-infrared silicon Schottky photodetectors. IEEE J. Quant. Electron. 2016;**52**:4000110.

[35] Sze SM, Kwok KN. Physics of Semiconductor Devices. New York: Wiley; 2006.

[36] Donati S. Photodetectors: Devices, Circuits and Applications. New Jersey: Prentice Hall PTR; 1999.

[37] Esper J, Panetta P, Ryschkewitsch M, Wiscombe W, Neeck S: NASA-GSFC Nano-satellite technology for earth science missions. Acta Astronaut. 2000;**46**:287–296.

[38] Daffara C, Pampaloni E, Pezzati L, Barucci M, Fontana R: Scanning multispectral IR reflectography SMIRR: an advanced tool for art diagnostics. Acc. Chem. Res. 2010;**43**:847.

[39] Kimata M, Denda M, Yutani N, Iwade S, Tsubouchi N: A 512/spl times/512-element PtSi Schottky-barrier infrared image sensor. IEEE J. Solid-State Circ. 1987;**22**:1124–1129.

[40] Casalino M, Sirleto L, Moretti L, Della Corte F, Rendina I: Design of a silicon resonant cavity enhanced photodetector based on the internal photoemission effect at 1.55 μm. J. Opt. A Pure Appl. Opt. 2006;**8**:909–913.

[41] Casalino M, Sirleto L, Moretti L, Gioffrè M, Coppola G, Iodice M, Rendina I: Back-illuminated silicon resonant cavity enhanced photodetector at 1550 nm. Physica E. 2009;**41**:1097–1101.

[42] Casalino M, Sirleto L, Moretti L, Gioffrè M, Coppola G, Rendina I: Silicon resonant cavity enhanced photodetector based on the internal photoemission effect at 1.55 micron: fabrication and characterization. Appl. Phys. Lett. 2008;**92**:251104.

[43] Casalino M, Coppola G, Gioffrè M, Iodice M, Moretti L, Rendina I, Sirleto L: Silicon technology compatible photodetectors at 1.55 μm. J. Lightwave Technol. 2010;**28**:3266.

[44] Casalino M, Coppola G, Iodice M, Rendina I, Sirleto L: Critically coupled silicon Fabry-Perot photodetectors based on the internal photoemission effect at 1550 nm. Opt. Express. 2012;**20**:12599–12609.

[45] Desiatov B, Goykhman I, Mazurski N, Shappir J, Khurgin JB, Levy U: Plasmonic enhanced silicon pyramids for internal photoemission Schottky detectors in the near-infrared regime. Optica. 2015;**2**:335–338.

[46] Zhu S, Yu MB, Lo GQ, Kwong DL: Near-infrared waveguide-based nickel silicide Schottky-barrier photodetector for optical communications. Appl. Phys. Lett. 2008; 92:081103.

[47] Zhu S, Lo GQ, Kwong DL: Low-cost and high-speed SOI waveguide-based silicide schottky-barrier MSM photodetectors for broadband optical communications. IEEE Photon. Technol. Lett. 2008;**20**:1396–1398.

[48] Casalino M, Sirleto L, Iodice M, Saffioti N, Gioffrè M, Rendina I, Coppola G:Cu/p-Si Schottky barrier-based near-infrared photodetector integrated with a silicon-on-insulator waveguide. Appl. Phys. Lett. 2010;**96**:241112.

[49] Casalino M, Iodice M, Sirleto L, Rendina I, Coppola G: Low dark current silicon-on-insulator waveguide metal-semiconductor-metal photodetector based on internal photoemission effect at 1550 nm. J. Appl. Phys. 2013;**114**:153103.

[50] Casalino M, Iodice M, Sirleto L, Rendina I, Coppola G: Asymmetric MSM sub-bandgap all-silicon photodetector with low dark current. Opt. Express. 2013;**21**: 28072–28082.

[51] Zali AR, Moravvej-Farshi MK: Nonidealities and dark current in IR photodetector based on silicide-nanolayer Schottky barrier integrated into a Si microring resonator. IEEE J. Quant. Electron. 2015;**51**:4000108.

[52] Hosseinifar M, Ahmadi V, Ebnali-Heidari M: Si-Schottky photodetector based on metal stripe in slot-waveguide microring resonator. IEEE Photon. Technol. Lett. 2016;28: 1363–1366.

[53] Huang YP, Wang LA: In-line silicon Schottky photodetectors on silicon cored fibers working in 1550 nm wavelength regimes. Appl. Phys. Lett. 2015;**106**:191106.

[54] Maier SA: Plasmonics: Fundamentals and Applications. New York: Springer; 2007.

[55] Akbari A, Berini P: Schottky contact surface-plasmon detector integrated with an asymmetric metal stripe waveguide. Appl. Phys. Lett. 2009;**95**:021104.

[56] Olivieri A, Akbari A, Berini P: Surface plasmon waveguide Schottky detectors operating near breakdown. Phys. Status Solidi RRL. 2010;**4**:283–285.

[57] Scales C, Breukelaar I, Charbonneu R, Berini P: Infrared performance of symmetric surface-plasmon waveguide Schottky detectors in Si. J. Lightwave Technol. 2011;29: 1852–1860.

[58] Zhu S, Lo GQ, Kwong DL: Theoretical investigation of silicide Schottky barrier detector integrated in horizontal metal-insulator-silicon-insulator-metal nanoplasmonic slot waveguide. Opt. Express. 2011;**19**:15843–15854.

[59] Knight MK, Wang Y, Urban AS, Sobhani A, Zheng BY, Nordlander P, Halas NJ: Embedding plasmonic nanostructure diodes enhances hot electron emission. Nanoletters. 2013;13:1687–1692.

[60] Goykhman I, Desiatov B, Khurgin J, Shappir J, Levy U: Waveguide based compact silicon Schottky photodetector with enhanced responsivity in the telecom spectral band. Opt. Express. 2012;**20**: 28594–28602.

[61] Goykman I, Desiatov B, Khurgin J, Shappir J, Levy U: Locally oxidized silicon surface-plasmon Schottky detector for telecom regime. Nano Lett. 2011;11:2219–2224.

[62] Fathauer RW, Iannelli JM, Nieh CW, Hashimoto S: Infrared response from metallic particles embedded in a single-crystal Si matrix: the layered internal photoemission sensor. Appl. Phys. Lett. 1990;**57**:1419–1421.

[63] Fathauer RW, Dejewski SM, George T, Jones EW, Krabach TN, Ksendzov A: Infrared photodetectors with tailorable response due to resonant plasmon absorption in epitaxial silicide particles embedded in silicon. Appl. Phys. Lett. 1993;**62**:1774–1776.

[64] Stuart RH, Hall DG: Island size effects in nanoparticle-enhanced photodetectors. Appl. Phys. Lett. 1998;**73**:3815–3817.

[65] Schaadt DM, Feng B, Yu ET: Enhanced semiconductor optical absorption via surface plasmon excitation in metal nanoparticles. Appl. Phys. Lett. 2005;**86**:063106.

[66] Raissi F: A possible explanation for high quantum efficiency of PtSi/porous Si Schottky detectors. IEEE Trans. Electron. Dev. 2003;**50**:1134.

[67] Fukuda M, Aihara T, Yamaguchi K, Ling YY, Miyaji K, Tohyama M: Light detection enhanced by surface plasmon resonance in metal film. Appl. Phys. Lett. 2010;**96**: 153107.

[68] Nazirzadeh MA, Atar FB, Turgut BB, Okyay AK: Random sized plasmonic nanoantennas on silicon for low-cost broad-band near-infrared photodetection. Sci. Rep. 2014;**4**:7103.

[69] Muehlbrandt S, Melikyan A, Harter T, Kohnle K, Muslija A, Vincze P, Wolf S, Jokobs P, Fedoryshyn Y, Freude W, Leuthold J, Koos C, Kohl M: Silicon-plasmonic internal-photoemission detector for 40 Gbit/s data reception. Optica. 2016;**3**:741–747.

[70] Novoselov KS, Fal'ko VI, ColomboL, Gellert PR, Schwab MG, Kim K: A roadmap for graphene. Nature. 2012;**490**:192–200.

[71] Dawlaty J, Shivaraman S, Strait J, George P, Chandrashekhar M, Rana F, Spencer MG, Veksler D, Chen Y: Measurement of the optical absorption spectra of epitaxial graphene from terahertz to visible. Appl. Phys. Lett. 2008;**93**:131905-1-13.

[72] Sensale-Rodriguez B, Yan R, Kelly MM, Fang T, Tahy K, Hwang WS, Jena D, Liu L, Xing HG: Efficient terahertz electro-absorption modulation employing graphene plasmonic structures. Appl. Phys. Lett. 2012;**101**:261115.

[73] Chen C-C, Aykol M, Chang C-C, Levi AFJ, Cronin SB: Graphene-silicon Schottky diodes. Nano Lett. 2011;**11**:1863–1867.

[74] Goykhman I, Sassi U, Desiatov B, Mazurski N, Milana S, de Fazio D, Eiden A, Khurgin J, Shappir J, Levy U, Ferrari AC: On-chip integrated, silicon–graphene plasmonic Schottky photodetector with high responsivity and avalanche photogain. Nano Lett. 2016;16:3005–3013.

[75] Levy U, Grajower M, Goncalves PAD, Mortensen NA, Khurgin JB: Plasmonic silicon Schottky photodetectors: the physics behind graphene enhanced internal photoemission. APL Photon. 2017;**2**:026103.

[76] Padovani FA, Stratton R: Field and thermionic-field emission in Schottky barriers. Solid State Electron. 1966;**9**:695–707.

[77] Vabbina P, Choudhary N, Chowdhury A-A, Sinha R, Karabiyik M, Das S, Choi W, Palaoi N, Pala N: Highly sensitive wide bandwidth photodetector based on internal photoemission in CVD grown p-type MoS2/graphene Schottky junction. ACS Appl. Mater. Interfaces. 2015;**7**:15206–15213.

Chapter 12

Advanced SiC/Oxide Interface Passivation

Yogesh Kumar Sharma

Additional information is available at the end of the chapter

http://dx.doi.org/10.5772/67867

Abstract

To save energy on an electric power grid, the idea of redesigned 'micro-grids' has been proposed. Implementation of this concept needs power devices that can operate at higher switching speeds and block voltages of up to 20 kV. Out of SiC and GaN wide band gap semiconductors, the former is more suitable for low- as well as high-voltage ranges. SiC exists in different polytypes 3C-, 4H- and 6H-. 4H-SiC due to its wider band gap, 3.26 eV has higher critical electric field of breakdown (E_c) and electron bulk mobility compared to 6H-SiC. Even with all these benefits 4H-SiC full potential has not yet been realized. This is due to high trap densities (D_{it}) at the interface. In addition to 4H-polytype, in recent years, there is a reignited interest on cubic silicon carbide (3C-SiC), which can be potentially grown heteroepitaxially on 12″ Si substrates, as it would result in a drastic cost reduction of semiconductor devices compared to the successful but exorbitantly expensive SiC hexagonal polytype technology (4H-SiC). In this chapter, we discuss and summarize all different interface passivation techniques or processes that have led to a vast improvement of these (4H- or 3C-SiC/SiO_2) interfaces electrically.

Keywords: silicon carbide, MOSFET, interface trap density, mobility, PSG, nitrogen plasma, NO passivation, BTS

1. Introduction

1.1. Progress in semiconductor devices

The first solid-state amplifier was manufactured by using germanium (Ge) which was seen as the semiconductor material of the future. With time, silicon (Si) turned out to be more appropriate for a plethora of reasons [1–4]. Silica, the source of Si, is commonly available and is easier to get high-purity Si from it. Silicon can easily be doped to produce n-type, p-type and semi-insulating material [5]. In addition to all these, a native oxide SiO_2 can be grown on Si using thermal oxidations at relatively low temperatures of around 900°C [6–8]. These properties make Si semiconductor industry favourite. At present, semiconductor industry worth is more than $300 billion [9]. Around 10% of this total is in smart integrated circuits and electronic power devices [10, 11]. In excess to more than 50% of our electricity is conditioned by electronic power devices [12, 13]. These devices are important because they determine the cost and efficiency of an electronic system. Hence, they have a greater influence on the economy of a country. The arrival of devices like the bipolar transistors in the 1950s led to the replacement of vacuum tubes [13, 14], and these improvements made possible the Second Electronic Revolution with Si as the material of choice. Power devices had a vital place in this revolution. In the 1970s, there were bipolar devices with a blocking voltage capacity of 500 V and high current capabilities. Also in 1970, International Rectifier Inc. launched the first metal-oxide-field-effect transistor (MOSFET) [15]. The idea was to switch bipolar devices with MOSFETs for high power use. The MOSFET is a unipolar device and thus can switch at a higher speed. Also, the MOSFET is a voltage-controlled device where the junction transistor is a current-controlled device. Higher switching speed means operation at higher frequencies where other system components such as inductors can be made smaller in size, and voltage control instead of current control means saving of internal energy in the device.

1.2. Need of a wide band-gap semiconductor device

To save energy on an electric power grid, the idea of redesigned 'micro-grids' has been proposed [16, 17]. Implementation of this concept needs power devices which can operate at higher switching speeds and block voltages of up to 20 kV [18]. A potential solution for this problem is to use wide band-gap semiconductor (e.g. SiC) power devices [19]. For a power device, the Baliga figure of merit (BFOM) [20] is given by

$$\mathrm{BFOM} = \mu_N \varepsilon_S E_C^3 \tag{1}$$

μ_N = bulk mobility of SiC, ε_S = permittivity of SiC, E_C = critical electric field of breakdown for SiC. Higher the BFOM, more suitable the semiconductor is for high power operation.

1.3. Unipolar devices

Power devices can be divided into two categories—bipolar and unipolar. Schottky diodes and MOSFETs are examples of unipolar devices. In a unipolar device, only one type of carrier (either a majority electron or a majority hole) is responsible for current flow. The device can operate at higher frequencies which results in lower switching losses [20]. There is a flow of both majority and minority carriers in bipolar devices. The slower minority carriers have to be injected and removed to get the device to turn on and off, so in bipolar devices there is power loss due to switching and leakage current. The n-channel Si-MOSFET is a better choice for low voltages (~100 V), and it can operate at high switching speed, 100 kHz. But as the blocking voltage increases, the on-state resistance increases drastically. The SiC-MOSFET enables us to go to higher operating voltages (order of kilo volts) with higher switching speed. This is possible because SiC has a high critical breakdown field, almost seven times that of Si. The specific on resistance (R_{ON}) of a MOSFET is given by [2]

$$R_{ON} = 4\,V_B^{\,2}/\mu_N\,\varepsilon_S\,E_C^3 \quad (2)$$

V_B is the desired blocking voltage, μ_N is the bulk electron mobility and ε_S is the semiconductor permittivity.

Bulk electron mobilities are similar for lightly doped Si and SiC (900–1200 cm^2/V s). However, $E_C^{SiC} \sim 7E_C^{Si}$, so that for a given blocking voltage, R_{ON} can be a factor of 343 (7^3) times lower for SiC. Another way to think of this advantage is that lower critical field of Si means a much thicker drift layer is needed to support the source-drain voltage in blocking state. A thicker drift layer means higher drift resistance and thus higher R_{ON} for Si but lower R_{ON} for SiC. Furthermore, due to unipolar nature of the device we do not have to deal with stored charge and hence a FET will have higher switching speed.

1.4. Oxidation

Oxidation of 4H-SiC is a very important processing step during the manufacturing of a device. The performance of a metal-oxide semiconductor (MOS) device is dependent on the quality of the gate oxide layer. Out of many oxidation processes, thermal oxidation is the process most commonly used to form the interface (4H-SiC/SiO_2). Thermal oxidation is typically carried out in an oxygen (O_2) atmosphere (500 sccm) at 1150°C. The thermal oxidation process has been investigated both experimentally and theoretically by researchers. First-principle calculations done by Di Ventra et al. have shown that during thermal oxidation, atomic oxygen diffused onto the surface of SiC and formed an advancing interface (4H-SiC/SiO_2) [21]. Tan et al. confirmed experimentally that the excess carbon atoms diffused out as carbon monoxide (CO) [22]. For thicker oxide layers, their simulations showed that CO may break up either in SiO_2 bulk or at the interface (4H-SiC/SiO_2). The released oxygen participates in another round of oxidation, and the carbon atoms may lead to the formation of carbon clusters. Di Ventra et al. also proposed the formation of carbon dioxide (CO_2) while CO was emitted out through a thick oxide layer. Kanup et al. developed theoretical predictions of the formation of stable carbon pairs and carbon interstitials [23]. These defects combined with silicon interstitials form near-interface traps (N_{IT}). Near-interface traps are more critical compared to bulk traps for the mobility of SiC MOSFETs. The oxidation process can also cause the injection of carbon into SiC substrate. This injected carbon can exist in different forms such as carbon interstitials (C_i) and carbon di-interstitials $(C_i)_2$ to further degrade the FET channel mobility [4–6, 24]. The oxidation rate of 4H-SiC depends upon the orientation of 4H-SiC wafers. This has been determined experimentally by Shenoy et al. [25]. The oxidation rate for C-face is three to five times faster than for the Si-face. Alumina-enhanced oxidation (AEO) is very fast due to the Na that is released from the alumina at the oxidation temperature. AEO on Si-face at 1050°C gives growth rate which is 10 times faster than normal thermal oxidation at 1150°C [26].

1.5. 4H-SiC/SiO_2 interface passivation

Silicon carbide exists in different polytypes 6H-, 4H- and 3C-. A MOSFET fabricated using 6H- polytype has field-effect inversion channel mobility which is much higher than that of 4H- polytype MOSFET (due to higher band gap of 4H-, $Eg \sim 3.26$ eV, most of the interface traps for 4H-SiC falls in the forbidden band, $Eg \sim 3.0$ eV for 6H- polytype). 6H-SiC due to its lower band gap, 3.0 eV, has lower critical electric field of breakdown (E_c) and bulk mobility compared to 4H-SiC. Also, in 4H- the field-effect mobility is more isotropic [5, 27] as compared to 6H-SiC. All these benefits make 4H- polytype a preferred choice among other polytypes of SiC for power devices. Although 4H- polytype has all these advantages, its full potential has not yet been exploited. This is due to high interface trap densities (D_{it}) at the interface (4H-SiC/SiO_2) 10^{13} $eV^{-1}cm^{-2}$. This value is much higher than the D_{it} of Si/SiO_2 interface (10^{10} $eV^{-1}cm^{-2}$). The cause of higher D_{it} is the formation of Si-dangling bonds, C-dangling bonds, O vacancies and C clusters. All this occurs during the oxidation of 4H-SiC [6–8]. Different post-oxidation annealing (called as passivation) techniques had been tried in the past to reduce the interface traps, for example, post-oxidation annealing in nitric oxide (NO), nitrous oxide (N_2O), ammonia (NH_3) and hydrogen (H_2) ambient [13, 29, 30]. NO/N_2O led to the incorporation of nitrogen (N) at the interface and forms different chemical species C-N, Si-N, Si-N-O and reduces the near D_{it} [6, 7]. The combined NO + H_2 passivation (NO passivation followed by H_2 passivation) gives improved interface which results in slightly better mobility ~40 cm^2/V s. Allerstan et al. have shown that the presence of sodium (Na^+) ion at the interface increases inversion channel mobility drastically up to 150 cm^2/V s, on the Si face of 4H-SiC MOSFET [11]. However, Na^+ moves under applied bias and hence destabilizes the device due to changing threshold voltage (V_T).The a-face (1120) of 4H- polytype with Al_2O_3/SiO_2 composite gate oxides leads to devices with channel mobilities of 100 cm^2/V s or more, but these MOSFETs showed higher leakage currents and lower breakdown voltages [13, 28]. Some groups have reported that the thermal oxidation at higher temperatures can lead to a better interface with a low-interface trap density value $\sim 2 \times 10^{11}$ $eV^{-1}cm^{-2}$ at 0.2 eV from the conduction band (CB) edge [15, 29]. In the following sections, different interface passivation processes will be discussed.

1.5.1. Phosphorous interface passivation

In phosphorous (P) passivation, the interface is treated with P source to get a gate oxide with P at the interface. This can be done by using either a gas mixture of $POCl_3$, N_2 and O_2 or SiP_2O_7 as a planar solid source to form phosphosilicate glass (PSG), P_2O_5, under high-temperature

http://dx.doi.org/10.5772/67867

(1000°C) annealing in nitrogen environment. With P passivation, we can have different process variations in order to obtain different results, which are discussed in the following sections.

1.5.1.1. Thick PSG process

In thick PSG device, a 70-nm thermal oxide is grown and then passivated by a 3-h P passivation process. During interface passivation, the following reaction takes place, which leads to the formation of phosphosilicate glass. The formation of this PSG layer leads to a high concentration of P at the interface. The P concentration near interface is ~2 × 10^{21} cm^{-3} using Secondary Ion Mass Spectroscopy (SIMS), data are not shown.

$$SiP_2O_7 \xrightarrow{N_2\text{ gas 1L/min, 1000°C}} SiO_2 + P_2O_5 \qquad (3)$$

The D_{it} is significantly lower as compared to NO-passivated device and is 2 × 10^{-11} $eV^{-1}cm^{-2}$ at 0.2 eV from the conduction band edge, $E_c - E = 0.2$ eV, **Figure 1**. In all the D_{it} measurements, high-low C-V technique is used to extract the values. MOS capacitor results show that P passivation is more effective than NO passivation. The field-effect mobility of an n-channel MOSFET after P passivation is two times higher compared to standard NO passivation (**Figure 2**).

Bias temperature stress (BTS) test was performed on metal-oxide semiconductor capacitors for a positive/negative bias and results are shown in **Figure 3(a)** [30]. For a positive BTS test, the electric field is ~+1.5 MV/cm. The value of 0 V for flatband voltage (V_{FB}) before BTS in the case of phosphorous passivation confirms fewer interface traps after this process. The V_{FB} is significantly higher (~2 V) after NO annealing. After the positive BTS test, V_{FB} increases drastically from 0 to −18 V. The reason behind this shift in V_{FB} is induced positive polarization charge at the interface due to the formation of PSG layer. Results of a negative BTS test are also shown in **Figure 3(a)**. Due to induced negative polarize charge at the interface, the V_{FB} shifts in positive direction. Thus, the PSG layer makes devices highly unreliable by shifting V_{FB} in opposite directions. Results of a positive BTS test for a thick PSG MOSFETs are shown in **Figure 3(b)**. A negative shift of −2 V in threshold voltage after a positive BTS test confirms

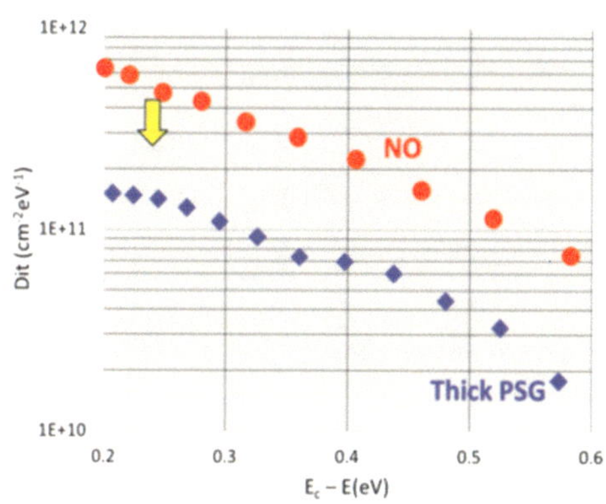

Figure 1. D_{it} after 3-h P passivation and comparison with NO passivation.

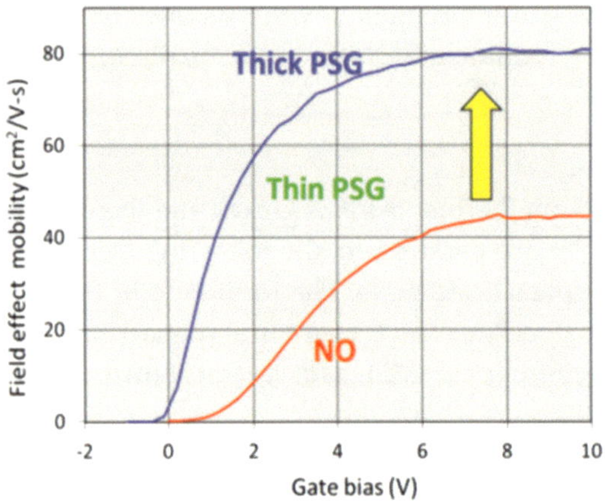

Figure 2. Field-effect mobility of a MOSFET after P passivation and comparison with NO passivation.

that all these devices are highly unstable and of no practical use. Again, the near zero values for V_T confirm the effectiveness of P passivation compared to NO passivation.

Phosphorous process leads to improved interface by making traps electrically inactive and hence leads to higher field-effect mobility in 4H-SiC MOSFETs [6, 30]. Although the values of diffusion coefficients of impurities in SiC are very low, the possibility of P diffusion into SiC cannot be neglected. Phosphorous in the SiC substrate could have two effects: (i) phosphorous in the SiC can passivate carbon di-interstitial clusters and the correlated dangling bonds and (ii) the presence of phosphorous in the substrate can increase the concentration of n-type dopants (P) in the 4H-SiC/SiO_2 interface region to produce a counter-doping effect. This phenomenon has been observed in nitrogen-implanted 4H-SiC MOSFET. Both these effects (reduction in interface trap and counter-doping) lead to a lower V_T and a higher channel mobility [31, 32].

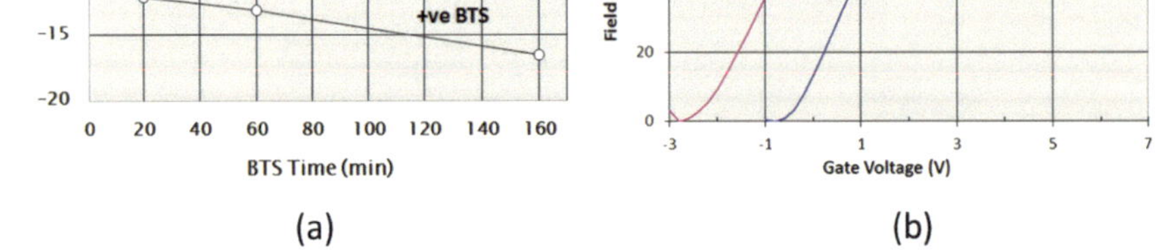

Figure 3. (a) V_{FB} after positive and negative BTS of a MOS capacitor. (b) Mobility data before and after positive BTS on a thick PSG MOSFET.

1.5.1.2. Etched PSG process

Annealing of an SiO_2 layer in a P_2O_5 ambient converts it into a phosphosilicate glass layer. PSG is a polar material [33], and if a positive/negative bias is applied at the gate terminal of MOSFET a positive/negative polarization sheet charge is induced at the 4H-SiC/SiO_2 interface. The effect of this induced charge is similar to the effect of Na^+ ions at the interface. The presence of either charge leads to an unsTable 4H-SiC MOSFET. For example, this polarize charge can change a "normally-off" device to a "normally-on" device. X-ray photoelectron spectroscopy (XPS) results [34, 35] reveal that PSG layers cannot be removed completely by etching in BOE if is grown on 4H-SiC while opposite is true for the layer grown on Si. After BOE etching in the case of SiC, a 2–3-nm Si-C-O-P interfacial layer can still be seen which is equivalent to a phosphorous areal density of 2×10^{14} atoms/cm^2 (approximately one-tenth of a monolayer). Before etching, the areal density of phosphorus is 10^{15} cm^{-2}. We lose P after etching which is reflected in higher trap density for the etched PSG sample, **Figure 4**. In order to understand this phenomenon, we need to address the following two questions: (i) Is the un-etched PSG layer responsible for the better interface trap and hence high field-effect mobility? and (ii) Is there any difference between the bulk PSG layer and the un-etched layer in terms of induced polarization charge? These two questions are very important in order to understand the field-effect mobility and threshold voltage stability of the devices. An etching experiment on P-passivated MOS capacitors was performed to answer these questions. On etched devices, a thick layer of deposited oxide is used to fabricate MOS capacitors by using a low-pressure chemical vapour deposition (LPCVD) system at a temperature of 650°C. The D_{it} and V_{FB} results of PSG-etched devices are shown in **Figures 4** and **5** and are also compared with etched NO-passivated devices.

The D_{it} for an etched NO device is like as-oxidized MOS capacitors. The values are significantly higher and similar to the D_{it} profile for an unpassivated MOS device. These results show that, after etching, there is a decrease in the areal densities of both P and N from the 4H-SiC/SiO_2 interface. As a result, we observed worst electrical properties of the interface

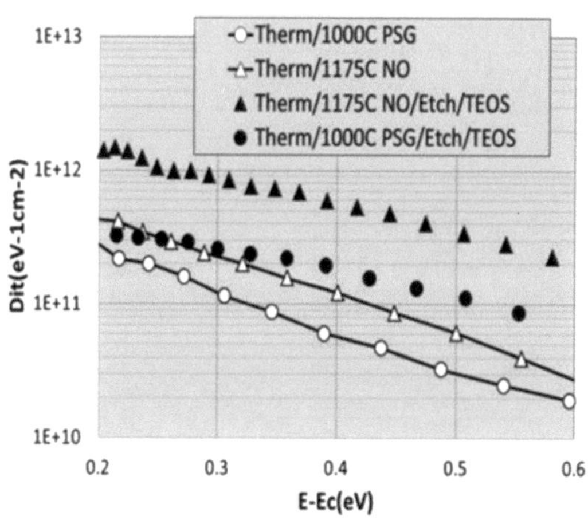

Figure 4. D_{it} before and after etching for NO and PSG MOS capacitors trap density increases after etching.

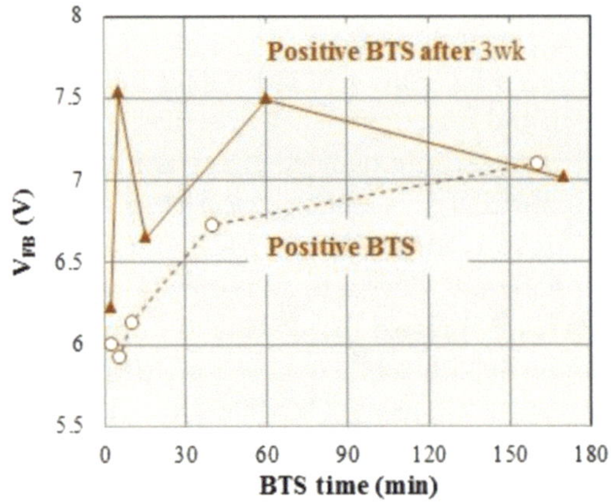

Figure 5. Positive BTS data of an etched PSG MOS capacitor device stability after etching.

caused by increased interface trap densities. The results of BTS tests for etched P-passivated samples are shown in **Figure 5**. The etched devices have flatband voltages, which remain constant at around 6 V and are caused by the electron injection phenomenon from the SiC into the polar (PSG) layer. For the etched PSG MOS capacitors, the values are reduced significantly showing the absence of high induced polarization charge at the interface. In the case of thick PSG samples, polarization charge induces a negative shift in V_{FB} which keeps on increasing with increasing BTS test time. The results of BTS tests for the etched PSG MOS capacitors show that, after etching, stability improves at the cost of higher interface trap density.

1.5.1.3. Thin PSG devices BTS

P passivation of the 4H-SiC/SiO_2 interface reduces the D_{it} but increases the threshold voltage instability due to the formation of a PSG layer during the process (**Figure 6**) [18]. In order to make the devices more reliable after P passivation, a different approach is used which involves a thin PSG layer (~10 nm). The D_{it} of a thin MOS capacitor is shown in **Figure 7**. D_{it} value for thin PSG MOS device is 3×10^{11} $eV^{-1}cm^{-2}$ at $E_c - E = 0.2$ eV which is two times lower than NO-passivated device. The corresponding lateral 4H-SiC MOSFET has a peak channel

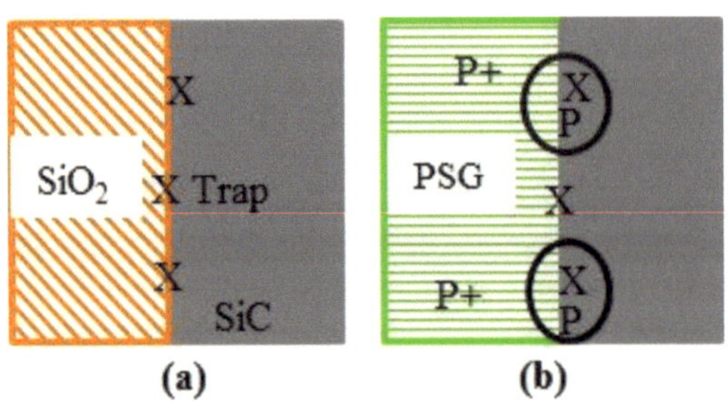

Figure 6. Gate oxide before (a) and after (b) P passivation. The P which passivates the traps at the interface also leads to instability by transforming SiO_2 into PSG.

http://dx.doi.org/10.5772/67867

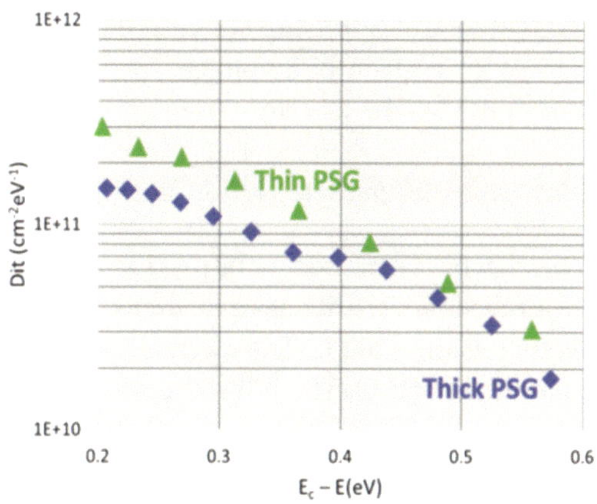

Figure 7. D_{it} for thin PSG MOS capacitor. D_{it} is lower than NO device, but higher than thick PSG device.

mobility of ~75 cm²/V s, **Figure 7**. This number is approximately two times higher compared to an NO device. Although the mobility of a thick PSG is higher (**Figure 2**), the device is plagued with threshold voltage instability. The breakdown field for thin PSG device is NO-like, but its leakage current is significantly higher (not shown). Results for positive BTS tests for thin PSG MOS capacitors and MOSFETs are shown in **Figure 8(a)** and **(b)**, respectively. An oxide field of ~+1.5 MV/cm was applied at 150°C during positive BTS tests for all the samples. The thin PSG MOS capacitors show an improved electrical interface which translates into stable flat-band voltage values of the MOS capacitors. **Figure 8(b)** shows the electric field-effect mobilities of the MOSFETs before and after positive BTS tests performed for 8 h. We can observe that there is a small right shift caused by electron injection, in the field-effect mobility curve. This once again proves that using the thin PSG process, the device stability can be improved. The instability of thick PSG devices is due to polarization charge which is negligible in the case of

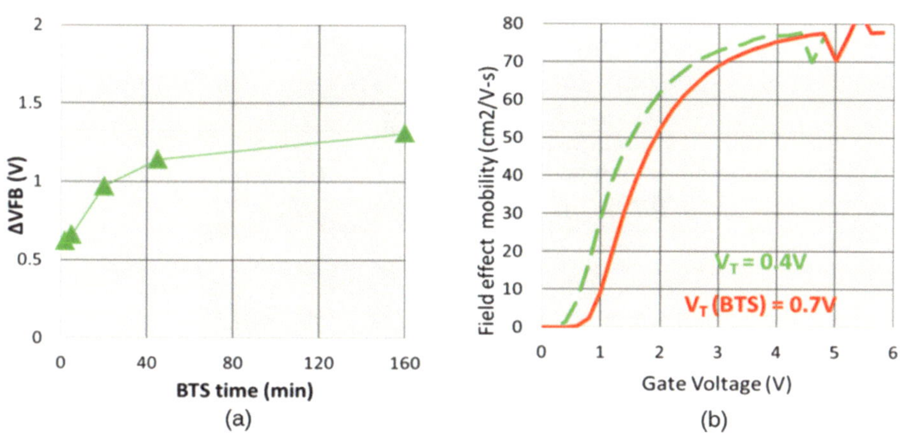

Figure 8. (a) Positive BTS data of a thin PSG MOS capacitor. Data show improved device stability. $\Delta V_{FB} = V_{FB}^{final} - V_{FB}^{initial}$ (b) Mobility data of a MOSFET with (dashed curve) and without 8-h positive BTS.

thin PSG devices. There is one-to-one relationship between the thickness of the PSG layer and shift in flatband voltage (ΔV_{FB}) for a given BTS voltage. As we increase the thickness of the gate oxide layer from 10 nm (thin PSG devices) to 70 nm, we get less stable device.

1.5.2. Nitrogen plasma (N-plasma) passivation of the 4H-SiC/SiO$_2$ interface

NO passivation is the process that has been used in the production of commercial SiC MOSFET. This process reduces the D_{it} of the interface, it has some limitations. This process is performed at high temperature between 1150 and 1200°C. The mechanism is that at such a high temperature, NO dissociates into atomic N and O. The N produced during the reaction reduces the number of the traps by passivating them, while the O reacts with the SiC layer forming an additional SiO_2 layer and hence a new set of interface traps [16]. Thus, there are two mechanisms going on simultaneously in competition. With N-plasma passivation where there is no atomic O, it is possible to eliminate additional oxidation and hence limiting the total number of traps caused during post-oxidation annealing. As a result during N-plasma passivation, the D_{it} decreases drastically. Also, we get some extra benefit from this process. It has been seen that after this process, the D_{it} continues to decrease with the increase of N plasma exposure time, which is not true in the case of NO passivation. In that process, the N concentration saturates at the interface after the annealing time of 2 h [17]. The D_{it} for 4-h N-plasma passivation is NO-like and results in NO-like mobility (~40 cm^2/V s). Eight-hour N-plasma-passivated MOS devices give D_{it} ~3 × 10^{11} eV^{-1}cm^{-2}. If mobility scales with D_{it}, MOSFETs that have undergone an 8-h N-plasma passivation should have a peak field-effect mobility of ~100 cm^2/V s.

For N-plasma passivation, ground-state atomic N is created in microwave plasma and a portion of these atoms recombine to emit at visible wavelengths. The set-up used to create N-plasma is shown in **Figure 9**. The snapshot of the spectrum formed during the N-plasma passivation is shown in **Figure 10**. In the spectrum, a peak is obtained at the wavelength of

Figure 9. High-temperature microwave plasma furnace used for nitrogen plasma passivation (N-plasma passivation).

http://dx.doi.org/10.5772/67867

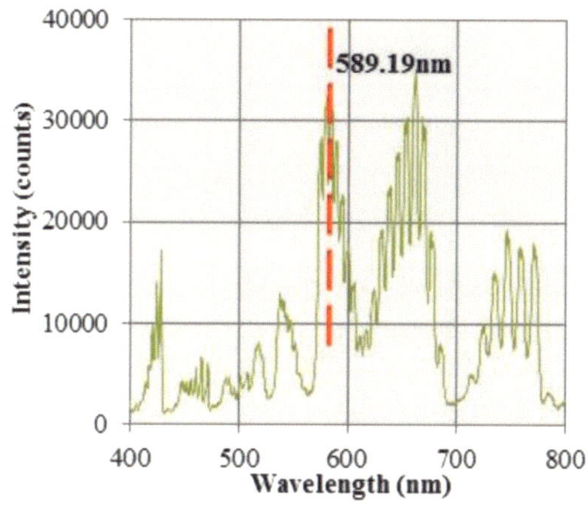

Figure 10. Typical spectrum of nitrogen during N-plasma passivation.

589.19 nm. This peak is obtained due to the recombination of active ground-state (^{4}S) N atoms. This recombination is followed by the decay of an excited state in the N_2 molecule and gives rise to yellow afterglow [19]. The mechanism, which causes the afterglow, takes place due to the following reaction.

$$\begin{aligned} &N(^4S) + N(^4S) \rightarrow N_2(^5\Sigma g+) + M \\ &N_2(^5\Sigma g+) + M \rightarrow N(^4S) + N(^4S) + M \\ &N_2(^5\Sigma g+) + M \rightarrow N_2\left(B^3\Pi_g\right) + M \\ &N_2\left(B^3\Pi_g\right) \rightarrow N_2(A^3\Sigma u+) + h\nu \end{aligned} \tag{4}$$

Also, the intensity of yellow afterglow of emission is proportional to the square of concentration of active ground-state atoms. The amount of radiation detected at 589.19 nm is therefore a measure of the atomic nitrogen concentration in the plasma [36, 37]. After N-plasma passivation for the desired time, the recovery step was performed at 1160°C in N_2 flow. This step is used to heal the damage caused during the nitrogen plasma exposure of oxide and helps to improve the breakdown characteristics of the devices.

1.5.2.1. Four-hour N-plasma passivation on thermal oxide

Initially, the N-plasma process was used on the interface grown using a standard thermal oxidation. Nitrogen plasma also causes damage to the interface so N-plasma process is followed by a recovery process. Results obtained for a 4-h N-plasma process followed by a 2-h recovery process (from plasma damage) are shown in **Figure 11(a)** and **(b)** [38]. The D_{it} and electric field-effect mobility obtained under these conditions are both "NO-like." The thickness of the oxide layer used is 70 nm. The peak value of field-effect mobility of MOSFET is ~45 cm^2/V s (which is like NO-passivated MOSFET). The V_T determined by using the linear portion of the drain current versus gate voltage curve is around 4 V (**Figure 12**). After N-plasma passivation, the oxide breakdown field of the MOS capacitor is ~4 MV/cm. This lower breakdown field is

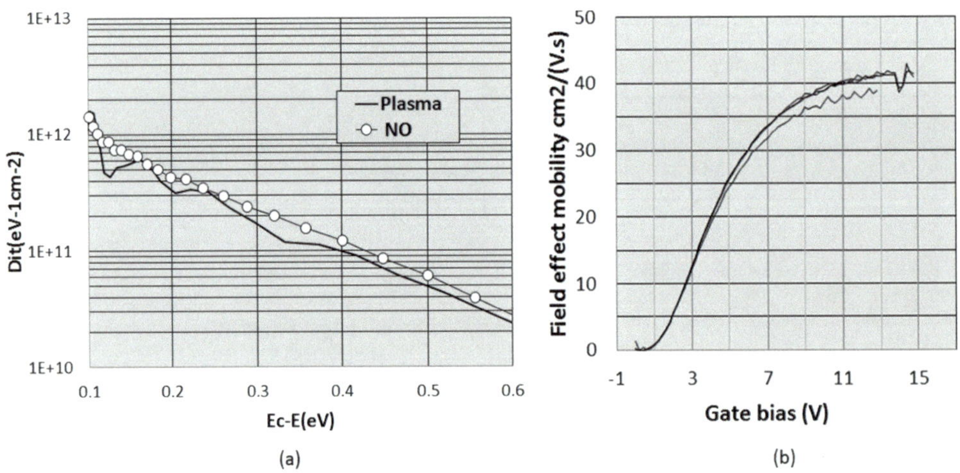

Figure 11. (a) Interface trap density (both NO and plasma) and (b) mobility for 4-h N-plasma passivation.

due to the damage caused during N-plasma passivation. This is the result of the gate leakage current and needs further optimization of the recovery process done after the passivation step. Also, we can conclude that the recovery anneal of 2 h is not enough.

1.5.2.2. Four-hour N-plasma passivation on deposited oxide

To limit carbon liberation (by minimizing the number of processing steps which causes oxidation) of the SiC, thermal oxide layers can be replaced by deposited oxides. The peak value of field-effect mobility for a companion MOSFET is ~50 cm²/V s (**Figure 12**), which is significantly higher than "NO-like" device, with threshold voltage again ~4 V (**Figure 13**). Note that this value is 25% higher than the one obtained with thermally grown oxide [38, 39].

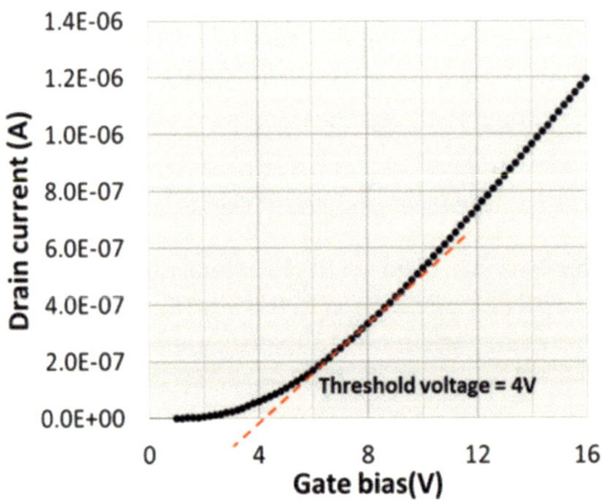

Figure 12. Threshold voltage (V_T) for a 4-h N-plasma passivation. The extrapolation in linear region method is used to extract the threshold voltage.

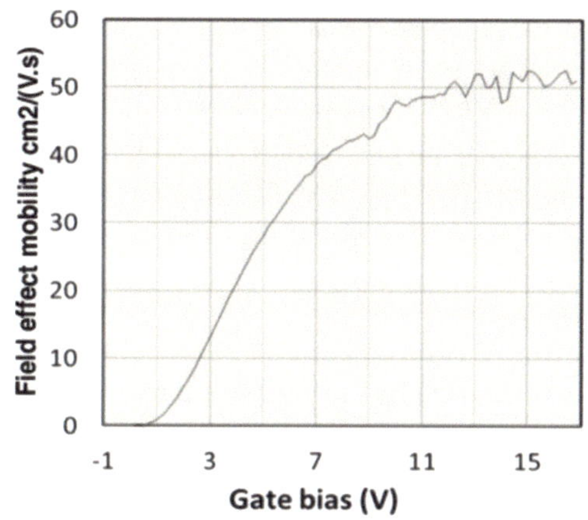

Figure 13. Mobility for a deposited oxide + 4-h N-plasma-passivated MOSFET.

1.5.2.3. Eight-hour N-plasma passivation

The interface trap density after 8-h N-plasma passivation with 6-h recovery is shown in **Figure 14(a)** for 62-nm thermal oxides [3, 39, 40]. The D_{it} at 0.2 eV is 2×10^{11} $cm^{-2}eV^{-1}$, which is like the MOS capacitors having a thick PSG layer. But as we discussed, these types of devices are highly unstable and of no practical use. With N-plasma passivation, we do not have this problem because the oxide layer is still SiO_2 and still non-polar in nature. The areal densities of N obtained after Secondary Ion Mass Spectroscopy measurements done on the MOS capacitors, which underwent 4-h and 8-h N-plasma passivation, are 6×10^{14} and 1.5×10^{15} cm^{-2}, respectively. The areal density for standard NO-passivated MOS capacitor is also 6×10^{-14} cm^{-2} and has been used as a reference sample. Also, XPS data for N-plasma-passivated devices have shown (not presented here) that the D_{it} for devices that have undergone this process is

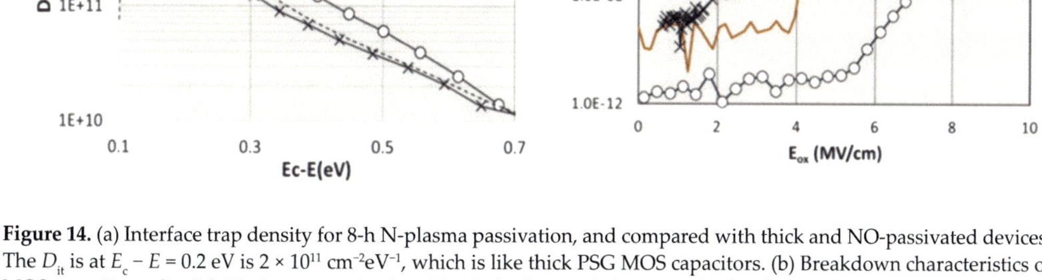

Figure 14. (a) Interface trap density for 8-h N-plasma passivation, and compared with thick and NO-passivated devices. The D_{it} is at $E_c - E = 0.2$ eV is 2×10^{11} $cm^{-2}eV^{-1}$, which is like thick PSG MOS capacitors. (b) Breakdown characteristics of MOS capacitors after 8-h N-plasma passivation. The dashed curve is for a standard NO device.

almost 2.5 lower than that for NO-passivated devices. **Figure 14(b)** shows breakdown characteristics after an 8-h N-plasma passivation. The oxide is leaky compared to NO, and the breakdown field is lower (~2 MV/cm for some devices). If mobility scales with D_{it}, MOSFETs that have undergone an 8-h N-plasma passivation should have a peak field-effect mobility of ~100 cm^2/V s. Further optimization of the process is desirable to achieve higher N concentration at the interface and to improve the yield of the process.

1.5.3. High-temperature oxidation

Lately, there has been a growing interest in high-temperature oxidation of 4H-SiC. Studies have shown that if the oxidation conditions are optimized, then the 4H-SiC/SiO_2 interface grown after the process has much better electrical properties to the ones grown under standard conditions (1100–1200°C). In the following sections, we see the progress made in this area. Also, we discuss the effect of performing P and N_2O post-oxidation annealing on the oxides grown at high temperatures. It has been observed that high-temperature oxidation performed at 1500°C can lead to a better interface with D_{it}. **Figure 15(a)** shows the effect of temperature on the D_{it}. We can see with the increase of temperature from 1200 to 1500°C that D_{it} reduces from 2×10^{12} to 5×10^{11} $cm^{-2}eV^{-1}$ at $E_c - E = 0.2$ eV. **Figure 15(b)** shows the linear transfer characteristics and field-effect mobilities of the devices measured at room temperature [36]. The V_T values for the devices without any passivation are typically around 5 [3, 49] which is due to the large number of traps at the interface. The net charge of these traps is negative in nature. This high-temperature process is effective in reducing in number of trapped and resulted in low threshold voltages observed in the MOSFETs. The reduction in trapped charge is reflected by the field-effect mobility of ~40 cm^2/V s. This value is much lower than the previously reported results, even the one obtained at 1400°C. All this is a direct reduction

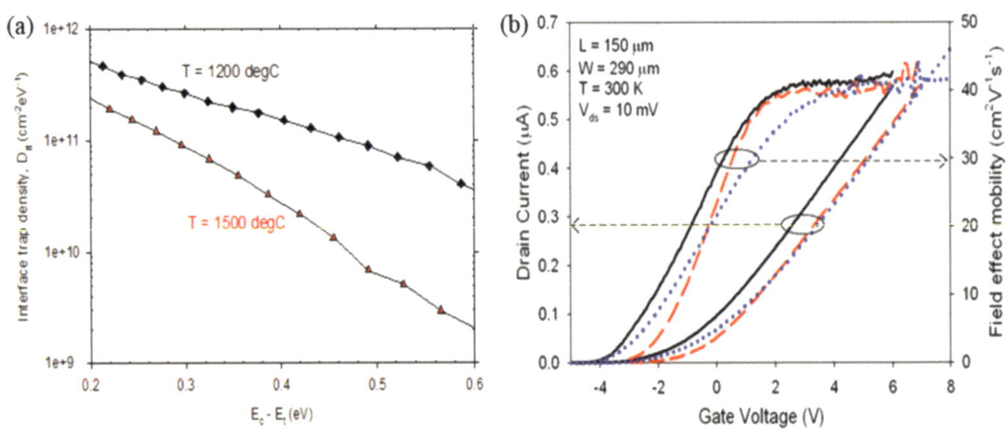

Figure 15. (a). D_{it} of thermal oxides grown at 1200 and 1500°C in a pure oxygen flow rate of 0.05 l/min. The increase in temperature resulted in approximately a twofold reduction in D_{it} from 5.3×10^{11} to 2.5×10^{11} $cm^{-2}eV^{-1}$, a $E_c - E = 0.2$ eV. (b). Linear transfer characteristic and field effect mobility of 3 lateral 4H-SiC MOSFETs. Whilst there is a variation in threshold voltage, the novel high temperature gate oxidation process yields a consistent maximum mobility of ~40 $cm^2V^{-1}s^{-1}$.

http://dx.doi.org/10.5772/67867

in Coulomb scattering. In this approach (high-temperature oxidation), there is no need of performing the post-oxidation annealing of the 4H-SiC/SiO_2 interfaces.

1.5.3.1. Combined N_2O and phosphorous passivations of the 4H-SiC/SiO_2 interface with oxide grown at 1400°C

Phosphorous (P) passivation is more effective than N_2O passivation in improving the 4H-SiC/SiO_2 interface by reducing the number of traps at the 4H-SiC/SiO_2 interface. There are some studies performed by Rong Hua et al. [41] to see the combined effect of high-temperature oxidation with either P or N_2O passivation. The MOS capacitor with 1400°C dry oxidation and without any post-oxidation passivation process has the highest D_{it} as shown in **Figure 16(a)**. The D_{it} results from the MOS capacitors have one-to-one relation with the field-effect mobility of the MOSFETs shown in **Figure 16(b)**, where the lowest D_{it} corresponds to the highest field-effect mobility. From **Figure 16(b)**, it can be seen that the P passivation can increase the peak field-effect mobility of a 4H-SiC MOSFET to about 66 cm^2/V s, which is five times higher than the value obtained after high-temperature N_2O-annealing process. Compared to low field-effect mobility, ~2 cm^2/V s, of the MOSFETs with gate oxide grown at 1400°C for 1 h without post-oxidation annealing process, the P-post-oxidation annealing (POA) process can dramatically increase the field-effect mobility.

However, the combined N_2O- and P-passivation processes have shown a slight decrease in the peak field-effect mobility value (60 cm^2/V s) compared to the P-only passivation. This value is still much higher than obtained using N_2O passivation (12 cm^2/V s). The only drawback for the combined N_2O- and P-passivation processes is that the MOSFET still has a negative threshold voltage (V_T) value (~−5 V) but better than the P-only passivation. The net positive charge at/or near the interface 4H-SiC/SiO_2 resulted in a negative V_T of the device. This value can be improved with further optimization of the process. The negative threshold voltage means that the device is normally on which is not an ideal choice from the application point

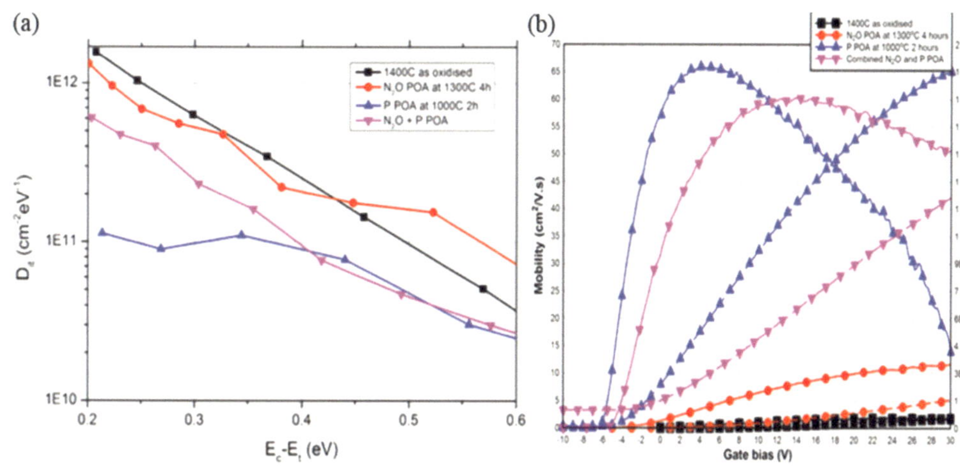

Figure 16. (a). Interface trap density extracted using high-low frequency capacitance method for MOS capacitors fabricated with different passivation conditions. (b). Field-effect mobility and the drain current against the gate bias for lateral MOSFETs with different passivation conditions.

of view. Potentially, this problem could be solved by using a thin layer (a few nm thick) of SiO_2 which afterward undergoes the suggested combined N_2O and P POA and then finally topped up by a thick(40-nm) deposited SiO_2 layer. There is no extra benefit of performing P passivation on the devices which have already had N_2O POA.

1.5.3.2. Impact of N_2O passivation on 4H-SiC/SiO_2 interfaces grown at high temperature

The results on high-temperature oxidation (1500°C) have shown a reduction in the interface trap density (D_{it}) without performing any kind of interface passivation. It would be interesting to see the impact of N_2O passivation on the oxides grown via high-temperature oxidation process. The interface is grown with high-temperature oxidation, and the results are shown in **Figure 1**. There is a decrease in the D_{it} for the high-temperature as-oxidized MOS capacitors as compared with the MOS capacitors grown using standard oxidation process. But there is no further improvement in the D_{it} with N_2O passivation performed at 1350°C, **Figure 17**. The electric field of breakdown for MOS capacitors with high-temperature oxidation with and without N_2O passivation is lower than 8 MV/cm. The reason for an early breakdown in the case of high-temperature as-oxidized MOS capacitors is not clear and further study is needed.

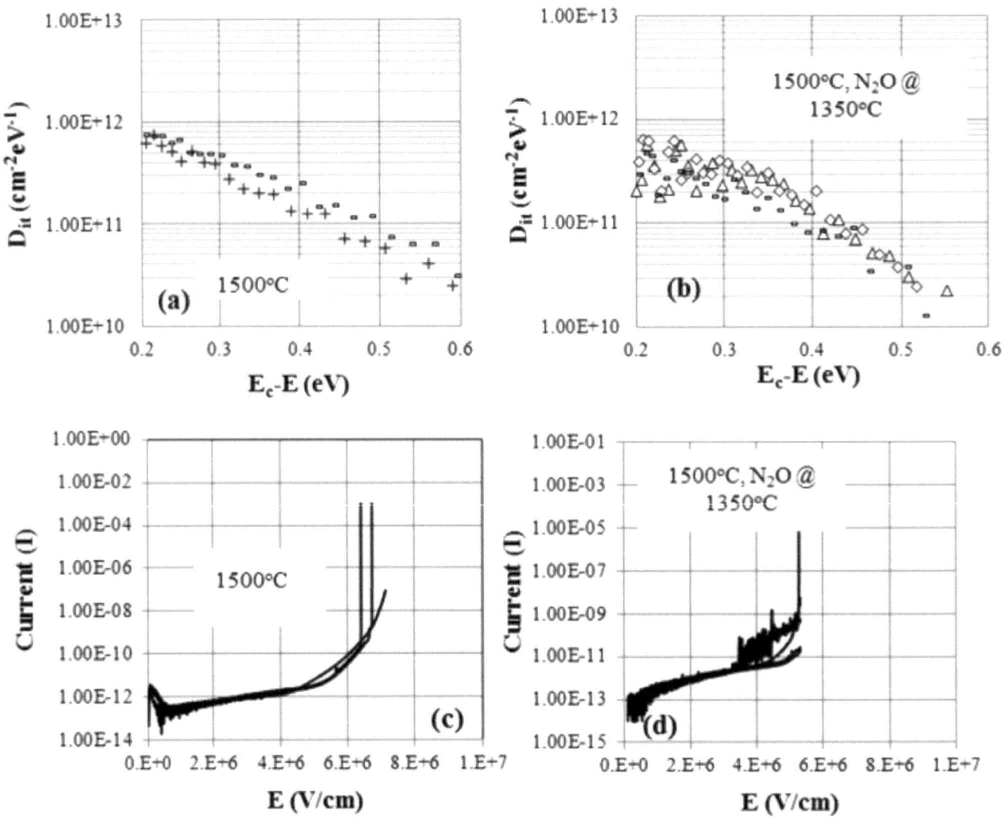

Figure 17. The electrical properties of the 4H-SiC MOS capacitors with a high-temperature (1500°C) thermal oxidation, interface trap density (a–b) and oxide breakdown (c–d).

In the case of passivated devices, this could be due to the incorporation of nitrogen throughout the bulk of the oxide, as previously reported on 4H-SiC MOS capacitors after ammonia passivation. We also performed N_2O passivation on the MOS capacitors grown via the standard oxidation process, and the results are shown in **Figure 18**. Again, the electric field of breakdown is lower for these devices. Also, there is no improvement in the D_{it} with the increase of temperature for N_2O passivation. The breakdown data for N_2O passivation at 1450°C could not be obtained due to a high increase in the oxide thickness after the passivation.

1.5.4. Other interface passivation processes

In addition to all these interface passivation processes, there are some studies done using boron (B) and Sb. Modic et al. have shown that Sb-doped surface channel in combination with nitric oxide post-oxidation annealing can increase the channel field-effect mobility to 100 cm²/V s [42]. Also, Okamoto et al. were able to increase the channel field-effect mobility to 102 cm²/V s by introducing boron atoms to the interface [43].

1.6. 3C-SiC/SiO_2 interface

There is a reignited interest on cubic silicon carbide (3C-SiC), which can be potentially grown heteroepitaxially on 12" Si substrates, as it would result in a drastic cost reduction of semiconductor devices compared to the successful but prohibitively expensive SiC hexagonal

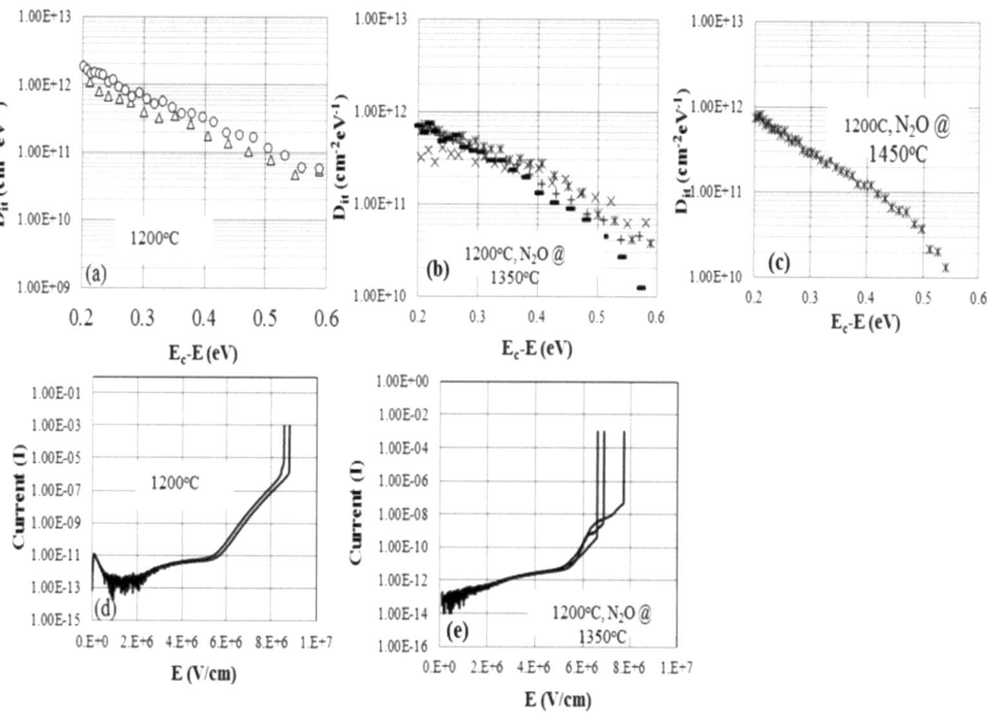

Figure 18. The electrical properties of the 4H-SiC MOS capacitors with a standard oxidation process (1200°C thermal oxidation) followed by N2O passivation at 1350 and 1450°C. From (a) – (c) D_{it} versus E_C-E and (d) – (e) show breakdown characteristics of various MOS capacitors with different oxidation/passivation processes.

polytype technology (4H-SiC). It has been demonstrated that lateral power transistors in 3C-SiC outperform Si and 4H-SiC devices up to 1200 V, and represent an alternative to gallium nitride (GaN) technology. Also, GaN transistors are normally on, and as a result, it is challenging to control them eclectically. The voltage ratings for which these 3C-SiC devices are targeted for make them useful in automotive and other domestic appliances. Thus, this 3C-SiC technology has a huge potential for reducing the global carbon footprint.

1.6.1. High-temperature dry/thermal oxidation (1200–1400°C) and N_2O passivation of the 3C-SiC/SiO_2 interface

Due to the smaller band gap of 3C- (2.2 eV) compared to 4H- (3.2 eV), a fewer number of traps lie within the energy band gap of 3C-SiC in a metal-oxide-semiconductor structure resulting in better field-effect mobility [44]. It is found that 3C-SiC has different oxidation chemistry compared with 4H-SiC; 70–80-nm oxide can be grown at 1100°C in 1 h, which is 10 times faster than the oxidation rate of 4H-SiC on the Si face [45]. Also for the Si face in 4H-SiC, it has been observed that mobility increases with decreasing D_{it}. For example, as we go from thermally grown gate oxide to post-oxidation-annealed (passivated) oxide using hydrogen (H_2) passivation, NO/N_2O passivation, nitrogen plasma (N-plasma) passivation or phosphorous (P) passivation, D_{it} decreases significantly [37, 46–48]. Thomas et al. have shown that as-grown oxidized at 1500°C resulted in a 4H-SiC MOSFET with maximum field-effect mobility of 40 cm^2/V s. Sharma et al. have studied high-temperature oxidation of 3C-SiC [49]. In that work, the highest temperature used for the oxidation of 3C-SiC is 1400°C, because of the limit dictated by the melting point of Si (1414°C). The oxidation is done in 100% oxygen ambient. In addition, a standard nitrous oxide post-oxidation annealing of the interface is performed to see its effect.

A 3C-SiC/Si wafer with heterostructure grown on on-axis p-type Si (001) substrate was used in the work. The 3C-SiC epilayer is n-type with a doping concentration of $N_D \sim 1 \times 10^{16}$ cm^{-3}. The thickness of epilayer was around 10 μm. Lateral MOS-C structure had been used to study the electrical properties of the interface (inset of **Figure 19(a)**). The oxidation was performed at high temperatures followed by 30-min argon (Ar) annealing at temperatures used for oxidation. Standard high-low C-V and conductance techniques were used to analyse the 3C-SiC/SiO_2 interface. In addition, lateral *n*-channel MOSFETs were fabricated to extract the field-effect mobilities (μ). Planar MOSFETs were fabricated on the same substrate which was used to fabricate the MOS capacitors. **Figure 19** shows the normalized C-V curves at 1 MHz for different MOS-Cs. There are three distinct features of these curves. The first is a large flatband voltage shift ($\Delta V_{FB} = V_{FB,i} - V_{FB,ideal}$) towards negative gate bias, the second is increased capacitance in depletion and the third is to observe the large stretch-out in C-V characteristics. The subscript '*i*' stands for different processing conditions used to grow the gate oxide, $i = 1$ for 1200°C oxidation, $i = 2$ for 1300°C oxidation, $i = 3$ for 1300°C oxidation and $i = 4$ for 1300°C oxidation followed by N_2O annealing. **Figure 19(b)** shows a comparison between the ideal and experimental C-V curves for each temperature. The ideal C-V curve is obtained (solving the electrostatic and Poisson equation) by assuming that the 3C-SiC/SiO_2 interface is perfect, having no defects near the interface and also in the bulk of the oxide and 3C-SiC. The doping value used in the computations for the high-temperature-oxidized samples was $N_D = 0.63 \times 10^{16}$ cm^{-3} for all the oxidation temperature range, that is, the 3C-SiC doping concentration

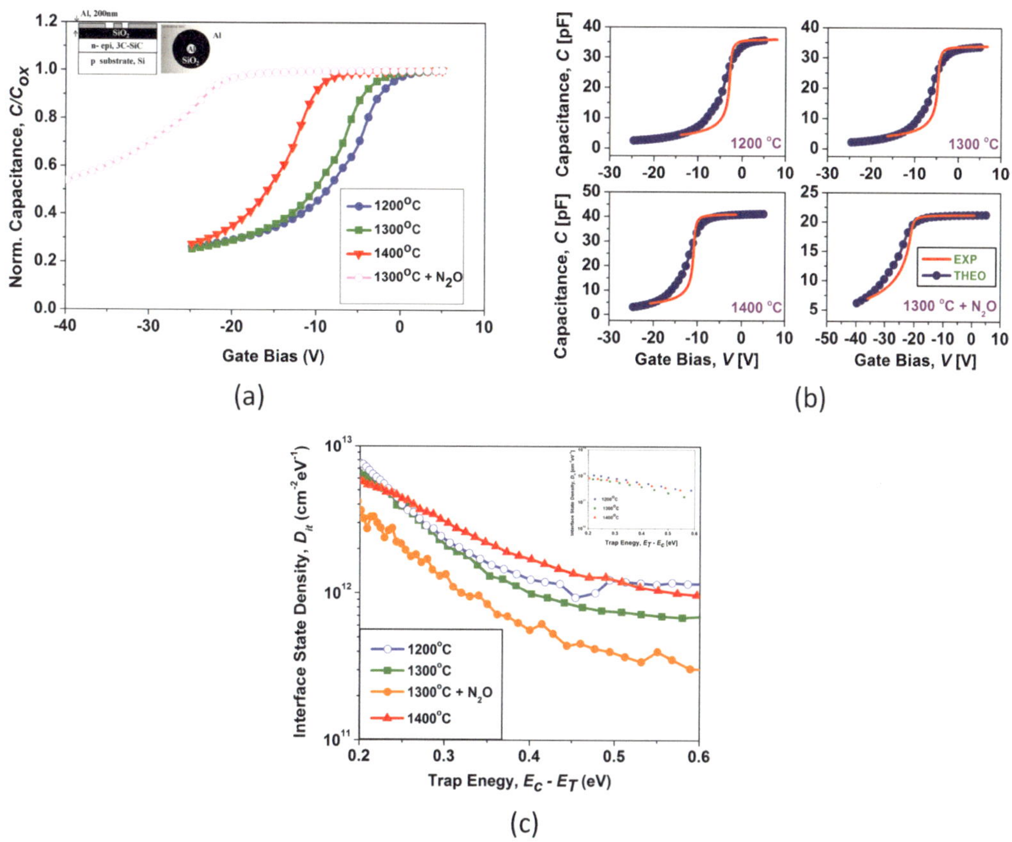

Figure 19. (a) Typical 1-MHz C-V curves for 3C-SiC MOS structures with oxides grown at 1200, 1300, 1400 and 1300°C+N_2O. (b) A comparison between the ideal and experimental C-V curves for each temperature. (c) Interface state densities (D_{it}) for 3C-SiC MOS structures with and without post-oxidation anneal (N_2O) process, calculated using Terman C-V and conductance (inset) methods. Due to a very wide Gaussian dispersion, it is not possible to extract D_{it} for the N_2O-annealed device using the conductance method.

seems to be stable even for 1400°C. Interestingly, a significant increase in the 3C-SiC doping was observed for the sample further annealed in N_2O. In this case, a value of $N_D = 2.45 \times 10^{16}$ cm^{-3} was used to fit the experimental capacitance. The flatband voltage shift shows the existence of net positive fixed charge at the interface and its origin is believed to be the presence of carbon clusters and dangling bonds formed after thermal oxidation [6] Carbon clusters and dangling bonds act as donor-like states and are positively charged when they are empty and as a result give rise to a negative shift in $V_{FB.}$ The calculations for effective oxide charge (Q_{EFF}) has been done (not shown) for all the MOS-Cs. It was found that Q_{EFF} increases with temperature, which is different from the results reported previously on 3C-SiC MOS-Cs and further research is required to explain this behaviour [50]. In addition to carbon clusters, the nitrogen-related complex in the case of 1300°C (oxidation + N_2O annealing) MOS-C causes a large shift in V_{FB}, likely due to its incorporation in the 3C-SiC surface and resulting in an increased doping concentration [51]. Also, as mentioned above, all the curves are significantly stretched out compared to the ideal curve, representing the presence of high interface traps. These traps could be near the conduction band edge and/or the presence of oxide near interface traps (also known as slow traps) [52]. Hysteresis (not shown) in the C-V curves indicates the slow nature of some of the traps [52]. In addition, less band banding is caused by increased net positive fixed charge at the interface resulting in increased capacitance in depletion region [53]. **Figure 19(c)** shows the density of interface traps (D_{it}) extracted from the Terman C-V method and compared with conductance methods (shown as inner graphic of **Figure 1**). Both these methods give consistent results with 1300°C (oxidation + N_2O) leading to the lowest D_{it} = 1.8 × 10^{12} $cm^{-2}eV^{-1}$ at $E_c - E_T$ = 0.25 eV. This value is two times smaller than the one obtained with the 1300°C oxidation process (D_{it} = 3.7 × 10^{12} $cm^{-2}eV^{-1}$). It is believed that these D_{it} values are rather high, primarily because of the contribution of slow states. As it can be seen in **Figure 19(b)** for the 1300°C (oxidation + N_2O) device, the plateau is quite ideal for voltages larger than the flatband (accumulation) but in depletion there is a relevant dispersion (ideal versus experimental), typical signature of the oxide near interface traps. The superiority of annealed device is also confirmed by the G-V plot as shown in **Figure 20(a)**, performed at 100 kHz. The area under the conductance peak is a measure for D_{it}. The position of the peak corresponds to the energy position in the band gap and as all these peaks occur at gate bias close to V_{FB}, it can be inferred that these peaks are a measure of D_{it} close to the CB edge of 3C-SiC [49]. Devices without the N_2O post-oxidation annealing have larger area under G-V curves when compared with the annealed device, implying higher D_{it} for non-annealed devices. The peak value of 1300°C (oxidation + N_2O) is much smaller than other devices and hence this device has the lowest D_{it} being consistent with Terman and conductance methods. Also, all of the admittance-voltage (*G-V*) curves have finite full width at half maximum (FWHM) signifying that these traps are dispersed over an energy range within the band gap and are not localized at a fixed energy value. Similar results have been reported on the MOS capacitors implanted with N [54]. As the 1300°C oxidation process with N_2O post-oxidation annealing gives the lowest D_{it}, this process was used to fabricate a lateral 3C-SiC MOSFET. Also, a lateral MOSFET was fabricated with the 1300°C oxidation process, as for a reference sample. The results of mobility

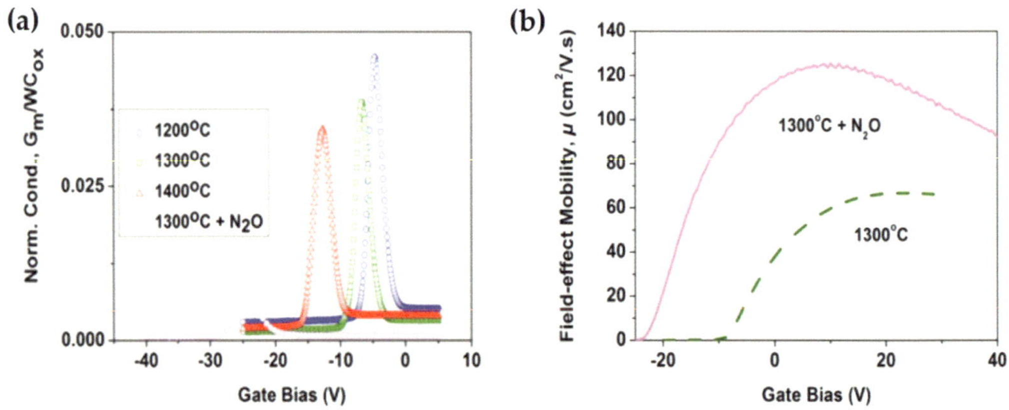

Figure 20. (a) Normalized conductance curves of 3C-SiC MOS structures taken for different gate voltages at probe frequency (ω) = 100 kHz. (b) Field-effect channel mobilities (μ) for N_2O-annealed and non-annealed 3C-SiC MOSFETs. Both the N_2O post-oxidation annealing and thermal oxidation were done at 1300°C.

measurements for lateral MOSFETs are shown in **Figure 20(b)**. The device with N_2O annealing has a peak channel mobility of approximately 120 cm^2/V s, and shows a ×2 improvement in peak mobility as compared with a 1300°C as-oxidized MOSFET which has a mobility of 60 cm^2/V s. This is consistent with the fact that the N_2O-annealed MOS-C has better D_{it} as compared with the as-oxidized 1300°C MOS-C. The MOSFET annealed in N_2O has a relatively sharp turn-on and a peak channel mobility of 125 cm^2/V s. At greater gate biases, the field-effect mobility decreases because of the increasing surface field, which may be an indication of the surface roughness to be a dominant scattering mechanism [40]. **Figure 21** shows atomic force microscopy (AFM) analysis that has been performed on all of the devices in order to see the effect of temperature on the surface morphology. The root-mean-square (RMS) values of roughness for all these MOSCs lie in the range of 0.54–0.60 nm, showing that high temperatures (≥1300°C) do not have effect of temperature on the surface morphology. The RMS values of roughness for all these MOSCs lie in the range of 0.54–60 nm, showing that high temperatures ($T \geq 1200$°C) do not have any significant detrimental effect on the oxide surface. The Secondary Ion Mass Spectrometry profiles are shown in **Figure 22**. The N_2O-annealed device shows the accumulation of nitrogen at the interface which is responsible for the improved D_{it} and field-effect mobility (μ) in this process. There is virtually no appreciable difference between the Si, O and C profiles for all the devices, and it could be due to no out-diffusion or diffusion of species from the interface. There is a possibility that the Si, O and C concentration

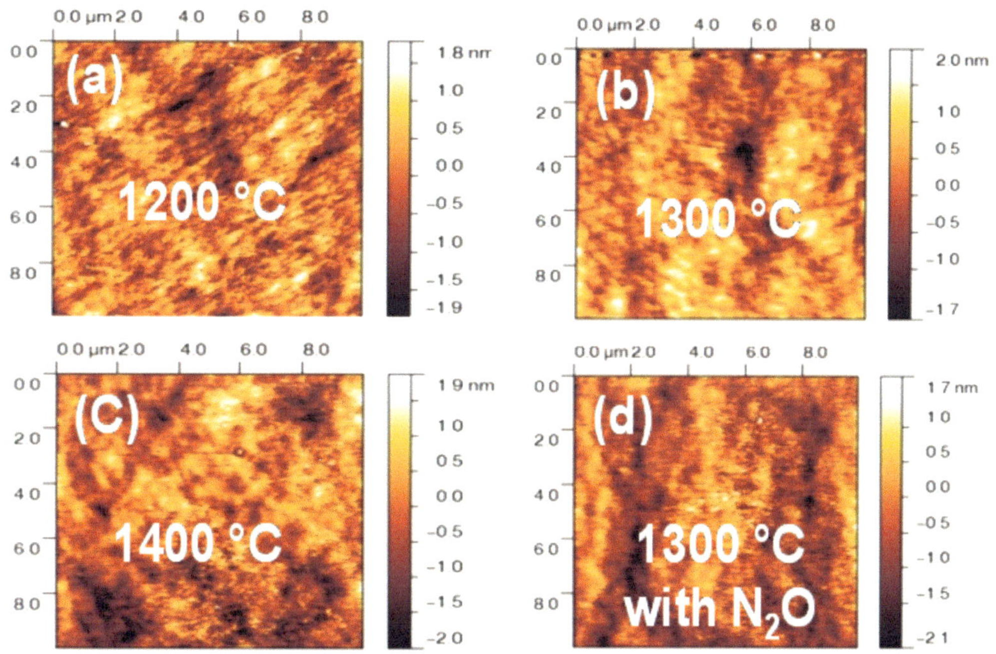

Figure 21. Atomic force microscopy (AFM) for results different processed MOS capacitors with as-grown thermal oxide (a) at T = 1200°C (b) T = 1300°C, (c) T = 1400°C and (d) T = 1300°C with N_2O post-oxidation annealing. The RMS values of roughness within a range of 0.54–0.60 nm, implying that high temperature does not have deteriorating effect on the surface morphology of devices.

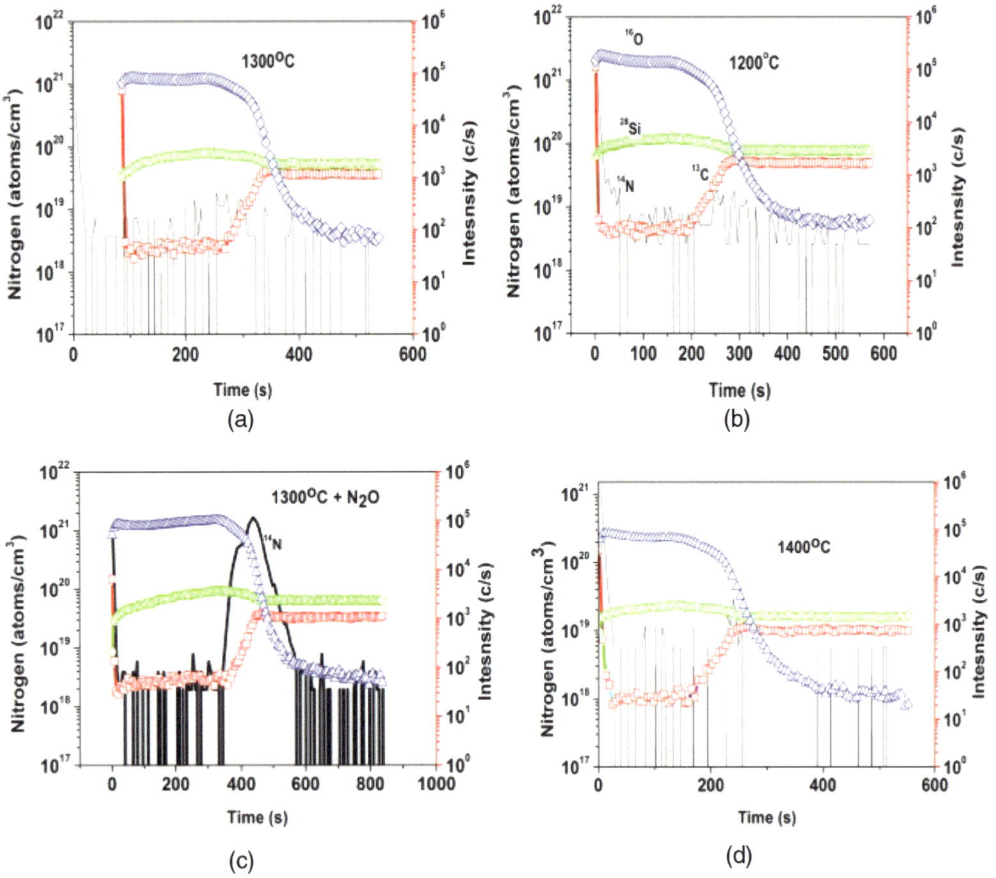

Figure 22. SIMS profiles for 3C-SiC MOS capacitors fabricated under different conditions with (a)–(c) for as-oxidized MOS capacitors. From plot (d), it is evident that most of the nitrogen has accumulated at the 3C-SiO_2/SiC interface in the case of 1300°C N_2O-annealed device. It is obvious that most of the nitrogen has accumulated at the SiO_2/SiC interface after an NO anneal.

decreases monotonically very slightly with temperature, which is not within the tolerances of SIMS measurement. The parallel equivalent conductance spectroscopy is used to further analyse the characteristics of the high-temperature oxidation. **Figure 23** shows G_p/ω versus probe frequency (f) curves for non-annealed and 1300°C (oxidation + N_2O) devices. By fitting the curves to a Gaussian fit, the interface state density, trap time extracted. As shown in **Figure 5**, the 1300°C (oxidation + N_2O) device could have no Gaussian dispersion or a very wide Gaussian dispersion, and as a result it, is not possible to fit the experimental data and it is not possible to extract D_{it}, trap time constant (t_p) and surface-potential fluctuations (σ_s) for the Gaussian dispersion, for the annealed device. The D_{it} extracted using this technique for all the devices is shown as the inset of **Figure 1**. Naik and Chow have reported a wide Gaussian dispersion for conductance curves on NO-treated 4H-SiC MOS-C [54], which potentially explains the results presented here. **Figure 24(a)** plots the trap time constant as a function of energy. As we can see from the figure that across the different energy levels in the band gap, t_p increases with increasing temperature. At a constant temperature, t_p increases as we go

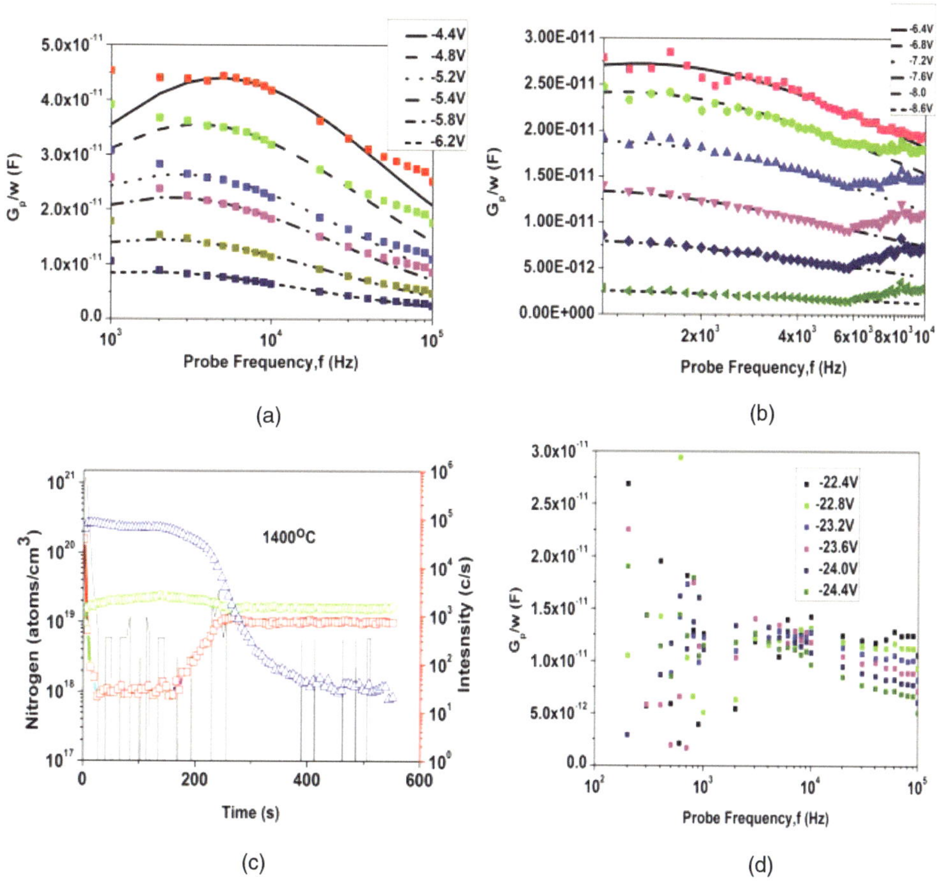

Figure 23. G_p/ω versus probe frequency (ω) at different biases in depletion region for MOS capacitors with as-grown thermal oxide (a) at $T = 1200°C$, (b) $T = 1300°C$, (c) $T = 1400°C$ and (d) $T = 1300°C$ with N_2O post-oxidation annealing.

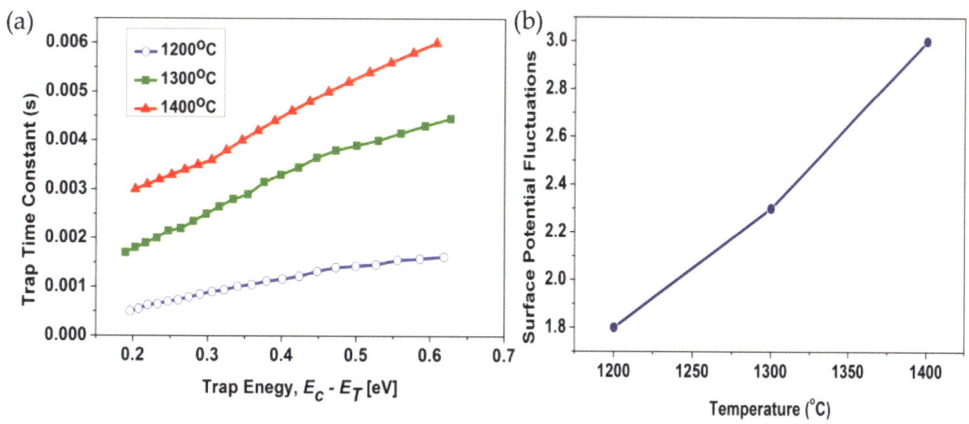

Figure 24. (a) The trap time constant as a function of energy and (b) the temperature dependence of standard deviation of the surface potential (σ_s).

deeper into the band gap. The surface-potential fluctuations can be extracted by performing the curve fitting to the *G-V* curves. In **Figure 24(b)**, we can see the temperature variation of the standard deviation of the surface potential (σ_s) which is caused by fluctuations of interface states for different MOS capacitors. The values lie between 2 and 3, which indicates that the 3C-SiC/SiO_2 interface is electrically better than 4H- polytype (σ_s = 4 for 4H- and around 2 for Si). This fairly low value of σ_s, coupled with the low value of D_{it} and the MOSFET field-decreasing large field-effect mobility, suggests that Coulombic interface-scattering-related effects should not limit transistor performance.

In conclusion, high-temperature oxidation (1200–1400°C) has been used to grow the 3C-SiC/SiO_2. Out of all the oxidation temperatures investigated, 1300°C was found to be the optimum temperature for oxidation. The interface can be improved further by performing the N_2O post-oxidation annealing again at 1300°C for 2 h, though this leads to high accumulation of N at the interface. The lateral MOSFET with N_2O-annealed oxide yielded a field-effect mobility of 125 cm^2/V s, which is twice the value of non-annealed MOSFET, with the gate oxide grown at 1300°C (60 cm^2/V s). The low values of σ_s and larger μ for as-oxidized MOS-Cs show that the 3C-SiC/SiO_2 interface is better than its 4H-SiC counterpart (at least in terms of interfacial fast traps). These findings have important implications for SiC-MOS technology as 3C-SiC/Si can provide a low-cost alternative even in the case of high temperature of processing.

Acknowledgement

I would like to thank all my old colleagues from Auburn University and University of Warwick where all the work had been done. This work was supported by the US Army Research Laboratory, the US National Science Foundation, the II–VI Foundation, and the Engineering and Physical Sciences Research Council. Finally, special thanks to Prof. John R. Williams and Prof. Minseo Park who introduced me to the world of wide band gap semiconductors (SiC and GaN).

Author details

Yogesh Kumar Sharma

Address all correspondence to: aceyogesh83@gmail.com

Dynex Semiconductor Ltd., Lincoln, UK

References

[1] Pfann WG, Scaff JH. The p-germanium transistor. Proceedings of the IRE. 1950 Oct;**38**(10):1151–1154.

[2] Valdes LB. Transistor forming effects in n-type germanium. Proceedings of the IRE. 1952 Apr;**40**(4):445–448.

[3] Brittain JE. The evolution of electrical and electronics engineering and the Proceedings of the IRE: 1938-62. Proceedings of the IEEE. 1990;**78**(1):5–30.

[4] Lécuyer C. Making Silicon Valley: Innovation and the Growth of High Tech, 1930-1970. MIT Press; 2006

[5] O'Mara W, Herring RB, Hunt LP. Handbook of Semiconductor Silicon Technology. Crest Publishing House; 2007 Aug 30

[6] Deal BE, Grove AS. General relationship for the thermal oxidation of silicon. Journal of Applied Physics. 1965 Dec;**36**(12):3770–3778.

[7] Mott NF, Rigo S, Rochet F, Stoneham AM. Oxidation of silicon. Philosophical Magazine Part B. 1989 Aug 1;**60**(2):189–212.

[8] Kao DB, Mcvittie JP, Nix WD, Saraswat KC. Two-dimensional thermal oxidation of silicon. II. Modeling stress effects in wet oxides. IEEE Transactions on Electron Devices. 1988 Jan;**35**(1):25–37.

[9] https://www.semiconductors.org/news/2017/02/02/global_sales_report_2015/global_semiconductor_sales_reach_339_billion_in_2016/.

[10] Baliga BJ. An overview of smart power technology. IEEE Transactions on Electron Devices. 1991 Jul;**38**(7):1568–1575.

[11] http://www.marketsandmarkets.com/PressReleases/power-electronics.asp

[12] http://www.power-mag.com/news.detail.php?NID=38.

[13] Baliga BJ. Power MOSFETs. In: Fundamentals of Power Semiconductor Devices. Springer US; 2008. pp. 276–503.

[14] Warner RM. Microelectronics: Its unusual origin and personality. IEEE Transactions on Electron Devices. 2001 Nov;**48**(11):2457–2467.

[15] Bose BK. Power electronics-a technology review. Proceedings of the IEEE. 1992 Aug;**80**(8):1303–1334.

[16] Lasseter RH, Paigi P. Microgrid: A conceptual solution. In: Power Electronics Specialists Conference, 2004. PESC 04. 2004 IEEE 35th Annual. Vol. 6. IEEE; 2004 Jun 20. pp. 4285–4290.

[17] Hatziargyriou N, Asano H, Iravani R, Marnay C. Microgrids. IEEE Power and Energy Magazine. 2007 Jul;**5**(4):78–94.

[18] Zhang Q, Callanan R, Das MK, Ryu SH, Agarwal AK, Palmour JW. SiC power devices for microgrids. IEEE Transactions on Power Electronics. 2010 Dec;**25**(12):2889–2896.

[19] Liu G, Tuttle BR, Dhar S. Silicon carbide: A unique platform for metal-oxide-semiconductor physics. Applied Physics Reviews. 2015 Jun;**2**(2):021307.

[20] Bhatnagar M, Baliga BJ. Comparison of 6H-SiC, 3C-SiC, and Si for power devices. IEEE Transactions on Electron Devices. 1993 Mar;**40**(3):645–655.

[21] Di Ventra M, Pantelides ST. Atomic-scale mechanisms of oxygen precipitation and thin film oxidation of SiC. Physical Review Letters. 1999 Aug 23;**83**(8):1624.

[22] Tan J, Das MK, Cooper Jr JA, Melloch MR. Metal–oxide–semiconductor capacitors formed by oxidation of polycrystalline silicon on SiC. Applied Physics Letters. 1997 Apr 28;**70**(17):2280–2281.

[23] Knaup JM, Deák P, Frauenheim T, Gali A, Hajnal Z, Choyke WJ. Defects in SiO_2 as the possible origin of near interface traps in the SiC/SiO_2 system: A systematic theoretical study. Physical Review B. 2005 Sep 20;**72**(11):115323.

[24] Wang S, Di Ventra M, Kim SG, Pantelides ST. Atomic-scale dynamics of the formation and dissolution of carbon clusters in SiO_2. Physical Review Letters. 2001 Jun 25;**86**(26):5946.

[25] Shenoy JN, Das MK, Cooper Jr JA, Melloch MR, Palmour JW. Effect of substrate orientation and crystal anisotropy on the thermally oxidized SiO_2/SiC interface. Journal of Applied Physics. 1996 Mar 15;**79**(6):3042–3045.

[26] Opila E. Influence of alumina reaction tube impurities on the oxidation of chemically-vapor-deposited silicon carbide. Journal of the American Ceramic Society. 1995 Apr 1;**78**(4):1107–1110.

[27] Schomer R, Friedrichs P, Peters D, Stephani D. Significantly improved performance of MOSFETs on silicon carbide using the 15R-SiC polytype. IEEE Electron Device Letters. 1999 May;**20**(5):241–244.

[28] Lichtenwalner DJ, Misra V, Dhar S, Ryu SH, Agarwal A. High-mobility enhancement-mode 4H-SiC lateral field-effect transistors utilizing atomic layer deposited Al_2O_3 gate dielectric. Applied Physics Letters. 2009 Oct 12;**95**(15):152113.

[29] Thomas SM, Sharma YK, Crouch MA, Fisher CA, Perez-Tomas A, Jennings MR, Mawby PA. Enhanced field effect mobility on 4H-SiC by oxidation at 1500°C. IEEE Journal of the Electron Devices Society. 2014 Sep;**2**(5):114–117.

[30] Sharma YK, Ahyi AC, Issacs-Smith T, Shen X, Pantelides ST, Zhu X, Feldman LC, Rozen J, Williams JR. Phosphorous passivation of the SiO_2/4H–SiC interface. Solid-State Electronics. 2012 Feb 29;**68**:103–107.

[31] Tuttle BR, Dhar S, Ryu SH, Zhu X, Williams JR, Feldman LC, Pantelides ST. High electron mobility due to sodium ions in the gate oxide of SiC-metal-oxide-semiconductor field-effect transistors. Journal of Applied Physics. 2011 Jan 15;**109**(2):023702.

[32] Liu G, Ahyi AC, Xu Y, Isaacs-Smith T, Sharma YK, Williams JR, Feldman LC, Dhar S. Enhanced inversion mobility on 4H-SiC (1120) using phosphorus and nitrogen interface passivation. IEEE Electron Device Letters. 2013 Feb;**34**(2):181–183.

[33] Snow EH, Deal BE. Polarization phenomena and other properties of phosphosilicate glass films on silicon. Journal of the Electrochemical Society. 1966 Mar 1;**113**(3):263–269.

[34] Xu Y, Xu C, Liu G, Lee HD, Shubeita SM, Jiao C, Modic A, Ahyi AC, Sharma Y, Wan A, Williams JR. Concentration, chemical bonding, and etching behavior of P and N at the SiO_2/SiC (0001) interface. Journal of Applied Physics. 2015 Dec 21;**118**(23):235303.

[35] Sharma YK, Xu Y, Jennings MR, Fisher C, Mawby P, Feldman LC, Williams JR. Improved stability of 4H SiC-MOS devices after phosphorous passivation with etching process. International Journal of Fundamental Physical Sciences. 2014 Jun;**4**(2):37–42.

[36] McCullough RW, Geddes J, Croucher JA, Woolsey JM, Higgins DP, Schlapp M, Gilbody HB. Atomic nitrogen production in a high efficiency microwave plasma source. Journal of Vacuum Science & Technology A: Vacuum, Surfaces, and Films. 1996 Jan;**14**(1):152–155.

[37] Williams JR, Isaacs-Smith T, Wang S, Ahyi C, Lawless RM, Tin CC, Dhar S, Franceschetti A, Pantelides ST, Feldman LC, Chung G. Passivation of oxide layers on 4H-SiC using sequential anneals in nitric oxide and hydrogen. In: MRS Proceedings. Vol. 786. Cambridge University Press; 2003. pp. E8–1.

[38] Sharma YK, Ahyi AC, Issacs-Smith T, Modic A, Xu Y, Garfunkel E, Jennings MR, Fisher C, Thomas SM, Fan L, Mawby P. Advancements in SiC power devices using novel interface passivation processes. In: Physics of Semiconductor Devices. Springer International Publishing; 2014. pp. 47–52.

[39] Sharma YK. Advanced SiO_2/SiC interface passivation [Doctoral dissertation]. Auburn University

[40] Pérez-Tomás A, Brosselard P, Godignon P, Millán J, Mestres N, Jennings MR, Covington JA, Mawby PA. Field-effect mobility temperature modeling of 4H-SiC metal-oxide-semiconductor transistors. Journal of Applied Physics. 2006 Dec 1;**100**(11):114508.

[41] Rong H, Sharma YK, Dai TX, Li F, Jennings MR, Russell S, Martin D, Mawby PA. High temperature nitridation of 4H-SiC MOSFETs. In: Materials Science Forum. Vol. 858. Trans Tech Publications Ltd.; 2016 Feb 29. pp. 623–626.

[42] Modic A, Liu G, Ahyi AC, Zhou Y, Xu P, Hamilton MC, Williams JR, Feldman LC, Dhar S. High channel mobility 4H-SiC MOSFETs by antimony counter-doping. IEEE Electron Device Letters. 2014 Sep;**35**(9):894–896.

[43] Okamoto D, Sometani M, Harada S, Kosugi R, Yonezawa Y, Yano H. Improved channel mobility in 4H-SiC MOSFETs by boron passivation. IEEE Electron Device Letters. 2014 Dec;**35**(12):1176–1178.

[44] Afanasev VV, Bassler M, Pensl G, Schulz M. Intrinsic SiC/SiO_2 interface states. Physica Status Solidi (A). 1997 Jul 1;**162**(1):321–337.

[45] Fung CD, Kopanski JJ. Thermal oxidation of 3C silicon carbide single-crystal layers on silicon. Applied Physics Letters. 1984 Oct 1;**45**(7):757–759.

[46] Li HF, Dimitrijev S, Harrison HB, Sweatman D. Interfacial characteristics of N_2O and NO nitrided SiO_2 grown on SiC by rapid thermal processing. Applied Physics Letters. 1997 Apr 14;**70**(15):2028–2030.

[47] Okamoto D, Yano H, Hirata K, Hatayama T, Fuyuki T. Improved inversion channel mobility in 4H-SiC MOSFETs on Si face utilizing phosphorus-doped gate oxide. IEEE Electron Device Letters. 2010 Jul;**31**(7):710–712.

[48] Modic A, Sharma YK, Xu Y, Liu G, Ahyi AC, Williams JR, Feldman LC, Dhar S. Nitrogen plasma processing of SiO_2/4H-SiC interfaces. Journal of Electronic Materials. 2014 Apr 1;**43**(4):857.

[49] Sharma YK, Li F, Jennings MR, Fisher CA, Pérez-Tomás A, Thomas S, Hamilton DP, Russell SA, Mawby PA. High-temperature (1200–1400°C) dry oxidation of 3C-SiC on silicon. Journal of Electronic Materials. 2015 Nov 1;**44**(11):4167–4174.

[50] Cohen NL, Paulsen RE, White MH. Observation and characterization of near-interface oxide traps with CV techniques. IEEE Transactions on Electron Devices. 1995 Nov;**42**(11):2004–2009.

[51] Kong FC, Dimitrijev S, Han J. Electrical characteristics of near-interface traps in 3C-SiC metal–oxide–semiconductor capacitors. IEEE Electron Device Letters. 2008 Sep;**29**(9):1021–1023.

[52] Nicollian EH, Brews JR, Nicollian EH. MOS (Metal Oxide Semiconductor) Physics and Technology. New York et al.: Wiley; 1982 Mar.

[53] Krieger M, Beljakowa S, Trapaidze L, Frank T, Weber HB, Pensl G, Hatta N, Abe M, Nagasawa H, Schöner A. Analysis of interface trap parameters from double-peak conductance spectra taken on N-implanted 3C-SiC MOS capacitors. Physica Status Solidi (B). 2008 Jul 1;**245**(7):1390–1395.

[54] Naik H, Chow TP. Comparison of inversion electron transport properties of (0001) 4H and 6H-SiC MOSFETs. In: Materials Science Forum. Vol. 679. Trans Tech Publications; 2011. pp. 678–681.